Systems to Support Health Policy Analysis

Theory, Models, and Uses

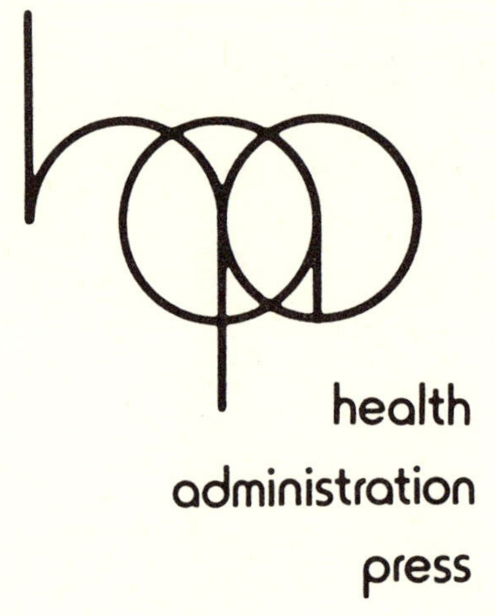

Health Administration Press is a division of the Foundation of the American College of Healthcare Executives. The Editorial Board is a cooperative activity of the Association for Health Services Research, The University of Michigan, and the Foundation of the American College of Healthcare Executives and selects publications for the Health Administration Press imprint. The Press was established in 1972 with the support of the W.K. Kellogg Foundation.

Systems to Support Health Policy Analysis

Theory, Models, and Uses

David H. Gustafson
William L. Cats-Baril
Farrokh Alemi

Health Administration Press
Ann Arbor, Michigan 1992

95 5 4 3 2

Library of Congress Cataloging-in-Publication Data

Gustafson, David H.
Systems to support health policy analysis : theory, models, and uses / David H. Gustafson, William L. Cats-Baril, Farrokh Alemi.
p. cm.
Includes bibliographical references and index.
ISBN 0-910701-73-3 (hardbound : alk. paper)
1. Medical policy. 2. Health planning. 3. Decision-making. 4. System analysis. 5. Expert systems (Computer science) I. Cats-Baril, William L. II. Alemi, Farrokh. III. Title.
[DNLM: 1. Decision Support Techniques. 2. Health Policy. 3. Information Systems. 4. Models, Theoretical. WA 525 G982s]
RA395.A3G87 1992 362.1—dc20
DNLM/DLC for Library of Congress 91-20880 CIP

The paper used in this publication meets the minimum requirements of American National Standard for Information Sciences—Permanence of Paper for Printed Library Materials, ANSI Z39.48-1984. ♾™

Health Administration Press
A division of the Foundation of the
American College of Healthcare Executives
1021 East Huron Street
Ann Arbor, Michigan 48104-9990
(313) 764-1380

Harold E. and Olive A. Gustafson
"We are a provider of our genes and environment."
—D.H.G.

Dave, this one's for you.
Thanks for the wisdom and inspiration.
—W.L.C.-B.

Mastaneh Badii and Roshan Badii Alemi
—F.A.

Contents

Foreword

We know of no other book quite like the one you are holding in your hand. It provides a rigorous and systematic approach to problem solving, combining quantitative techniques and methods for guiding human behavior toward objectives. It takes the reader from basic concepts through to their application, offering practical and realistic examples that make the book readable and useful to students as well as health professionals.

Many academicians are like the proverbial child with a hammer. They treat every problem as a nail, trying to make the hammer work rather than finding the right tool for the task. This book opens up a larger tool kit with a selection of tools for a variety of problems. The authors draw on a special combination of knowledge of industrial engineering, human behavior, and the delivery of medical care. The information they present is useful not only for addressing public policy but also for health care management and medical decisions.

To understand the applications of this book, consider for a moment the problem of low back pain. This common and painful twentieth-century symptom is rarely caused by a clear-cut disease or condition. Because it is so common, however, it is associated with high medical care costs as well as high costs to employers and employees due to loss of work. This is the kind of health care problem that needs fundamental rethinking at the health policy level, well beyond the bounds of the single institution.

Organizing sensible care for back pain requires the involvement of providers of care, people with back pain, and the companies and insurers who pay for this care. How rational will their decision-making process be? How can they share information about this problem? What can be done to treat back pain, and at what cost? Nearly all treatments for back pain are of unproven value, many may be worthless, and some (like prolonged bed rest) are probably even harmful. So where do we begin in developing an appropriate policy for approaches to treatment?

Systems to Support Health Policy Analysis offers the right combination of tools to begin untangling the answers to these questions. Decision analysis and utility (value) analysis may be very useful in analyzing the relationship between treatment and outcomes. Decision analysis, in particular, helps in developing clinical guidelines for appropriate care. In the case of back pain, perhaps it will become clear that back pain is often precipitated by poor working conditions, including unhappiness on the job. Prevention might call for job redesign, education about body motion, and improved management of the work environment. A randomized trial has shown that work loss from back pain can even be reduced by having the direct supervisor call the employee after a three-day absence to express concern and to ask when the employee plans to be back on the job. It is this type of integrated solution that the authors of this book are in a unique position to address.

If one sees the solution to better back pain care (or any other kind of care) as requiring an ongoing partnership between payers, patients, and providers, then one needs to know how to bring these people together to improve this process. This requires an understanding of group process, conflict resolution, and setting priorities. The methods described in this book reflect that understanding. In addition, the proposed process improvement for treating back pain will need to be evaluated. Evaluation requires forecasting and information collection to relate expectations to reality. The methods in this book will support that endeavor. In short, the methods described in this book will help the reader find the right tools to think through new solutions to ongoing health policy issues.

Duncan Neuhauser, Ph.D.
Xi Peng Yin, M.D.
Department of Epidemiology and Biostatistics
Case Western Reserve University
School of Medicine

Acknowledgments

The authors wish to acknowledge the National Institute of Mental Health for its early support of the research leading to this text. We also acknowledge the writing skills of David Tenenbaum and Laura Gustafson, who helped transform a series of coherent but rough chapters into a finished work, and Myrna Kasdorf, who helped prepare the final draft.

1

Introduction

As costs rise, planning, policy analysis, and evaluation are gaining importance in health and social service systems. Evaluations of such programs are commonly mandated at both the federal and local levels. Health and social service systems are also feeling pressures for change from other sources. The courts have addressed the issues of the right to treatment, the right to refuse treatment, and the right to appropriate treatment. The issue of national health insurance and the role of mental health services have been debated in Washington for years; the debacle over Medicare catastrophic insurance was only one chapter in that debate. The emergence of health maintenance and preferred provider organizations has powerful implications not only for the private sector but for the public sector as well. The issues of cost containment and efficiency, long discussed in the abstract, have recently become critical considerations in providing health and social services. And attempts at cost containment, in their turn, raise concerns about the quality of care.

Thus, health and social service systems are facing complex challenges just as they are undergoing substantial change. With the heightened frequency and importance of these challenges, one would hope that rigorous planning and policy analysis would be playing a greater role in the policy process. Yet interviews with health policymakers and planners in nine states suggest that evaluation, data, and policy analyses have little influence on policy. Our research on mental health data bases found that many state organizations spontaneously report dissatisfaction with the realities and/or operation of the management information system. Our experience suggests that policy analysis, planning, and evaluation has minimal influence on the process of forming policy. It is almost as if the large data bases and rigorous analyses that have been conducted for the health and social service fields did not exist.

This conclusion does not appear to have altered the emphasis in health information systems development, which often develops large data bases containing "minimum" data sets collected according to uniform definitions and protocols. Too little emphasis is placed on learning how the vast stores of data could help decision makers, and even less emphasis is put on finding what kind of help decision makers need before designing systems that supposedly respond to them. One might say most information systems are obsessed with providing data for everybody but information for no one.

No single institution can claim the credit or bear the blame for the continued emphasis on creating data bases. The philosophy of large data bases not directly linked to applications is shared by nationwide federally supported data collection efforts such as the National Ambulatory Medical Care Survey (1979); cooperative health manpower research efforts such as the Functional Task Analysis Cooperative Study (1981); and federally funded local and statewide efforts such as the Cooperative Federal-State-Local Statistics System (1982).

There is no agreement about why these elaborate information systems have been so lacking in influence. Some suggest that potential users are at fault because they are unwilling to learn the skills needed to use the new systems or that a general resistance to change prevents the use of a powerful tool. Others suggest that disuse is at least partly the result of the current organization of the system. Ida Hoos, an early critic of computers and quantitative methods policy, observed in 1971: "Crucial to the input-output models and cost benefit ratios that characterize planning, the data bank has a solid sound and significance far exceeding its intrinsic worth, for somewhere in its generation, the items plugged in become 'hard facts' and are accorded a reliability that is more wishful than actual."

One hopeful development has been the support for medical treatment effectiveness programs by the new Federal Agency for Health Care Policy and Research (Salive et al. 1990), which combines development of data bases with creation of clinical guidelines and dissemination of results. It is hoped the results will include the design of applications-oriented and integrated data bases that will efficiently describe variations in medical practice. In fact, Connell et al. (1987) report on how large data bases have recently been used to describe variations in medical practice and to provide longitudinal analysis of outcomes of care. If successful, these efforts may provide a model for designing administrative systems that influence a wide variety of health and social services policy.

If data bases are to support decision makers in their work, we, with Williamson (1990), see the need to put first things first. Instead of gathering statistics and seeking applications for them, we prefer (1) to determine what information potential users need, (2) to obtain relevant and valid information to meet those needs, and (3) to provide that information to policymakers in such a way that it will be used. The system we envision is called a policy information management system (PIMS). This system differs from a management information system (MIS) in many ways, including its use of subjective judgments as well as empirical data, its use of decision science models to model expert thinking, its access to expertise via computer-mediated communication, its offer of service as well as data to policymakers, and its access to relevant literature. All these services are seldom elements of an MIS.

The task of building a PIMS is critical and difficult. A serious obstacle to widespread acceptance of PIMS is a widespread doubt that computers, or, for that matter, systematic analysis, can enhance our intuitive understanding of complex social and political issues. It is difficult to convince people that analytical models can improve on their judgment, yet there is increasing evidence that analytical approaches to complex policy issues can enhance the quality and effectiveness of social policy. This book provides an opportunity for students of the policy process to examine how quantitative models can be used to analyze and enhance human intuition and judgment.

We have observed that the actual process of formulating, implementing and evaluating health and social service policy (which we call the policy process) is largely based on the values, judgments, and uncertainties of health care managers and policymakers. What distinguishes a successful policy process is often the integration of those values, uncertainties, and judgments. As in any field, a few policymakers stand out for their ability to grasp a problem, select a solution, and successfully implement it in a political environment. We believe the overall quality of the policy process could be improved if we could model how the best experts think and then make such models widely available in the form of support systems. This book provides methodologies for and examples of such support systems. These support systems, based on models of the judgmental processes of outstanding policymakers, are built upon three disciplines: decision analysis, computer technology, and group processes.

This book is divided into three sections: Section I, Issues in the Policy Process; Section II, Methodological Foundations; and Section III, Policy Support Tools.

Section I: Issues in the Policy Process

Chapter 2, Rationality and Policymaking, summarizes the literature on how managers and policymakers make decisions. Several models of how decisions are or should be made in organizations are presented along with our model of organizational and individual change. Psychological, physiological, and environmental factors can limit the quality of decisions. This chapter explains various limitations that we believe a good policy support system can surmount.

Chapter 3, The Issue Life Cycle, suggests that issues move through a predictable life cycle with identifiable stages and specific characteristics. Information needs change according to the stage of the life cycle at which an issue is located. These insights should help players in the policy process anticipate how issues will develop and prepare their analyses to fit the issue's current state.

Chapter 4, Individual Differences, deals with differences among policymakers, which are a key challenge to a support system. Using the variety of the tools discussed in this book, a PIMS can respond to individual differences among policymakers. The system should provide easy access to resources for policymakers when they are wanted. The support system must be simple to use and designed for policymakers. A system can have components dealing with information, analysis, communication, and referral. An example of a policy information system for use in quality improvement efforts in hospitals is given.

Section II: Methodological Foundations

Chapter 5, The Integrative Group Process, explains the important roles groups play in the policy process, reviews literature on group dynamics to show the strengths and weaknesses of groups, and discusses important issues related to group facilitation and leadership. A variety of processes for leading problem-solving groups are presented, as are hints about choosing the best process. We introduce the integrative group process, which we developed to support problem-solving groups and groups developing decision models.

Chapter 6, Introduction to Decision Analysis, shows the importance of decision analysis as a tool for policy support, by reference to a particular health policy issue. In this context we introduce the essential elements of decision analysis (options, outcomes, values, and uncertainty). Methods are available to structure policy issues within the decision analysis framework and to develop models to estimate value and uncertainty. Some of these models are designed to replicate the approach of recognized experts; others try to replicate how experts wish they could view a problem. Different aspects of decision analysis can be emphasized for different problems.

Chapter 7, Value Models, presents this essential aspect of decision analysis in detail. After introducing the concept of values, we present a brief history of multiattribute value models and suggest how they can be useful in the policy process. Next we explain a five-step process for developing these models and a way to evaluate them.

Chapter 8, Forecasting without Real Data, introduces subjective probability and Bayes' theorem and gives a structure for applying those concepts to the policy process. Subjective probability, the estimate of uncertainty obtained from expert judgments instead of empirical data, allows us to formally incorporate opinions of policymakers and power brokers into a policy analysis and to analyze policy issues in which uncertainty plays a large role but the empirical data is inadequate. Bayes' theorem, a tool to formally revise estimates of uncertainty as new information accumulates, allows us to account for the evolving nature of uncertainty. In other words, it lets us enhance our understanding of a problem as we receive further information.

Chapter 9, Option Generation, presents two approaches to generating options that build on decision theory, creativity theory, and change theory. Option generation is one of the underemphasized aspects of policy analysis. Often the options under consideration represent the extremes of the political spectrum (for example, parental notification if teenagers use family planning services is a yes-or-no situation). In reality, we need not a compromise but a creative alternative that allows both sides to win. Our approach to generating options should help policy analysts find creative alternatives even when time is pressing.

Chapter 10, Expected Utility and Decision Trees, examines a technique that combines aspects of values modeling and uncertainty into a powerful tool to examine the relative benefits of alternative solutions. Decision trees are particularly useful for representing an analysis that depends on values and uncertainty. This chapter teaches how to represent a policy analysis in decision tree form and how to combine the results of uncertainty analysis and value modeling into summary analyses of the expected benefits of alternative policies. We also introduce the concept of risk, which allows us to formally incorporate into an analysis the policymaker's preference for or against risk.

Chapter 11, Decision-Oriented Program Evaluation, argues that program evaluation efforts (by policy analysts and others) often provide information that does not help policy formation. Program evaluation efforts have focused on making information more accurate rather than more relevant. Emphasizing quasi-experimental design has provided powerful tools for explaining system behavior. However, little emphasis has been placed on identifying the decision or policy problem that justifies the evaluation. Often the policymaker thinks evaluation will help make those decisions but is frustrated by irrelevant information. We present a decision-analytic approach to program evaluation that forces the evaluator to identify the perceptions and values of affected parties and define their importance.

Section III: Policy Support Tools

Chapter 12, Determining Information Requirements, is a tool for policy analysis used to find out exactly what must and must not be known to solve a problem. Empirical data can play a vital role in the policy process, but more often data are useful in supporting the wisdom of a chosen policy. Too frequently, data are unrewarding because they are overly general or irrelevant. This chapter presents a methodology to identify needs and establish four categories of data: (1) data that should be routinely collected and analyzed to give immediate answers to key questions; (2) data for which we want a quick-response collection and analysis system; (3) infrequently needed data that can be supplied by a vendor; and (4) data too trivial to justify any planning or resource allocation.

Chapter 13, Setting Priorities for Policy Analysis, presents a model to identify policy issues for analysis to help us overcome time constraints. Policy analysts need time to understand, analyze, and report, so issues must be identified as early as possible. The ability to predict the policymaker's needs distinguishes successful policy staff units. We believe issues become important as a result of intrinsic importance or political ripeness. We present a decision analysis model of factors that influence political ripeness, suggest ways of measuring each factor, and present a Bayesian model to aggregate those measures into a prediction of issue ripeness. Finally, we exemplify how the Bayesian methodology can identify crucial issues and set priorities to reflect ripeness and intrinsic importance.

Chapter 14, Conflict Analysis, describes how to manage conflict, an essential ingredient and critical barrier to policy formation and implementation. Conflict, which can cause delays and fruitless compromises, often stems from misunderstanding the value systems of the parties and viewing a situation in win-lose terms. This modeling process identifies the value systems of key parties, helps spot opportunities for compromise and trade-offs, and evaluates the acceptability of various solutions. We explain a model incorporated in a policy information management technology that has helped individuals and groups deal with a variety of conflicts.

Chapter 15, Using Bayesian and MAV Models to Analyze Implementation, explains the use of policy analyses to help with implementation, a critical aspect of the policy process. Implementation rarely receives formal analysis, and little attempt has been made to translate the change literature into an effective tool for planning and managing implementation. We present such a tool for predicting and explaining implementation success, based on synthesis of available data and judgments of experts. The chapter synthesizes literature on implementation, describes the development of a decision-analytic model and a computer-based policy information management system, and shows by example how the model can help evaluate a program to improve nursing home care.

Chapter 16, Using MAV Models to Test the Effectiveness of Hospital Categorization, describes how decision-analytic methodology can be used to evaluate the cost-benefit ratio of categorizing hospitals into several levels of sophistication regarding burn treatment. This was one of the earliest studies we know of to demonstrate how severity indexes using multiattribute value modeling can strengthen program evaluation. Like Chapters 14 and 17, this chapter presents a case study of the application of decision analysis, group processes, and (to a lesser extent) policy information management systems to the policy process.

Chapter 17, Using Subjective Probabilities to Predict the Impact of National Health Insurance on Low-Income People, is a case study using expert subjective probability estimates to predict the effect of several national health insurance programs on low-income populations. This study demonstrates how powerful quantitative tools can be adapted to situations where empirical data are unavailable or inadequate.

References

Arkes, H., and K. Hammond. 1986. *Judgment and Decision Making*. Cambridge University Press.

Connell, F. A., P. Diehr, and L. G. Hart. 1987. "The Use of Large Data Bases in Health Care Studies." *Annual Review of Public Health* 8: 51–74.

Gustafson, D., and G. Huber. 1977. "Behavioral Decision Theory and the Health Delivery System." In *Human Judgment and Decision Processes: Applications in Problem Settings,* edited by M. Kaplan and S. Schwartz. New York: Academic Press.

Gustafson, D., W. Edwards, and L. Phillips. 1969. "Subjective Probabilities in Medical Diagnosis." *IEEE Trans. on Man-Machine Sys.* 10 (3): 61–65.

Gustafson, D., J. Greist, J. Kestly, and D. Jensen. 1971. "Initial Evaluation of a Subjective Bayesian Diagnostic System." *Health Services Research* (Fall).

Hoos, I. 1972. *Systems Analysis in Public Policy: A Critique*. Berkeley: University of California Press.

Sainfort, F., D. Gustafson, K. Bosworth, and R. Hawkins. 1989. "An Experimental Evaluation of a Computer-Based Conflict Resolution Program." *Organizational Behavior and Human Decision Processes* 42 (1).

Salive, M. E., J. A. Mayfield, and N. W. Weissman. 1990. "Patient Outcomes Research Teams and the Agency for Health Care Policy and Research." *Health Services Research* 25 (5) (December): 697–708.

Von Winterfeldt, D., and W. Edwards. 1986. *Decision Analysis and Behavioral Research*. New York: Cambridge University Press.

Williamson, J. 1990. "Education in Science Information Management." *Quality Assurance and Utilization Review* 5 (4) (November): 121–26.

Section I

Issues in the Policy Process

2

Rationality and Policymaking

Ideally, policymakers would behave perfectly rationally. They would frame questions accurately, obtain all necessary information, discuss the problem with interested parties, and weigh all factors carefully when making a wise and fair decision. But in real life, this is more the exception than the rule. Everyone can cite irrational decisions by policymakers in federal, state, and local governments, in corporations and nonprofit agencies.

This chapter is devoted to exploring the ideal of rational policymaking, the impediments that stand in the way of reaching that ideal, and alternative means of making decisions that retain the positive parts of rationality while sidestepping the negative ones. In terms of making decisions, rationality implies consistency, clear purpose, focusing on the needed information (content rationality), and combining that information logically (process rationality). Rationality is essential to unbiased and comprehensive processing of information, both of which are required to make good decisions.

Specifically, this chapter addresses three questions: How do real policymakers behave? What biases impede their information processing? How can policy information management systems help policymakers behave more rationally? We also explore why decision makers cannot fulfill the demands of rationality, alternative conceptions of rationality in organizations, physiological limitations, cognitive biases, psychological characteristics, and organizational pathologies. In the entire discussion, the goal is to learn how policy information management tools can increase decision makers' rationality. The obvious starting point is to examine the concept of rationality itself.

The literature offers several definitions of rational decision making. Chaffee (1980) summarizes these definitions to say it is characterized by

1. Accumulating information before making a decision
2. Collecting information that is problem-centered and goal-directed
3. Documenting the existence of a problem, the need to solve it, and what benefits will arise from its solution
4. Considering several alternatives for reaching the goal or solving the problem
5. Positing logical and consistent cause-effect relationships
6. Assessing the value of the costs and benefits of the various alternatives
7. Determining the individual and organizational values behind the data and interpretations being used

Research in decision theory has shown that individuals cannot meet the requirements of rationality because of common limitations that cause real policymakers to behave erratically. Most people

- Are incapable of integrating information, particularly if it is not perfectly coherent
- Collect the wrong kind or amount of information
- Develop too few hypotheses about the cause of the problem
- Fail to gather contradictory information
- Underestimate the diagnostic value of the information and prematurely commit themselves to action
- Are misled by apparent correlations
- Do not define problems adequately
- Have difficulty articulating their values and beliefs

Other problems arise if decision makers fail to recognize dissatisfaction with the present regime, are sloppy in marshaling social and political support, lack skill in implementing a decision, or fail to develop necessary skills among their associates. Decision makers may also be too busy to monitor the success or failure of their actions.

Realities of Decision Making

When theoreticians in decision analysis, behavioral decision theory, and cognitive psychology warn about human shortcomings in policymaking, some people dismiss the cautions as academic extrapolations from studies of undergraduates doing irrelevant tasks in unrealistic settings. But enough evidence has been gained in real settings for policymakers and systems designers to accept the pervasiveness of cognitive limitations—and the need to compensate by using decision aids.

It could be argued that poor decision making was specific to the samples used in those studies, but some studies have found frequent occurrence of less than optimal behavior. The following observations are from longtime observers of policy decision making.

Kepner and Tregoe (1981) conclude from their extensive consulting experience that policymakers do a poor job of getting information because they do not

- Consciously explore what is known and unknown
- Organize available information into a coherent picture
- Try to generate a problem statement

Kepner and Tregoe further assert that policymakers have difficulty establishing causation because they collect arguments to support pet theories and become convinced they have a complete explanation of the problem, even if only some of the data confirm their hypothesis. These decision makers commonly resist other explanations.

Watson (1976) identified three common concerns about problem definition, perhaps the most critical phase in decision making. Decision makers frequently

1. Do not believe a problem exists at all. Perhaps this is so because objectives are not clearly identified or they are insulated from complaints about the status quo.
2. Identify the wrong problem, the wrong causes, or both. This failing can be caused by defining problems simply around attitudes and beliefs without using empirical data to test the validity of those opinions and beliefs, oversimplifying the causes, or mistaking symptoms for causes.
3. Apply old solutions without checking whether they fit the problem. Creative approaches are shunned for various reasons: "This has worked before"; "There is only one way to do this"; or "That may have worked elsewhere, but it won't work here."

Other scholars have pointed out that management education trains policymakers to see problems as neatly structured and focused entities. This approach leads managers to be solution-minded rather than problem-minded and inclines them to select narrow, short-term solutions. Cohen et al. (1972) say most policymakers already know the answers to the questions they ask—or at least where to find the answers. Better, the authors suggest, is to ask questions that enhance the problem analysis phase even if they are tough enough to delay the solution phase.

Janis and Mann (1977) summarize their observations on actual policymaking by listing seven basic mistakes that afflict decision makers and interfere with rational decision making. Decision makers do not

1. Thoroughly examine a wide range of alternative courses of action
2. Survey the full range of objectives to be fulfilled
3. Carefully weigh what they know about the risks and costs of negative consequences
4. Weigh the benefits of each alternative
5. Intensively search for new information needed to evaluate the alternatives
6. Correctly assimilate new information or expert judgment, particularly when it does not support their preference
7. Reexamine the positive and negative consequences of the alternatives, including those they originally considered unacceptable, before making a choice
8. Make detailed provisions to implement or execute the chosen action
9. Give special attention to contingency plans to deal with known risks

One reason for this irrational behavior is a lack of adequate strategy on the part of decision makers. To put it another way, perhaps they do not know how to implement the solution they choose. Perhaps the policy information management is weak because it does not give feedback on previous actions or devise a reasonable set of alternatives. Perhaps the work environment is not conducive to rational decisions. For example, Mintzberg (1973) observed policymakers and concluded that their activities are characterized by:

1. *Brevity.* They rarely spend more than 30 minutes on an activity.
2. *Fragmentation.* They deal with a number of problems in a day and deal with a particular problem sporadically over a long period.
3. *Variety.* They deal with many types of problems and decisions.

These observations undermine the validity of the rational decision model. Clearly, policymakers do not devote unlimited time and resources to collecting perfect information on every problem and solution. These observations also buttress the notion that policymakers make extensive use of simplifying strategies—shortcuts—to reduce the time burden of decisions.

Evidence for this simplifying phenomenon was offered by Downs (1966). After studying several government agencies, he concluded that time limitations cause policymakers to consider a limited (and often insufficient) amount of information at one time. Job realities being what they are, decision makers must be involved in several activities simultaneously; thus, they only attend to one problem while another is latent. Also, the amount of information initially available is seldom sufficient to solve the problem. Sometimes additional information is available at a cost; other times it is totally unavailable. This forces policymakers to make one more decision: whether to spend the time and money to collect additional information.

Huber (1980a) summarizes the psychological and situational difficulties that policymakers encounter while attempting to identify and use information for decisions. He concludes that "limits on human rationality" lead to the use of simplistic strategies and inadequate models. In the short run these strategies and models save time and other resources, but they tend to produce solutions with less than maximum quality.[1] The use of poor strategies and models increases when stress increases or when time and other resources decrease (Huber 1981, p. 29).

This last conclusion is crucial since, as previously noted, policymakers tend to be short of time and resources. We next review

[1]While we generally agree with Huber, we don't expect policy information management systems to promote "maximum" quality decisions. Like Phillips (1983), we seek decision support systems that structure and analyze the situation well enough to solve the problem.

several of the biases that interfere with rationality and decision making among policymakers.

On the Limits of Rationality

Human beings need help processing information, and we must first understand these weaknesses in order to design aids to overcome them.

> The first consequence of the principle of bounded rationality is that the intended rationality of an actor requires him to construct a simplified model of the real situation in order to deal with it. He behaves rationally with respect to this model, and such behavior is not even approximately optimal with respect to the real world. To predict his behavior we must understand the way in which the simplified model is constructed, and its construction will certainly be related to his psychological properties as a perceiving, thinking, and learning animal. (Simon 1957, p. 198)

The first limit on rationality stems from the fact that the human perceptual system is not all-powerful.

The second limit on rationality is the extensive use of overly simplistic strategies and the interference of personal biases. The literature reports a plethora of biases from human information processing, and most are pervasive, consistent, and systematic. Though there is debate about whether cognitive biases actually impair decision making as much as was first thought, there is no doubt that some biases decrease decision quality.

Hogarth (1980), in his excellent summary of empirical findings on biases, classified them into four categories: (1) biases in acquiring information, (2) biases in processing information, (3) biases in transmitting information, and (4) biases in receiving and storing feedback. The likelihood that these biases will be present depends on two factors: the level of psychological and environmental stress and the schema (the frame of reference and problem-solving approach) the policymaker uses.[2]

---

[2]It should be noted that some investigators disagree with these findings, arguing that the difference between subject behavior and "optimal behavior" occurs not because of heuristics and biases but because the subjects are responding rationally to a problem structure that is different from the structure perceived by the experimenter.

The third reason why people need help processing information is limitations caused by personal characteristics. Cognitive style (the method of collecting and evaluating information), dogmatism (the resistance to dissonant information), risk propensity (the willingness to decide with partial information), creativity (the ability to generate new perspectives), and integrative complexity (the ability to absorb input from many sources) all affect the efficiency and effectiveness of information processing.

Finally, organizational behaviorists have found that organizational structure along with the content and direction of the message all affect the way the message is processed and transmitted.

Now we will examine these causes of distortion and faulty information processing so we can later propose intelligent decision aids to overcome them.

Our basic equipment for observing the world

Experiments show that certain characteristics of the human information processing system are fixed across individuals and tasks. These characteristics include the size and access speed of the different types of memory, the mode of processing, and the usual overload of environmental stimuli.

Briefly, we know that humans can assimilate and process information relatively slowly and that they compute and remember poorly. Our information processing system is limited by low capacity in short-term memory, slow storage in long-term memory, the slow serial (as opposed to parallel) processing, and the use of sometimes inappropriate but familiar patterns when analyzing new information.

Allport (1955) was one of the first to suggest that human information processing is limited by other factors. He suggested that physical needs tend to determine what is perceived; that the type of reinforcement influences which stimuli are perceived, their apparent size, and how fast they are recognized; and that personality characteristics bias perception. Allport also found that we are slow to recognize stimuli that are emotionally disturbing and that these stimuli undermine our usual analytical capabilities, perhaps because they arouse emotional reactions even before being recognized.

Understanding how policymakers see problems is crucial for designing effective decision aids. Several researchers have proposed models to explain how people perceive problems and proceed to solve them. Newell and Simon (1972) have suggested that experience is critical in problem solving. MacCrimmon and Taylor (1976) said perceptions are more important than facts in determining the degree of uncertainty, complexity, and conflict a policymaker sees in a situation, so perceptions are a major determinant of the strategies chosen to solve the problem. Piaget (1981), Varela (1989), and others have suggested that people create their own reality based on a series of emotional, psychological, and political considerations.

Brunswick (1956) developed a model (shown in Figure 2-1) that has been helpful in conceptualizing how people make judgments. Suppose you want to predict demand for nursing home beds next year. You, the policymaker, would mentally review variables or cues that could affect demand and conclude that cues p_1, p_3 and p_i (see Figure 2-1) are important. The true relationship between these cues and next year's demand are lines $r_1, r_2 \ldots r_n$. Note that not every cue you consider relevant is indeed relevant, and vice versa. The accuracy of your judgment depends on the correspondence between relationships ($p_1 \ldots p_n$) and relationships ($r_1 \ldots r_n$). If we accept this model as a reasonable description of how people make judgments, then we can easily see several sources of error.

First, how do you decide what to judge? Why do you want to predict demand for beds instead of demand for long-term institutional care? Second, how do you choose which cues predict demand, and how can you be sure those cues are valid? Third, how do you assess and combine the values of various cues? Fourth, how do you know you were right or wrong in selecting those cues and assigning those weights? Finally, what effect will any errors have on your decision?

Before we discuss biases, let us return to the bottom line of this section: human beings are sequential information processors who create images and models of the world based as much on belief as on reality. Since those beliefs are often flawed, designers of decision aids must help policymakers distinguish between useless and useful information. Computer-augmented decision processes can help overcome the physiological bottlenecks and shortcomings of human information processing. Cognitive problems are another story, and they are our next topic.

Biases affecting the decision areas

Decision makers have a lot of tasks when they process information: they must integrate contradictory opinions, assess the causalities and

Figure 2-1 The Lens Model

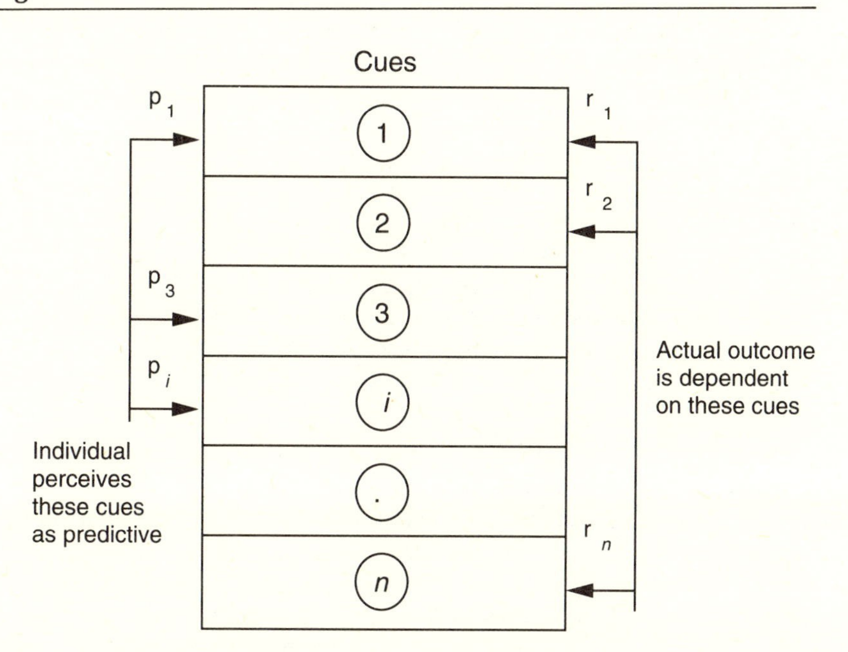

interconnections of events, evaluate the value of outcomes, and integrate information from several sources before selecting a course of action. Unfortunately, research shows that people do all these tasks less than optimally. Pervasive biases interfere with tasks like making inferences, predictions, diagnoses, evaluations, or choices.

We can postulate that the generic process of decision making has four phases: (1) information acquisition, (2) information processing, (3) information output, and (4) feedback. Decision makers are impaired by biases and processing limitations in all four phases.

Policymakers are affected by several biases in acquiring information. Tversky and Kahneman (1974) document the *availability* bias, in which the ease of recalling specific instances affects judgments of frequency. Dearborn and Simon (1967), and many others since, found that people's training colors the way they see and structure problems. Such *selective perception* also predisposes people to seek information consistent with their views and to downplay contradictory evidence. Hogarth suggests that anticipation of what one expects to see biases what actually one sees.

The *frequency* bias (Einhorn and Hogarth 1979) leads people to remember rare but impressive events and assume they occur more frequently than they really do, while underestimating the frequency of more common but rather mundane events. This bias causes people to base their judgments on observed frequency instead of the more accurate relative frequency.

The *illusory correlation* (Einhorn and Hogarth 1979) is the belief that two variables co-vary even though they do not. This bias leads policymakers to choose undiagnostic variables that lead to false predictions.

As a person becomes an expert in a subject area, these biases may occur less frequently. However, most policymakers are not experts in one subject and thus need help in compensating for such biases: policy information management systems need to play the compensatory role.

Report format influences how policymakers use data. For example, the order in which data are presented may affect processing. Biases induced by order or presentation include *recency* effects (the alternative reviewed most recently "bleeds" onto the one reviewed next, making it appear better or worse than it is). Another bias related to data presentation is *overstressing quantitative information* as opposed to qualitative information. Policymakers also tend to ignore data from large samples in the face of concrete data based on incidents (e.g., a friend telling you a certain car is a lemon may have more weight than a large contradictory survey in *Consumer Reports*). This bias has important implications on what type of information a decision aid should be able to manipulate.

Finally, individuals tend to use only information that is displayed, and then only in the form in which it is displayed. This causes us to discount or ignore information that must be remembered, inferred, or transformed from the display. The moral of the story is that one must consider very carefully how to display information so that it easily and quickly communicates the intended meaning.

Tversky and Kahneman (1974) have described heuristics that are commonly used to reduce mental effort but cause systematic errors. One such heuristic is anchoring and adjustment, a means of making predictions by starting with an initial opinion about which event is most likely to occur (anchoring) and then moving away from that opinion as new information dictates (adjustment). This may seem to be a reasonable strategy. But unfortunately these authors also detected insensitivities to statistical rules; their subjects incorrectly extrapolated from small samples and thought extreme values of a variable to be unrealistically common.

Some biases affect the ways in which people must produce answers. For example, the scale on which responses are recorded seems to impair responses.

Hogarth (1980) mentions another output-phase bias. He points out that a person who is taking action on something that has an uncertain outcome can become convinced that he or she has control over something that is actually uncertain. This illusion of control leads to overconfident assessments.

The biases that may have the greatest impact are associated with feedback, perhaps because these biases interfere with learning. For example, the success/failure attributions bias impairs how people store information about successes and failures. This leads decision makers to attribute their successes to skill and their failures to poor luck. This bias impairs policymakers' learning because it leads them to falsely attribute certain outcomes to previous actions and allows them to forget failures.

Hindsight is a feedback bias that differs conceptually from success/failure attributions. Hindsight is the phenomenon by which people, in retrospect, are not surprised by what actually happened (Fischoff 1977). They easily find plausible explanations for the event and say it was obvious and could have been predicted. Logical fallacy is a hindsight bias (Hogarth 1980) that consists of being unable to recall details of how an event came to pass and reconstructing the event in a "logical"—but inaccurate—fashion.

Why this somewhat lengthy review of cognitive biases? Biases in processing information give us opportunities to improve decision quality with fairly simple decision aids. Indeed, these findings on the suboptimality of human information processing launched an enormous amount of research comparing the performance of people versus models of people (Slovic and Lichtenstein 1971; Dawes 1971; Brightman and Urban 1970; Smith 1968; Edwards et al. 1968; and many others). These regression models show that bootstrapping—the phenomenon in which models built with the decision maker's decision rules actually outperform the decision maker—is possible because humans are unable to apply those criteria consistently. Another example of a technique that has outperformed people is the probabilistic information processing (PIP) system that uses Bayes' theorem to revise the likelihoods of events. This result arises because people suffer from conservatism—after receiving new information, they are not able to revise their predictions as well as Bayes' theorem.

While biases are widespread and systematic, certain environments make them more or less acute. Stressful environments tend to magnify biases and promote the use of simplistic heuristics. As mentioned before, managerial work is characterized by time pressures, frequent interruptions, attention to numerous diverse tasks, limited information, and high stakes (Downs 1966; Mintzberg 1973). Each distraction and pressure has been shown to impair information handling. Now we shall discuss the relation between stress and information processing.

Stress

Four sources of stress can reduce policymaker efficiency: (1) environmental, (2) psychological, (3) emotional, and (4) the nature of the task. Environmental stress can be induced by time pressure, information overload, noises, or distraction. Psychological stress is created by a risky decision with high stakes. Emotional stress can be caused by fatigue and depression, among many other factors. Unfamiliar tasks, or those with no "good" solution, generate stress, as do tasks at which the policymaker is not confident. Table 2-1 (adapted from Hogarth 1980) summarizes how stress can cause bias.

Psychological stress that impairs decision making is in contrast to anticipatory psychological stress (shown in table) which is beneficial to decision making. Different forms of stress bring out various biases in policymakers. Time deadlines and other stresses have been associated with a diminished ability to use available information, decreased search for new information, and the adoption of suboptimal comparison strategies (Janis and Mann 1977; Payne 1976; Wright 1974). Heavy environmental stress induces people to decide with too little information and to use noncompensatory and other simplistic strategies (Wright 1974). Emotional stress also causes people to use less information in making decisions and to make inconsistent decisions (Hogarth 1980). Psychological stress seems to cause people to quit searching for information prematurely, and this allows visceral impulses (fears, hopes, or anger) to interfere. Psychological stress also seems to exaggerate favorable consequences of decisions and minimize negative ones and to make the decision seem less controversial than it actually is. The use of decision aids in cases of high psychological stress is almost imperative (but these are unfortunately

Table 2-1 Potential for Bias and Use of Simplistic Heuristics under Different Conditions of Stress

	Causes of Stress							
	Environmental (Complexity)		*Task (Procedural Uncertainty)*		*Emotional (e.g., Fatigue)*		*Anticipatory (Regret)*	
Potential for Bias	*Low*	*High*	*Low*	*High*	*Low*	*High*	*Low*	*High*
Small	X		*		X			X
Large		X		X		X	X	

*Depends on how good (in the normative sense) the known procedure is.
Adapted and modified from Hogarth (1980).

just the types of situation in which policymakers do not want to take time to use these aids).

While many stresses impair the ability to make decisions, certain types of stress are beneficial. Anticipatory regret is a beneficial form of psychological stress (see Table 2-1). If this stress is absent, the policymaker is prone to bias. A policymaker concerned by anticipatory regret is likely to be more vigilant about and have more invested in the search for a good solution than one with nothing to lose. Regret is present when (1) the preferred choice is no better than the alternative, (2) negative consequences could materialize almost immediately after the decision, (3) significant people in the policymaker's social network consider the decision important and expect him or her to adhere to it, (4) available information about potential gains and losses is not used, and/or (5) significant people in the policymaker's social network allow action to be postponed until all alternatives are evaluated (Janis and Mann 1977). Under these circumstances a decision maker is likely to be more vigilant and involved in the decision.

Motivation is another form of stress that encourages active search for a solution. A moderate degree of stress will induce a policymaker to scrutinize alternative courses of action and work for a good solution, provided one seems available. This latter point is crucial: decisional stress is related to the aspiration level (the desire to do well) of the policymaker.

While stress per se is not bad, the amount and type of it are significant. Janis and Mann (1977) correlate the amount of stress to the number of goals a person expects to remain unsatisfied—and the importance of the needs underlying those goals. That is, stress increases along with the number of goals expected to be unfulfilled. Also, when a person encounters threats or opportunities that motivate him or her to consider a new course of action, the degree of stress is a function of commitment to the present course of action. Furthermore, when the policymaker anticipates not having sufficient time to find a way to prevent serious losses, the level of stress will remain high.

The relevance of this discussion for designing policy information management systems is that under certain circumstances decision makers become more erratic and bias-prone. The nature of these circumstances can help us determine what sort of support is needed.

Now we turn to the mechanisms used to choose a strategy for solving a problem—and the potential biases involved in that decision.

Schema

A schema is the series of steps by which a decision maker chooses a strategy to solve a problem. How a policymaker perceives a problem is a major determinant of which strategies he or she considers to solve it (MacCrimmon and Taylor 1976; Newell and Simon 1972). If we study how and when experts invoke schema, we can develop a training device for decision makers; this is the design strategy for expert systems.

A schema is developed and modified through experience, as we remember how our previous attempts have worked. Policymakers are predisposed to apply a successful schema to other, similar problems (Taylor 1975). Experience can facilitate or inhibit the selection of schema. Functional fixedness—the tendency to consider a tested schema applicable to every kind of problem—inhibits retrieving other, more appropriate problem-solving approaches. However, schemas are not permanent; schemas policymakers remember as ineffective with particular types of problems may be reserved for use only in certain situations or may be rejected altogether.

A schema, relative to a particular task, can be characterized on four dimensions:

1. *Veridicality*. Is there an objectively superior schema?
2. *Stability*. How subject is the schema to change?

3. *Generality.* How widely does the schema apply to the phenomena?
4. *Accessibility.* How easily can the schema be recalled from memory?

Gettys et al. (1980) and Hogarth (1980) suggest that a schema that lacks any of these dimensions can lead to suboptimal information processing and poor decisions.

Veridicality is crucial because it determines whether the policymaker is representing the task environment realistically (Simon 1977). The phenomenon of illusory correlation is a schema in which people falsely believe that variables are related. Such schemas cause policymakers to include inappropriate information in predictions and to use irrelevant and misleading decision rules.

Stability is important because a person with a stable schema is likely to approach similar problems in similar manners, and this promotes consistent information processing. It should be pointed out, however, that stability alone does not eliminate bias. A schema could be stable but nonveridical, which would clearly lead to systematic bias.

Generality is important because the extent of generalization is inversely related to bias in information processing (all other characteristics being equal). Abelson (1976) has distinguished among levels of schema ranging from concrete/episodic (based on one experience) to abstract/general (based on many experiences).

Accessibility is important because accessible schemas are used more often, regardless of their appropriateness. Gettys et al. (1980) reported that subjects tend to use accessible schema spontaneously but must be prompted to use inaccessible ones. Accessibility is determined by how frequently a schema is used and by how recently it was last used.

Table 2-2 summarizes the potential for bias under different schema characteristics. Although veridicality of the schema may be the single key determinant, schema that are unstable, specific, and inaccessible are problematic. Finally, we should add that stress substantially increases the chance of using the inappropriate schema.

Information about schemas is important for designing policy information management systems. Indeed, while most policy information management systems and decision aids provide computational support, they should also offer procedural support. Such support could be more important than computational support be-

Table 2-2 Potential for Bias in Judgment under Different Schema Characteristics

	Characteristics of Schema							
	Veridicality		*Stability*		*Generality*		*Accessibility*	
Potential for Bias	*Low*	*High*	*Low*	*High*	*Low*	*High*	*Low*	*High*
Small		X		*		*		*
Large	X		X		X		X	

*If schema is veridical.
Adapted and modified from Hogarth (1980).

cause computing correctly within the wrong schema is extremely misleading.

With the exception of stress, so far we have only considered sources of bias and distortion that are "internal" to the policymaker. We now turn our attention to pressures within the organization that increase the potential for biased and distorted information processing. In the next section we will discuss how organizational structures restrict information flow, how standard operating procedures bias people toward certain types of information, how reward systems encourage us to transmit some messages and suppress others, and how power and status relationships among senders and receivers affect the content of messages and the direction in which they flow.

Understanding how information can be distorted within the organization can identify ways in which policy information management systems can minimize their occurrence.

Distortion in organizational communication

The quality of decisions made in organizations is affected not only by biases and beliefs but also by the validity of the information coming through the organization's communications system. By this system, we mean the processes through which requests for information reach the point of collection and the information is returned to the requester.

Unfortunately, the process of transmission is rough. For example, Wiksell (1960) claims that 70 percent of communication is bound to be distorted, misunderstood, rejected, forgotten, or disliked. Mac-

Crimmon (1974), in studying team decision making by policymakers, found that 90 percent of individuals showed inefficient and mutually inconsistent communication heuristics.

What factors have been proven to increase distortion? The literature offers strong evidence that information processing is dependent on the characteristics of the processor (such as background and personality), the processor's relative status and location (internal or external to the organization), the structure of the organization, the credibility of information sources, and the content and relevance of the message (O'Reilly and Pondy 1980). Here we emphasize three sources of distortion: the organizational structure, interpersonal variables between sender and receiver, and the content of the message.

Several studies have linked structural variables to dysfunctional communication (Athanassiades 1971; Hage, Aiken, and Marrett 1971; Wilensky 1967; among others). Specialization and differentiation create problems of coordination and information transfer; formalization adds links to the communication chain and thus increases the probability of distortion; centralization impedes information flows from the top down and the bottom up, thus inhibiting and delaying the delivery of large amounts of information.

The interests and concerns of the sender-receiver dyad determine whether information is added, modified, or eliminated—in a word, distorted—before transmission. In general, if the sender perceives a high material or psychological cost, the probability of distortion is high. Kelley (1951) and Read (1962), for example, found the subordinate's accuracy in upward communication to be inversely related to desires for advancement, leading to distortion of any matter that could harm the subordinate's career. Athanassiades (1971) investigated the relationships among distortion of upward communication, subordinate's needs, and organizational climate and found that distortion of upward communication was negatively related to the communicator's level of security and positively related to achievement needs and degree of hierarchy in the organization. Allen and Cohen (1969) found in studies of communication patterns that high-status individuals communicate more frequently with each other than low-status people do and that individuals with low status communicate more with those with high status (though often without reciprocation) than with each other.

O'Reilly (1978) showed that a sender's mistrust of the receiver leads to the suppression of unfavorable but relevant information and an increased flow of favorable but possibly irrelevant information. Functional location in an organization also fosters distortion. For example, Redfield (1958), Zander and Wolfe (1965), Strauss (1962), Dearborn and Simon (1967), and others have found that depending on the area they work in, people have different perceptions of the same information, even if it has the same source—and they react differently to this information as well.

Rosen and Tesser (1970) tested the common-sense notion that people are more reluctant to convey negative information than positive information. Interestingly, the study was structured to rule out certain variables as determinants of this difference: the recipient's prior behavior to the communicator, their relationship, and obvious rewards and punishments resulting from the transmission. The subjects transmitted information regardless of whether they saw it as good or bad; however, the desire to transmit good information was directly related to the urgency of the message. When the news was bad, the relationship was inverse, and senders showed clear signs of stress. O'Reilly and Roberts (1974) found that repression of information differs according to the direction of flow; that favorable information moves upward much faster than unfavorable information; that trust has a bigger impact on upward transmission than on downward or lateral transmission; and that the amount of favorable information passed is directly correlated to the sender's belief that the receiver has a lot of influence over the sender's future.

Finally, Huber (1980b, 1980c) has extensively reviewed the literature and suggested a set of propositions on the logistical determinants of routings, delays, modifications, and summarizations in organizational information processing (see Table 2-3). These four categories of characteristics were established: (1) the sending unit, (2) the receiving unit, (3) the sender-receiver dyad, and (4) the message.

From this review of the literature, we can conclude that distortion is the rule rather than the exception. Blockage and distortion may be intentional (fostered by aspirations to climb the hierarchy) or unintentional (caused by cognitive limitations), but they are present in all communications, upward, downward, and lateral.

Is there a way to eliminate distortion? While total eradication may not be possible or even desirable, two strategies can alleviate the problem: (1) creating an organizational climate that promotes free and accurate communication and (2) developing an organization-wide information system with complex routing capabilities that depend exclusively on message content (Galbraith 1973).

Table 2-3 Huber's Propositions on the Logical Determinants of Organizational Information Processing

	Probability or Extent of the Sending Unit			
	Routing Message	*Delaying Message*	*Modifying Message*	*Summarizing Message*
Sending unit characteristics				
Goal attainment (increase)			+	
Stress (decrease)			+	
Work load	–	+	+	
Receiving unit characteristic				
Work load (perceived by sending unit)				+
Units relationship				
Discretion allowed			+	
Power and status	+			
Past history, patterns	+			
Distance (number of links)		+	+	+
Message characteristics				
Timeliness		–		
Actual vs. desired content			+	
Ambiguity			+	
Relevance (perceived)	+			
Adverse effects (bad news)	–			
Cost of transmission	–		–	–
Savings in transmission			+	+

+ Means that the probability or extent is positively related to the factor.
– Means that the probability or extent is negatively related to the factor.

Models of Decision Making

Now that we have sampled the biases and limitations on rational behavior, we turn to several models of behavior on the part of decision makers.

In order to support the decision process of policymakers, we must understand how they construct or perceive reality. Several models have been proposed for this, each with its own definition of rational behavior. The classic concept of rationality includes these factors: consistency, objectivity, and transitivity (if A is better than B and B is better than C, then A is better than C). We distinguish relative from absolute rationality in behavior because an act may make sense in its context but not in another. Another important distinction is made between global rationality (strategies needed to achieve the objectives of the entire organization) and local rationality (strategies to achieve the goals and objectives of an individual or unit of the organization).

When we speak of rational behavior, we should remember that our actual goal is not to make decisions but rather to support the process of making decisions. Policymakers are change agents, not just decision makers, so the steps before and after a decision are as important as the actual choice of action. Preparatory steps include creating tension for change, understanding the positions of the constituencies, and developing social support for a chosen action. Steps after the decision include naming the change monitor and identifying the monitoring methods. Therefore, the mission of a good policy information management system is broader than just supporting a choice. Analysts must understand not only how policymakers think but also how the decision process will be implemented in their environment. This is why we are reviewing three models of organizational decision making—rational, administrative, and political.

The models of rationality are described according to their position on a spectrum from normative to descriptive. Normative models prescribe behavior, while descriptive models merely represent what is found in the world. Normative models may or may not be based on evidence.

The implications for policymaking of each model are also discussed to elucidate them. These models are compared in Table 2-4. Later in this chapter we discuss the implications of each model for the design of a policy information management system. This section concludes by discussing a compromise among the rational, administrative, and political views and giving an example of how this strategy would apply to making a choice.

The rational model

The rational policy model of decision making is based on the logic of optimal choice—choices that would maximize value for the organization. The model is highly normative and, in most situations, highly idealistic. The policymaker is assumed to be an objective, totally in-

Table 2-4 Comparison of Organizational Decision-Making Models

	Rational (Economic)	*Administrative (Bureaucratic)*	*Political (Incrementalism)*
Goals and references	Consistent across organizational units	Reasonably consistent	Pluralistic inconsistent
Objectives	Efficiency and growth	Stability and predictability	Power; survival; avoid costly mistakes
Decision process	Orderly; systematic; analytical; proactive	Embodied in programs, SOP; simplifying heuristics; minimizing effort	Disorderly; push and pull of interests; judgmental; reactive
Solution approach*	Optimization	Precedent; habit	Incrementalism; bargaining
Choice criterion*	Maximization of value	Good enough	Compromise
Number of constraints*	Large, static	Few, dynamic	Number depends on the number of interest groups
Number of alternatives*	Exhaustive	Limited by aspiration levels	Restricted by political plausibility
Examination of alternatives*	Comprehensive	Sequential	Haphazard

*Information processing requirements.

formed person who would select the most efficient alternative, maximizing whatever amount and type of output he or she values. Simon (1976) summarizes the rational choice process:

1. An individual is confronted with a number of known alternative courses of action.
2. A set of possible consequences is attached to each alternative.

3. The individual has a system of preferences or utilities that permits him or her to rank the consequences and choose an alternative.

There is no empirical support that these three phases are actually used. In reality, policymakers seldom have the time or money to analyze all alternatives or envision all consequences. If rationality were pervasive among members of an organization, it would appear as a coherent and rational policymaking entity that maximizes its attainment of a unique set of goals and has no internal conflict. In other words, a rational decision process implies a rational organization, meaning it has (1) centralized power, (2) harmony and consistency of goals across boundaries, and (3) members who are objective, fully informed, and inclined to choose alternatives that maximize the common good of the organization.

The rational model represents a sanitized vision of how organizations could make decisions. In contrast, organizations seem more like complex groups of coalitions fighting for shares of limited resources and using multiple sources of information with varying reliability. Individuals within them have widely divergent perceptions and goals and act to maximize their own gains, not necessarily those of the organization. Because of this disparity, we prefer to accept the rational model primarily as a benchmark for comparing the remaining two organizational decision-making processes. In searching for a more realistic description of how organizations make decisions, we turn to the "satisficing," or administrative, model.

The administrative model

The quest for a more realistic description of organizational decision making produced a mutant called the administrative model. This model sees decision makers as people with varying degrees of motivation who are besieged by demands but have little time to make decisions and thus seek shortcuts to acceptable solutions. The decision maker is seen as "rationally bounded," meaning the ability to behave rationally (in the classic sense discussed above) is limited by the task environment and personal motivation.

Simon (1957) proposed that under the administrative model a decision maker does not try to optimize but instead "satisfices"—treats objectives as loose constraints that can tighten if there are many acceptable alternatives. While optimization would require choosing the alternative with the highest value, satisficing requires finding the first alternative with an acceptable value. For example, if you listed your house for sale for $250,000 and had 35 offers, you could choose with either method. With the rational method, you would determine which offer had the highest value in terms of conditions and price. With the satisficing model, you would accept the first offer that met your lowest acceptable price. Satisficing may lead to a reduced decision quality, but it saves time and effort. Satisficing is a dynamic construct: the aspiration levels of the policymaker and the number of alternatives determine what is a "feasible, good enough solution." In explaining the concept, Simon (1957) wrote:

> In the real world we usually do not have a choice between satisfactory and optimal solutions for we only rarely have a method of finding the optimum. . . . We cannot within practicable computational limits, generate all the admissible alternatives and compare the relative merits. Nor can we recognize the best alternative, even if we are fortunate enough to generate it early, until we have seen all of them. We satisfice by looking for alternatives in such a way that we can generally find an acceptable one after only moderate search.

Miller and Starr (1967) point out that satisficing is an appropriate (i.e., rational) strategy when the cost of delaying a decision or searching for further alternatives is high in relation to the expected payoff of the supposedly superior alternative. They suggest "it is always questionable whether the optimum procedure is to search for the optimum value."

Satisficing does not preclude contemplating many alternatives, but it implies examining them sequentially with no attempt at comparison. Using a few basic principles that reduce complex decisions into a series of "go/no go" judgments, the policymaker stops considering actions after finding an acceptable one. This strategy obviously requires much less effort than the rational model.

Satisficing is usually based on heuristics, which are rules of thumb derived from experience. Under the rational model, every problem is solved with the same process, doing comprehensive information processing and then maximizing value. In contrast, under the administrative model, choice is always exercised on partial information processing and defining a good enough solution. The definition of *good enough* depends primarily on the policymaker's experience.

March and Simon (1958) summarize how bounded rationality forces simplifications in the decision process:

1. Alternatives and consequences of action are discovered sequentially through searching.
2. Existing problem-solving strategies are the strategies of choice in recurring situation.
3. Each program deals with a narrow range of situations and consequences.

In other words, policymakers deal with problems in piecemeal fashion, tending to handle one issue at a time and using an established repertory of problem-solving strategies.

March and Simon (1958) also suggest that the search for programs to reach an organizational goal does not follow objective logic but rather is limited by "psycho-logic." The policymaker attends to variables that are under his or her control. If that attempt fails, he or she directs attention to variables outside organizational control. If this fails, he or she waters down the criteria to allow a satisfactory strategy to emerge. Using the above example of selling a house, if you don't get any calls, you are likely to lower your price (dilute your criterion). People search more thoroughly for acceptable alternatives as each successive effort is frustrated. While you are lowering your expectations, you are also likely to intensify your search efforts.

The implication of having rationally bounded decision makers in organizations is that organizations cannot be seen as entities. Rather, problems are broken down and assigned to specialized units within the organization that develop their own priorities and goals. These goals, sometimes termed "subgoals," may not agree with the organization's overall goals. Cyert and March (1963) call this phenomenon "local rationality." Allison (1971) suggests that organizations are constellations of loosely allied units, each having a set of standard operating procedures and programs to deal with its piece of the problem. As time passes, these units become more distinct and their subgoals more entrenched. These divergences are enhanced by increasingly distinct perceptions of priorities, information, and uncertainty and reinforced by recruitment, rewards, and tenure. When these tendencies are very strong, the "loose alliance" of organizational units breaks down into "organized anarchies" (Cohen, March, and Olsen 1972; March 1978). In the extreme case, coalitions are created with conflicting interests. This leads us to the political model of rationality.

The political model

In contrast to the rational model, players in the political model (often referred to as incrementalists) do not focus on a single issue but on many intraorganizational problems that reflect their personal goals. In contrast to the administrative model, the political model does not assume that decisions result from applying existing programs and routines. Decisions result from bargaining among coalitions (which may depend on the issue). Unlike in the previous models, power is decentralized. This concept of decision making as a political process emphasizes the natural multiplicity of goals, values, and interests in a complex environment. The political model views decision as a process of conflict and consensus building and decisions as products of compromise. The adage "Scratch my back and I'll scratch yours" is the dominant decision-making strategy.

When a problem requires a change in policy, the political policymaker considers a few alternatives, all of them similar to existing policy. Janis and Mann (1977) note that "the incrementalist shows preference for the sin of omission over the sin of confusion."

This perspective was first introduced in the incrementalism theory presented by Lindblom (1959). He pointed out that decisions tend to be incremental—that policymakers make small changes in response to immediate pressures instead of working out a clear set of plans and a comprehensive program. While we distinguish incremental adjustment from the administrative model, incrementalism can be seen as the simplest or most extreme form of satisficing.

The incremental approach allows policymakers to simplify the search and definition stages. Incremental decision making is geared to alleviate shortcomings in present policy rather than consider a superior, but novel, course of action. Since no effort is made to specify major goals or find the best means of attaining them, ends chosen are appropriate to means that are more or less available.

Lindblom (1959) labels incremental change "the art of the possible." His "muddling through" approach to decision making argues against the appropriateness as well as practicality of the rational model. He argues that problems are not solved in a single step but are repeatedly attacked with partial solutions.

In the political model, the actors have different perceptions, priorities, and solutions. Because actors have the power to veto some proposals, no policy that harms a powerful actor is likely to triumph even if it is objectively "optimal." Indeed, the incrementalist approach adheres to the first adage of conservative medicine: First do no harm. Incremental theory says that policymakers attempt to avoid mistakes in the following ways—rather than achieve grandiose goals:

1. Choices are made in a given political universe at the margin of the status quo.
2. A restricted variety of policy alternatives is considered, and these alternatives represent incremental changes.
3. A restricted number of consequences are considered for any given policy.
4. Analysis and evaluation are undertaken by several interest groups, and therefore the locus of these activities is fragmented and disjointed.

Studies by Wildavsky (1967) and Hoos (1973), among others, have found that incrementalism is the actual form of decision making in many government decisions. However, some researchers have argued that in crises, where transcendental decisions are required, incrementalism is inappropriate, and policymakers revert to a process that has features of the rational process. We present this point of view next.

A compromise: The mixed scanning model

The mixed scanning strategy is based on the observation that policymakers recognize that important decisions deserve relatively great amounts of care and attention. The basic presumption behind this strategy is that policymakers do not or cannot always use the rational model but they sometimes wish to do so. Etzioni (1967) proposes that when policymakers operate as "mixed scanners," they overcome their limited capacity to process information by classifying decisions as "fundamental" or "minor." This conserves time and energy by allowing the policymaker to examine only fundamental decisions and treat other choices with the incremental approach.

For comparison, we can say that all decisions under the rational model are "fundamental" and subject to comprehensive analysis. Under the satisficing model, all decisions are minor, and policymakers search for the first alternative that satisfies the existing criteria. Incrementalism is a form of satisficing that also deals with every decision in the same way. The mixed scanning strategy, on the other hand, is a contingency approach; since decisions have different natures and importance, they should be approached differently. Quasi-rational approaches are used only to select alternative solutions for fundamental decisions, such as dramatic policy change, and a satisficing approach is used thereafter. The point of the mixed scanning strategy is to focus scarce decision-making resources on decisions that are crucial to the policymaker.

While Etizioni proposes mixed scanning primarily as a normative strategy, he suggests that it has some descriptive power as well. As an example, he points to changes in the U.S. defense budget before and after the Korean War, and alterations in the U.S. space budget before and after *Sputnik*. Major changes were instituted in the funding immediately after the event, then marginal change became the rule.

In summary, the choice of modes in mixed scanning depends on the gravity of the situation. For critical decisions, a quasi-rational approach is used; otherwise, an incremental decision process is used.

We have said that these models of decision making have different definitions of rational behavior. To underscore these differences, we now discuss how each model would approach a multicriteria choice decision.

An example: Simplistic strategies of choice

Choice is an appropriate task for comparing models of rationality because it usually requires us to combine large amounts of information while evaluating alternatives. Comparing alternatives over several dimensions requires substantial cognitive effort. It is in precisely such situations that each model of decision making proposes a different strategy.

For choices, the rational model dictates the use of a compensatory strategy.[3] Compensatory models require us to evaluate every alternative on every relevant attribute. A common compensatory model is the linear one:

$$Y_i = (W_1)(X_1) + (W_2)(X_2) + \ldots + (W_i)(X_i) + \ldots + (W_n)(X_n)$$

where Y_i represents the score of alternative i, X_i represents attribute i, and W_i represents the relative importance of each attribute. Compensatory models need a great deal of data and a substantial cogni-

[3]In this model, alternatives that score poorly on one criterion are given a chance to compensate by scoring well on others.

tive effort by the policymaker. Because of these stringent requirements, individuals often opt to evaluate alternatives with simpler strategies. Indeed, the most common heuristics used by individuals in choice situations are noncompensatory, of this general form:

$$Y_i = (W_1)(X_1) \times (W_2)(X_2) \times \ldots \times (W_i)(X_i) \times \ldots \times (W_n)(X_n)$$

where Y_i, W_i, and X_i are as before. The reason such a model is simpler is that noncompensatory strategies do not allow trade-offs among criteria for alternatives. For example, if an alternative scores 0 on any criterion in the noncompensatory model, it would get a total score of 0 and be discarded from consideration. This is a severe limitation since an alternative can be disqualified on the basis of a single, trivial criterion. However, noncompensatory models require much less effort than compensatory ones because we simply compare alternatives on the basis of one criterion at a time and eliminate alternatives that lack a minimum value on that criterion.

There are four main types of noncompensatory strategies: (1) conjunctive, (2) disjunctive, (3) elimination-by-aspects, and (4) lexicographic.

In the *conjunctive* heuristic, the policymaker sets cutoff points for each criteria; any alternative falling below the cutoff is eliminated. In the *disjunctive* strategy, alternatives are evaluated on all attributes, and there are no cutoff points. That is, the policymaker accepts a low score on any one dimension provided it is balanced by another dimension with a very high score. If it is not balanced, the alternative is eliminated. The conjunctive and disjunctive models are satisficing choice strategies. With these strategies, decision makers can concentrate on a handful of dimensions that have worked in the past to rapidly discard alternatives from consideration.

The *elimination-by-aspects* strategy, which can be the most erratic of the noncompensatory strategies, requires the least effort. At each stage of the comparison, a criterion for all alternatives is selected randomly. The best alternatives on that criterion are retained; the others are discarded. The process continues until the best alternative remains. If all criteria are used and several alternatives remain, they can be considered equal in value (Tversky 1972).

The *lexicographic* strategy is also a satisficing, sequential strategy, but it requires a somewhat greater cognitive effort by the decision maker. The policymaker first orders the set of relevant criteria by relative importance, then compares the alternatives on the basis of the most important criterion. One alternative is chosen if it is superior on the first criterion. If two alternatives are considered nearly equal on the most important criterion, they are compared on the second-most important criterion, and so on.

The mixed scanner would use a compensatory strategy to make a fundamental choice and would use noncompensatory strategies afterward to make the minor decisions on the same topic.

Organizational Change and Continuous Quality Improvement[4]

In recent years, a number of important changes have come upon the health care field. Examples include practice guidelines, which have been offered as a way to reduce inappropriate care and total quality management, which has created substantial interest because of its potential to improve quality and, at the same time, control costs. The implementation of these changes has often been approached from a rational perspective. If we just give physicians information about how medical care should be practiced, they will change their behavior. If leaders simply announce their acceptance of continuous quality improvement (CQI) and train their employees in CQI principles, the organization will change. Unfortunately, the adoption of both innovations has been spotty at best.

This section will introduce a model of organizational change that presents an alternative to the rational philosophy and apply it to the implementation of continuous quality improvement in a health care organization. First, we will provide a brief introduction to total quality management, the umbrella philosophy within which CQI resides. Next we will review the primary principles of CQI. The change model will then be introduced and applied to the implementation of CQI.

TQM is composed of three components: continuous quality improvement (CQI), quality planning, and quality of daily work life (CDWL). Continuous quality improvement is typically the first component to be implemented in organizations and will be the focus of this discussion. CQI is intended to help an organization adopt an ongoing commitment to making small yet significant improvements.

---

[4]Kris Bosworth (a colleague of Dr. Gustafson at Indiana University) has been an important contributor to the conceptualization of this model of change. Her insights are most appreciated.

The argument is made that Americans are obsessed with innovation. They love to develop and implement new ideas but they don't "mind the store" to be sure the innovations work as intended and are maintained once implemented. As a result there is an initial step function improvement in performance and then a deterioration as people go back to the old way or just don't continue to use the innovation as intended. Sometimes things end up worse than before because there is no ongoing commitment to make sure the innovation is fully adopted and the bugs worked out. CQI provides the climate in which continued small improvements on existing processes, as well as innovations, are ensured.

Quality planning is often the next addition to a CQI organization and is aimed at, among other things, instituting innovation so that big improvements can be made, but with CQI being present to be sure the organization can hold the resultant gains. QDWL is the final phase of CQI where all employees naturally operate within the concepts of CQI. It becomes a way of life to use the tools and concepts of CQI in everything they do, and not just as part of quality improvement teams.

Our focus will be on how to use the change model to plan the implementation of CQI. But first we will introduce CQI principles.

CQI principles

In this historical context, health care is a late comer to CQI, with the earliest transformations beginning less than ten years ago. Organizations such as NKC and HCA began to adopt CQI in the mid-1980s. But in the last few years, interest has grown rapidly to the point where networks of hospitals, clinics, and HMOs have formed to facilitate the transformation to CQI. Partnerships among several hospitals and several major employers in single communities, such as in Kingsport, Tennessee, have begun to apply CQI to the objectives of promoting communitywide improvement in health. Educational programs have developed to introduce health care providers to CQI and to promote continuing learning of CQI principles and tools. Research programs to foster the development of improved methods for transforming organizations to CQI and monitoring performance can now be found in academia and in health delivery organizations. These and other signs signal the ascendency of CQI to an important position in U.S. health care.

However, while many now espouse CQI as the foundation for

managing their organizations, few health care organizations can legitimately claim to have made that transformation. It is this lack of good models that may ultimately cause the failure of efforts to make CQI a long-term management strategy rather than a fad. The reasons for the scarcity of CQI organizations are many, but it comes down to the length of time and enormous effort required to successfully transform an organization. A review of some of the key CQI principles may help explain why. These principles have been presented by many people including Deming, Juran, and others. We will use the translation of Deming's principles suggested by Batalden (1988).

Quality mindedness. This is probably more a cornerstone of CQI than a principle. In essence it argues that an organization must place quality ahead of everything else it does and the bottom line will take care of itself. As an example of how difficult it is to fully transform to CQI, consider just a few of the implications. A CQI organization should be willing to deemphasize short-term financial performance in favor of the long-term, to design services that work right the first time and every time, and to be obsessed with designing services that will delight its customers.

Customer mindedness. A CQI organization knows why it exists, who its customers are, and what they need. On the surface this sounds simple. But when an organization is obsessed with serving customers that it has carefully chosen to fulfill its mission, it doesn't just ask a breast cancer patient, "How did you like our nurses?" It also asks: "Do you need help telling your daughter that you had breast cancer? Do you need help selecting a surgeon? . . . Understanding complex terms? . . . Coping with the possibility of death? . . . Understanding what was happening to you? What needs were most important? How did we do?" A CQI organization commits itself to not only identifying those needs, but also to meeting those needs.

A CQI organization also recognizes that its efforts to understand its customers extends not only to patients, but also to families, employers, physicians, and employees (its internal customers). The effort to identify customer needs is a complex, comprehensive, ongoing process with enormous rewards.

Leadership constancy. A CQI organization sees this as a long-term pursuit that can be successful only when top management views quality as their only job. It means creating an environment where everyone within the organization can learn and grow, including the leadership, an environment where an obsession with quality permeates the daily work life of the organization. This doesn't mean that a CEO should immediately announce the CQI transformation to the whole organization. (Though this might work in a rational world.) Rather, the CEO begins to create a climate where CQI can thrive, and provides the resources to carefully test and document its effect.

Process mindedness. A CQI organization recognizes that 85% of the problems it has are caused not by people but by the processes within which they function. Thus, throughout the organization there is a commitment to understand the way work gets done, to understand and reduce what leads to variation in the way service is delivered. A natural implication of a process focus is seeing employees and suppliers as partners, not problems. Things begin to change. For example, a contract is awarded not on the basis of price alone and long-term partnerships with a small number of suppliers are established.

Employee mindedness. A CQI organization believes its employees understand processes better than anyone else and, given the chance and training, are in the best position to improve those processes. Employees throughout a CQI organization learn the principles and tools of CQI: statistical process control, group facilitating, flow process charting, etc. Everyone has the right and responsibility to improve processes, and everyone is equipped to do so. And because employees are provided the freedom to improve and the tools to do it, they have the intrinsic motivation to work for quality.

Statistical thinking. In a CQI organization, wise collection, analysis, and use of data become a way of life. Data is used to reduce uncertainty, prove the existence of an improvement opportunity, monitor the success of an improvement attempt, and so on. This means understanding of variation, separating out those causes that are due to the existing process(es) (common causes) from those due to individual events (special causes), and methods for using facts to guide decisions.

Plan–do–check–act driven. A CQI organization believes it must operate scientifically—that when improvements are considered, they are carefully tested to understand their effect. When the validity of the improvements has been demonstrated, steps are taken to ensure that these improvements are fully implemented and processes are not allowed to slide back into previous, less-effective styles of operation. To check and hold the gains, data is collected and statistical process control is used to monitor performance over time.

Innovativeness. CQI organizations are committed to being the best in their field and they are committed to doing it one process at a time. This implies a strategy to identify organizations both within and outside the industry that operate with outstanding versions of the processes similiar to those they are seeking to improve. CQI organizations study those processes to understand what makes them so successful and build those features into their own processes. They continue to compare their processes against the best and are happy to share with others who also benchmark.

Regulatory proactiveness. CQI organizations are regulated just like anyone else. Very often they find the regulatory environment to be a hostile one, but they view their role as being a leader in promoting improvements in that environment. They see CQI serving as the basis for a regulatory system that promotes improving quality. As a result they advocate the inclusion of CQI principles as the foundation of the regulatory system. Regulators demand evidence that organizations are becoming progressively more customer-minded, process-minded, and so on.

Any one of these principles would be difficult to implement one at a time, even in a small organization. When one attempts implementation of all at once, the task becomes daunting. Yet this adoption of all principles is exactly what is required of a CQI organization. It is not surprising then that the most common question asked at a CQI training program is not what is CQI, but how it is implemented. How do I transform my organization to CQI? How do I change my organization? To begin to understand how to answer that question, one needs to examine the theory of organization change.

A change model for implementing CQI

The implementation of CQI is similar to implementing any innovation. The principles may be sound but the implementation is very difficult and far from a rational process. A number of theorists have studied change in individuals (Bandura 1977; Fishbein et al. 1980;

McQueen 1984; Strecher 1986) as well as in organizations (Utterbach 1971; Freeman 1979; Cooper 1982; Maidique 1983; Delbecq and Mills 1985). They find that successful change occurs when a combination of political, social administrative, and rational processes are put in place. We have summarized them in the model shown in Figure 2-2.

Creating a tension for change. In this model, a program of change begins with the identification of two sets of key actors who will need to get involved if successful transformation is to take place. The first group is the innovators (those people who have a history of trying new things and a personality consistent with that required to lead CQI—a penchant for involving others, networking, and a commitment to quality). The innovators will likely be quick to see the merit of CQI and will want to try it.

The second set of actors are the opinion leaders; those people who are open to change, approach it cautiously, and who are highly respected by most people in the organization. While they are not likely to adopt CQI immediately, they need to be involved from the start because their opinions will strongly influence the rest of the organization.

Both of these groups need to *believe* the current status of the organization must change in order for it to survive and prosper. There are a variety of ways to create this tension, including some rational approaches, such as providing literature on problems with the health care (e.g., small area analyses and articles by major employers criticizing the existing system). But few people will become sufficiently dissatisfied unless they are already experiencing some problems with the current situation. In fact, some people would argue that CQI is best adopted when an organization is on its knees, when the tension for change is already highly focused. Thus, the change agents (the people who want to see CQI become the prevailing management philosophy of the organization) must take the time to understand the issues that are already keeping the innovators and opinion leaders awake at night. The stressors that can be alleviated with CQI are the tensions for change that need to be addressed by the change agent. It may seem a bit Machiavellian to approach the tension issue this way, but the fact is that people are already worried about a lot of things. Change agents need to understand those worries because CQI must help relieve those tensions and not create more. Innovators and opinion leaders need to see how their existing

Figure 2-2 A Model of Organizational and Individual Change

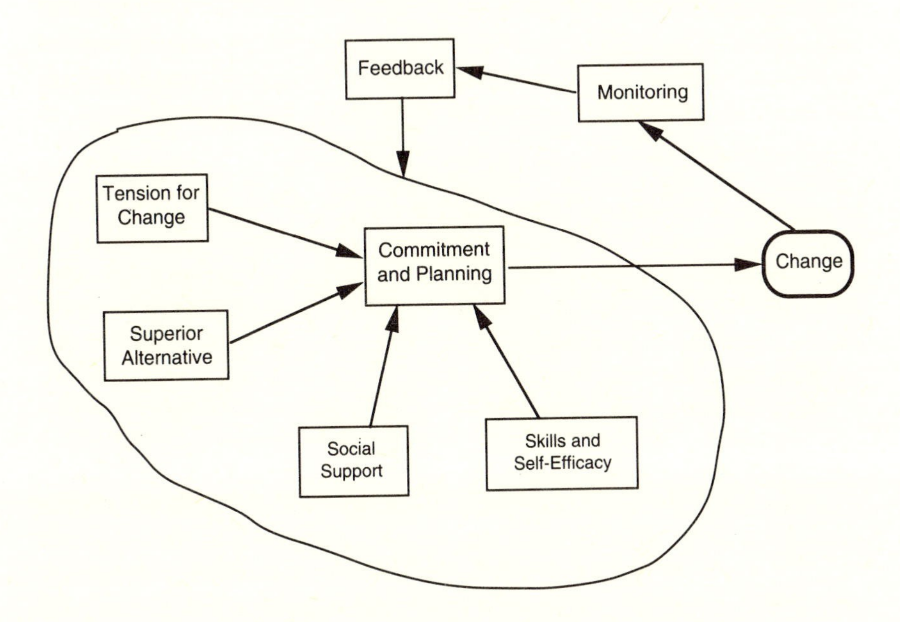

tensions can be reduced by the proposed change. And (as is addressed in the next paragraph) to see that CQI offers them hope for the future.

Seeing CQI as a uniquely superior alternative. Successful change is more likely to occur when key actors see the new way as both superior to the current way and unique in identifiable ways. The transformation to CQI thus requires that innovators and opinion leaders believe that CQI will not only correct important weaknesses and open new opportunities for the organization, but also meet their own professional needs.

People can be rationally told about the superiority of CQI and there are many training programs, texts, papers, and video tapes to do just that. But like the proverbial Missourian, most people need to be shown that it works. A small pilot study carried out quietly in one corner of the organization, using CQI principles and leading people to say, "Wow, how did you do that?" will have a powerful effect on opinion leaders. That is why choosing an innovator to apply the principles in one part of the organization can be such a powerful force for demonstrating the superiority of CQI.

Surveys to discover customer needs can be particularly effective as a means of demonstrating the superiority of CQI, but they must address needs, not wants or expectations. People *expect* a new car to start and a doctor to be qualified to deliver good care. People *want* good gas mileage and minimum waiting time for a nurse to respond to their call. People *need* help in living with the prospect of death from breast cancer and with coping with the side effects of treatment, even though they may not expect the health system to provide it. If the change agent were to show how CQI naturally leads to a new view of customers expectations, wants, and needs, the innovators and opinion leaders would begin to see how CQI offers new insights into management that are not provided with the traditional management culture.

CQI is a very complex philosophy of management, although it may not appear so on the surface. It would be nice if people were to take the time to slowly study the process and rationally reach a conclusion about its merits. In fact, most people don't have the time to come to such an appreciation of a new concept. If it is that complex, they will reject it just because they are so filled with things to do already. There needs to be a very simple way to communicate the concepts; a way that lets the key ideas jump out at the person. CQI principles provide that vehicle because their ideas are so graphic: Concepts such as "commit to quality and the bottom line will take care of itself," "reduce the number of suppliers and don't award contracts on the basis of price alone," "quality is identifying customer needs, designing services to meet those needs in ways that will delight the customer, and producing those services right the first time and every time; nothing more and nothing less," are so intriguing that they capture the attention and help people remember the ideas.

Each of the main principles of CQI serves to distinguish CQI from traditional management and are easy to grasp at a surface level. When innovators and opinion leaders are able to understand and communicate the uniqueness of CQI and see how it is superior, they are more likely to seek its implementation (in the case of innovators) or at least be prepared to observe and evaluate the effort to decide whether to support it (in the case of opinion leaders).

Building social support. Successful change is more likely to occur when those attempting the change believe the people they respect are also changing, want them to change, and will help them succeed. Moreover, they need to believe that if they fail, they fail not only themselves but also their social support group. Social ties with people committed to the change need to be established within and between organizations. Ties can extend even beyond their own industry.

In a rational world there would be no need for social support. People would implement the things they believe in. But in reality, social support is essential to continued success. The implementation of CQI is very difficult. There will be many who question its staying power and who believe it is a naive way to manage. The support of others is needed, even in a completely friendly environment, because each CQI principle seems simple at first but is very complex. (For example, what does it really mean to be data driven? How do you know what data to collect and when to collect it? How much can you continue to collect without becoming overwhelmed in data?) Textbooks have only so many answers. People need others to turn to for answers or at least to talk out the issues.

Fortunately, it is possible to develop social support systems to meet those needs. They range from study groups within the organization, to meetings of people within a community, to the participation in professional CQI organizations and networks. Some networks contain just hospitals and other health care organizations who meet quarterly to discuss issues of common concern. Others are networks of health care providers and manufacturers meeting to share experiences. All serve the function of helping each other through difficult times.

Other types of social support promote the implementation of a particular CQI principle by an individual or group. For instance, if people are trying to become more statistically minded in their work, others who want that to happen could agree to simply ask these people questions such as, "How do you know? What evidence do you have?" The social support group can take the responsibility for offering such challenges.

There is an important side effect of social support efforts. Organizations (and individuals) no longer see themselves as alone, but as one of many. This reinforces their belief in the practicality and success potential of the transformation.

Skills and self-efficacy. It is one thing to develop the skills needed to implement the change. Rationally that is all that should be needed. But Bandura (1977) and others have shown that it is also essential that one believes they are capable of implementing those skills needed to carry out the steps of the change.

A CQI transformation requires that employees be trained in the concepts, tools, and facilitation skills of CQI. Programs to do this are available from a number of outside sources. Over time, this training capability needs to be developed within the institution. In fact, one of the most important may be just-in-time training—the training of quality improvement teams just when they reach the stage where they need to employ a certain aspect of CQI. Management engineers need to see this as their important role in CQI. They need to abdicate their role as the problem solver to the quality improvement team and take on the important role of training and coaching both the teams and the organization in the CQI tools and philosophy. (One of the most important research objectives for industrial engineering in the next few years should be to transform progressively more of its tools to formats that will allow every employee to appropriately use them in quality improvement efforts. This suggests the need to develop very easy-to-use software that will expand the IE tools that are part of CQI beyond flowcharts, or cause-and-effect diagrams, to tools such as simulation, utility modeling, and workplace design, to name just a few.)

The opportunity for practice and feedback must be provided along with the training. People applying CQI for the first time must know what they are doing right and wrong. This feedback can come as part of the training, but it also calls for demonstration projects where people can see how CQI has been applied and how well it has worked. The innovators first trying CQI need to have their efforts observed and documented so others may learn and gain confidence in this innovation.

Commitment and plan to change. When a tension for change, superior alternative, social support system, skills and self-efficacy (through pilot studies and practice) are in place, the organization can begin to think about a wide-scale implementation of the change. A commitment to change can be made and a plan put in place to do so. The role of the opinion leaders observing the pilot test being run by the innovators is to help reach the decision that is now appropriate to extend CQI implementation and to help formulate the strategy to do so.

The pilot tests have several other benefits. One is to be sure the bugs are worked out before wide-scale implementation takes place.

This means that the observers should be considering how to modify the implementation of CQI as they watch the pilot test. They need to think about how the training of the workers should take place, how to select the few projects they will charter at a time and how to present CQI in a way that can be quickly understood and differentiated from, yet integrated with, current practice. This integration is particularly important in the health field where quality assurance and management engineering activities can play such important roles in identifying the quality improvement projects and in training quality improvement teams. What roles will management engineering and quality assurance play in CQI implementation?

A management strategy and schedule will be needed to

1. Examine the corporate mission to be sure it is consistent with CQI
2. Select the customers the organization wishes to serve, discover their needs, and put in place a program to meet those needs
3. Develop a vision showing what the new organization will be like and contrasting it with the current version
4. Decide what training will be given to which employees and what training will be ahead of time and what will be just-in-time
5. Decide how projects will be selected, how specific to be in directing the team, what feedback will be required (and when) from the teams to the management committee
6. Decide what resources will be offered and what expectations will be put in place about benchmarking with other organizations on specific projects and what networking will be initiated with other organizations on CQI

While these plans should be developed for the CQI in general, it will be important to develop similar plans for ensuring each CQI principle is properly implemented. In the "customer-mindedness" principle, how will a *tension for change* be created around the way customer needs are considered? How will CQI strategies for involving customers be shown to be *superior*? Who will provide *social support* to those using CQI to involve customers and how will it be provided? What *skills training* will be provided and how will *self-efficacy* in the ways of involving customers be developed? What specific *plans* need to be put in place for involving customers?

Beginning the change. Organizational transformation takes time. It is important to begin with a few carefully chosen pilot projects in which the innovators are given the resources, skills, responsibility, and authority to try out CQI in their segment of the organization with opinion leaders watching. Once a record of success has been established, the transformation can slowly spread to the opinion leaders and finally be implemented corporate-wide. A few pilot projects, in which early success is demonstrated and the bugs in the rollout are identified and corrected, leads more directly to success than an immediate wide-scale transformation with no prior testing.

Monitoring and feedback. Successful change is marked by an intentional program to monitor the attempts to change, learn from the mistakes, and improve the rollout process. One aspect that seems essential is a process to allow those who are expected to implement the change to give feedback on the strengths and weaknesses of the effort without fear of retribution. Processes need to be in place that will allow anonymity and ensure that the feedback influences future improvements in the process. Surveys, interviews, and external observation all provide means for learning how to improve. The resulting feedback can lead to changes in mechanisms by which each of the other aspects of the change model will be carried out. Tension for change, presentation of CQI as a uniquely superior product, creation of social support, development of skills and self-efficacy, commitment, planning, and feedback must all be continuously improved as the CQI transformation moves ahead.

Conclusion

We have established that policymakers' decision making does not meet the standards of classical rationality, and given the typical environment of tight deadlines, multiple interest groups, and many issues, we can understand the need for shortcuts.

Because the classical model of rational decision making is not useful in explicating how policymakers think, it is an unreliable guide for designing decision aids. The administrative and political models describe decision makers' behavior more accurately and should serve as a basis for understanding how policymakers think, choose, and solve problems. The mixed scanning model probably most approaches reality in its preference for spending time on the important issues.

Significant obstacles confront policymakers who want to behave rationally. The multiple demands for their attention and time, the shifting coalitions, the constraints on memory, the biases in information processing, and the sources of distortion in communication make the classical model of rationality an almost unreachable ideal. These limitations on rational behavior are the basis for creating policy information management systems.

References

Abelson, R. P. 1976. "Script Processing in Attitude Formation and Decision Making." In *Cognition and Social Behavior,* edited by J. S. Carroll and J. W. Payne. Hillsdale, NJ: Erlbaum.

Allen, T. J., and S. I. Cohen. 1969. "Information Flow in Research and Development Laboratories." *Administrative Science Quarterly* 14, no. 1.

Allison, G. T. 1971. *Essence of Decision.* Boston: Little, Brown.

Allport, G. W. 1955. *Becoming.* New Haven: Yale University Press.

Athanassiades, J. C. 1971. "The Distortion of Upward Communication in Hierarchical Organizations." *Academy of Management* 16, no. 2.

Bandura, A. 1977. *Social Learning Theory.* Englewood Cliffs, NJ: Prentice-Hall.

Batalden, P. 1990. "Organization Wide Quality Improvement in Health Care." *Topics in Health Record Management (*May 15).

Brunswick, E. 1956. *Perception and the Representative Design of Experiments.* Berkeley: University of California Press.

Chaffee, E. 1980. *Decision Models and University Budgeting.* Unpublished Ph.D. dissertation, Stanford University.

Cobb, R. W., and C. D. Elder. 1972. *Participation in American Politics: The Dynamics of Agenda Building.* Boston: Allyn & Bacon.

Cohen, M., J. G. March, and P. Olsen. 1972. "A Garbage Can Model of Organizational Choice." *Administrative Science Quarterly* 17, no. 1.

Cooper, R. 1984. "The Performance of New Product Innovation Strategies." *European Marketing* 18 (5): 3–54.

Cyert, R. M., and J. G. March. 1963. *A Behavioral Theory of the Firm.* Englewood Cliffs: Prentice-Hall.

Dawes, R. M. 1971. "A Case Study of Graduate Admissions—Applications of Three Principles of Human Decision Making." *American Psychologist* 26, no. 3.

Dearborn, D., and H. Simon. 1967. "Selective Perception: A Note on the Departmental Identification of Executive." In *Organizational Decision Making,* edited by M. Alexis and C. Wilson. Englewood Cliffs: Prentice-Hall.

Delbecq, A., and P. Mills. 1985. "Managerial Practices that Enhance Innovation." *Organizational Dynamics* 14 (1): 24–34.
Deming, W. E. 1986. *Out of the Crisis.* Boston: MIT Press.
Downs, A. 1966. *Inside Bureaucracy.* Boston: Little, Brown.
Dunker, K. 1945. "On Problem Solving." *Psychological Monographs* 58, no. 5.
Edwards, W. et al. 1968. "PIP Systems: Design and Evaluation." *IEEE Transactions* SSC-4, no. 3.
Einhorn, H. J., and R. M. Hogarth. 1979. "Behavioral Decision Theory: Process of Judgment and Choice." *Annual Review of Psychology* 32.
Etzioni, A. 1967. "Mixed Scanning: A Third Approach to Decision Making." *Public Administration Review* 27.
Fishbein, M., I. Aizen, and J. McArdle. 1980. *Understanding Attitudes and Predicting Social Behavior.* Englewood Cliffs, NJ: Prentice-Hall.
Fischoff, B. 1977. "Perceived Informativeness of Facts." *Journal of Experimental Psychology* 3.
Freeman, C. 1982. *The Economics of Industrial Innovation.* Cambridge, MA: MIT Press.
Galbraith, J. R. 1973. *Designing Complex Organizations.* Reading, MA: Addison-Wesley.
Gettys, C. et al. 1980. "Hypothesis Generation: A Final Report of Three Years of Research." Decision Process Laboratory, University of Oklahoma, Technical Report 15-10–80.
Hage, J., M. Aiken, and C. Marrett. 1971. "Organizational Structure and Communication." *American Sociological Review* 36, no. 4.
Hogarth, R. M. 1980. *Judgment and Choice.* New York: John Wiley.
Hoos, I. 1973. "Information Systems and Public Planning." *Management Science* 17, no. 1.
Huber, G. P. 1980a. *Managerial Decision Making.* Glenview, IL: Scott Foresman.
———. 1980b. "The Process Model of Organizational Decision Making: A Documentation and Further Explication." Wisconsin Working Paper 2–80–5.
———. 1981. "Organizational Decision Making and the Design of Decision Support Systems." *MIS Quarterly* 5, no. 2.
Janis, I. L., and L. Mann. 1977. *Decision Making: A Psychological Analysis of Conflict, Choice and Commitment.* New York: Free Press.
Juran, J. 1988. *Juran on Planning for Quality.* New York: Free Press.
Kelley, H. 1951. "Communication in Experimentally Created Hierarchies." *Human Relations* 4, no. 3.
Kepner, C. L., and B. B. Tregoe. 1981. *The New Rational Manager.* Princeton, NJ: Princeton Research Press.
Lindblom, C. E. 1959. "The Science of Muddling Through." *Public Administration Review* 19, no. 1.
MacCrimmon, K. R. 1974. "Descriptive Theory of Team Theory: Observation, Communication and Decision Heuristics in Information Systems." *Management Science* 20, no. 10.

MacCrimmon, K. R., and R. N. Taylor. 1976. "Decision Making and Problem Solving." In *Handbook of Industrial and Organizational Psychology,* edited by M. D. Dunnette. Chicago: Rand McNally.
McQueen, D. 1988. "The Development of a Framework for Health Promotion Research." Working Paper #26. Research Unit in Health and Behavioral Change, University of Edinburgh, Scotland.
Maidique, M., and B. Zirger. 1984. "A Study of Success and Failure in Product Innovation: The Case of the U.S. Electronics Industry." *IEEE Transactions in Engineering Management* EM-31 (4): 192–203.
March, J. G. 1978. "Bounded Rationality, Ambiguity, and the Engineering of Choice." *Bell Journal of Economics* 10, no. 6.
March, J. G., and H. Simon. 1958. *Organizations.* New York: Wiley.
Miller, D. W., and M. R. Starr. 1967. *The Structure of Human Decisions.* Englewood Cliffs, NJ: Prentice-Hall.
Mintzberg, H. 1973. *The Nature of Managerial Work.* New York: Harper & Row.
Newell, A., and H. A. Simon. 1972. *Human Problem Solving.* Englewood Cliffs, NJ: Prentice-Hall.
O'Reilly, C. A. 1978. "The Intention of Distortion of Information in Organizational Communication: A Laboratory and Field Investigation." *Human Relations* 32, no. 2.
O'Reilly, C. A., and L. Pondy. 1980."Organizational Communication." In *Organizational Behavior,* edited by S. Kerr. Columbus, OH: Grid.
O'Reilly, C. A., and K. H. Roberts. 1974. "Information Filtration in Organizations: 3 Experiments." *Organizational Behavior and Human Performance* 11.
Payne, J. W. 1976. "Task Complexity and Contingent Processing in Decision Making: An Information Search and Protocol Analysis." *Organizational Behavior and Human Performance* 16, no. 3.
Phillips, L. D. 1982. "Requisite Decision Modelling: A Case Study." *Journal of Operational Research Society* 33: 303–11.
———. 1983. "A Theoretical Perspective on Heuristics and Biases in Probabilistic Thinking." In *Analysing and Aiding Decision Processes,* edited by P. C. Humphreys, O. Sven, and A. Vari. Amsterdam: North-Holland.
Piaget, J. 1981. *Le possible et le nécessaire. L'évolution des possibles chez l'enfant.* Paris: PUF.
Read, W. H. 1962. "Upward Communication in Industrial Hierarchies." *Human Relations* 15, no. 5.
Redfield, C. F. 1958. *Communication Management.* Chicago: University of Chicago Press.
Rosen, S., and A. Tesser. 1970. "On Reluctance to Communicate Negative Information: The MUM Effect." *Sociometry* 33, no. 3.
Simon, H. A. 1957. *Administrative Behavior.* New York: Free Press.
———. 1976. *Administration Behavior,* 3rd Edition. New York: MacMillan.
———. 1977. "What Computers Mean to Man and Society." *Science* 195, no. 3.

Slovic, P., and S. Lichtenstein. 1971. "Comparison of Bayesian and Regression Approaches to the Study of Information Processing in Judgment." *Organizational Behavior and Human Performance* 6, no. 4.

Smith, R. D. 1968. "Heuristic Simulation of Psychological Decision Processes." *Journal of Applied Psychology* 52.

Strauss, G. 1962. "Tactics of Lateral Relationships." *Administrative Science Quarterly* 7, no. 3.

Strecher, V. 1986. "The Role of Self Efficacy in Achieving Behavior Change." *Health Education Quarterly* 13 (1): 73–91.

Strecher, V., B. McEvoy-DeVellis, M. Becker, and I.

Rosenstock. 1986. "The Role of Self-Efficacy in Achieving Health Behavior Change." *Health Education Quarterly* 13 (1): 73–91.

Taylor, R. N. 1975. "Perceptions of Problem Constraints." *Management Science* 22, no. 1.

Tversky, A. 1972. "Elimination By Aspects: A Theory of Choice." *Psychological Review* 19, no. 2.

Tversky, A., and D. Kahneman. 1974. "Judgment under Uncertainty: Heuristics and Biases." *Science* 185.

Utterbach, J. 1971. "The Process of Technological Innovation within the Firm." *Academy of Management Journal* (March): 76–88.

Varela, F. 1989. *Autonomie et connaissance. Essai sur le vivant.* Paris: Seuil.

Watson, C. 1976. "The Problem of Problem Solving." *Business Horizons* 19, no. 2.

Wiksell, W. 1960. *Do They Understand You?* New York: Macmillan.

Wildavsky, A. 1967. *The Politics of the Budgetary Process.* Boston: Little, Brown.

Wilensky, H. G. 1967. *Organizational Intelligence.* New York: Basic Books.

Wright, P. 1974. "The Harassed Decision Maker: Time Pressures, Distractions, and the Use of Evidence." *Journal of Applied Psychology* 59, no. 5.

Zander, A., and D. Wolfe. 1965. "Administrative Rewards and Committee Members." *Administrative Science Quarterly* 9.

3

The Issue Life Cycle: How to Plan for Evolving Information Needs

An adept policymaker seems to predict the future. Just as an issue gains importance, this person is developing a report or a plan of action to address something the press and public are only beginning to notice. This type of acumen may stem from the intuitive sense of the public mind that is found in all "natural" politicians. But we can also learn to analyze and predict the temporal patterns of public issues. This chapter presents one method of doing just that. We call it the issue life cycle—in other words, the natural history of a public issue. This concept is based on our belief that issues follow a certain pattern that indicates when they will require a policymaker to take a stand or make a decision. This is not to say that all issues behave similarly, only that they should be treated differently according to the stage in which they happen to be. The stages of the life cycle are dormancy, awareness, recognition, resolution, and alignment. An issue may move from any stage into dormancy and vice versa. Movement among the other stages is more or less linear.

The issue life cycle helps us study and analyze issues before they enter the political arena and become harder to handle. This model of issue evolution can serve as the infrastructure to support the policymaker throughout the resolution of an issue. From a policy information management systems point of view, knowing the cycle helps identify what type and format of information will be most useful at each stage. Knowing how issues evolve allows us to identify and monitor the key factors—catalysts and inhibitors—that shape issue evolution.

In Chapter 2, we discussed a model of political decision making for an incremental decision rule. For a variety of reasons, policymakers confronting a problem often revert to a rule of thumb: "Add 5 percent to what we are doing now, but don't rock the boat." This approach may be chosen because issues are hard to analyze, pressures are intense, and/or time is short. Under such circumstances, policymakers try above all to avoid mistakes and therefore think that caution is best, considering it dangerous to disturb the status quo. The incrementalist approach says that if the desired direction is unclear, small steps are preferable until a consensus emerges.

We believe the fields of policy analysis and policy information management have evolved to the point where they can be synthesized to allow a more aggressive approach to policy formation. The conception of the issue life cycle offers a framework for understanding the public policy approach which is described in the balance of the book.

In this chapter, we first discuss the model of the issue life cycle. We then describe how the model can be used as a master plan for designing policy information management systems.

The PIMS provides one source of policy support for decision makers, but it will never replace trusted advisers, good intuition, or common sense. A PIMS simply provides one way to capture and merge the advice of trusted advisers with empirical data, intuitive strategies, and common sense into a formal structure that promotes objectivity, completeness, and ease of communication. With existing group process techniques and computer programs (e.g., decision conferencing), these support services can be provided in a matter of hours or a few days.

While some political scientists have addressed how issues are placed onto agendas (Cobb and Elder 1976), few have looked at their life cycle. Most of us have an intuitive notion of how issues develop. However, few models have been proposed to describe and explain the processes that convert concerns into problems and problems into issues (and issues into calamities!). The issue life cycle is one paradigm to describe this process.

Why should we be interested in the issue life cycle? First, if we understand how issues evolve and enter the political arena, then we can anticipate them and be better prepared to analyze them. Second, understanding which factors are critical in the evolution and dissolution of an issue allows a policymaker to manipulate the timing—if not the resolution—of an issue. Third, evidence suggests that decision makers require different information at different stages in the life cycle. For example, information on how alternative solutions have worked elsewhere is important early in the process (while we are generating an understanding of the issue), while information on political constituencies is generally more important in the resolution stage. So, as analysts or policymakers, our interest in the issue life cycle reflects the desirability of having an information system that supports whatever analysis is appropriate at each stage.

The life cycle concept is important for other reasons. For example, it can facilitate communication between analysts and the policymaker in developing a strategy to deal with a particular issue. When confronting an issue in the recognition stage, we should begin asking the right questions to help a policymaker pick a position. Also, the life cycle provides a convenient framework for a quick overview and update of each issue of interest to a policymaker. It also provides a framework to focus on whatever element is critical at each life cycle stage.

The Life Cycle

The issue life cycle contains five stages: four are stages of development (the active stages), and one is a stage of stagnation (the inactive stage) which an issue can enter from any active stage. Issues pass through all active stages; the transition between stages may be gradual or sudden.

The five stages have specific characteristics:

1. *Issue awareness.* In this stage, groups and individuals inside and outside an organization start to express concerns as they become aware of a problem, perceived flaw, or injustice. The resulting tensions start shaping something that might eventually become an issue. We call the problem a "proto-issue" at this stage. The policymaker usually takes note of the concerns but does nothing more than superficially analyze them. The policymaker may assess his or her values and preferences and perform a preliminary survey of strategies (e.g., analyze, stonewall, recognize, or act) for dealing with the proto-issue if it ever becomes a full-fledged issue.

2. *Issue recognition.* A proto-issue becomes an actual, legitimate, officially recognized issue when a policymaker decides, either on his or her own initiative or through coercion from outside, that the issue must be resolved—or at least addressed. At this point, potential solutions must be evaluated and strategies analyzed.
3. *Issue resolution.* At this stage, policy decisions are made in an attempt to address the needs exposed by the issue. This resolution can occur with the policymaker in control or not, depending on circumstances.
4. *Issue realignment.* A realignment of coalitions, constituencies, or individuals usually occurs as a result of the policy developed to address the issue. How these individuals view the desirability and impact of the policy is a key factor in determining to which stage the issue moves next.
5. *Issue dormancy/pemission.* An issue may move into this inactive stage from any active stage. A dormant issue is in a latent stage of development; it can then reenter the life cycle at any active stage. An issue may enter dormancy for internal reasons (e.g., another crisis demands a policymaker's complete attention) or external forces (e.g., the legislator who pushed the issue loses an election). If the policymaker has control, he or she may try to force the issue into dormancy to defuse a conflict or allow more time to examine solutions. Dormancy should not be mistaken for inactivity, because customer needs may change during dormancy, the delivery of services may become more or less equitable, and the level of waste and duplication may change. At this stage, however, these changes are not recognized by policymakers or constituents.

Issue awareness stage

In the awareness stage, an issue is only one of a seemingly endless category of "things to worry about," and certainly not the most compelling one. Four broad categories of factors affect the emergence of an issue from this stage into the recognition stage: (1) issue catalysts, (2) issue characteristics, (3) interest group characteristics, and (4) the time in which the recognition process occurs. These four

Table 3-1 Factors Influencing Issue Emergence

Issue Catalysts	*Issue Characteristics*	*Interest Group Characteristics*	*Time Considerations*
Media, politicians, unexpected events	Specificity Social significance Long-term implications Complexity Routineness	Perceptions (ideology, norms and values) Power Intensity	Rate at which the issue emerges

categories interact to force an issue to the attention of the public and the policymaker. (See Table 3-1.)

Issue catalysts. Issue catalysts are forces that are instrumental in building the pressure needed to move a latent issue into the public arena. The primary catalysts are the media, politicians, advocacy groups, and unexpected events that focus interest in an issue.

Issue characteristics. Cobb and Elder define five important characteristics of an issue that influence its emergence:

1. *Specificity.* The more broadly an issue is defined, the more likely it is to expand and enter the public arena, because more groups can identify with it. An example is the Bakke case (in which a white man charged reverse discrimination at a California medical school, which had an admissions program favoring minorities). The issue could have been interpreted as simply concerning university admissions procedures. But more broadly defined, it affected affirmative action and equal employment opportunities in public and private sectors. Defined in this way, the case affected a much larger group, including unions, employers, and civil rights groups.
2. *Social significance.* The greater the potential impact of an issue (defined by the nature of the proposed change and the

number of people affected), the larger the issue is likely to grow. This was clearly demonstrated when the Food and Drug Administration attempted to prohibit saccharine because tests linked it to cancer. Since this proposal had significant impact on numerous saccharine producers and consumers, the issue expanded quickly.

3. *Long-term implications.* Issues are more likely to grow if they have long-term implications. This is apparent in the growing concern for workers' health and safety. As awareness that cancer, respiratory disease, and birth defects are associated with working in certain industries, the issue of inadequate health protection has grown as well. But long-term implications can also send an issue into dormancy. For example, preventive programs have relatively low success rates because their rewards are evident only in the long term.
4. *Complexity.* The more complex an issue, the less interest it draws. People who lack the time needed to understand an issue are likely to be apathetic. This is apparent in many defense projects for which military equipment funds are appropriated with little debate. These potential issue areas are too technical for the average person and legislator to comprehend, and the subject remains in the proto-issue stage.
5. *Routineness.* The fewer precedents an issue has, the more likely it is to develop. Examples can be found in the legal system, where issues with few precedents move to higher courts, gathering steam in the process.

Interest group characteristics. Almost by definition, interest groups have a direct stake in issue resolution because they will be directly affected by any resulting structural changes. Three interest group characteristics are vital in determining how much expansion and conflict will arise about a potential issue:

1. *Group perceptions (ideology, norms, and values).* The perceptual filters through which a group views a proto-issue determine whether sufficient tension will be generated to convert it into an issue.
2. *Power of the group.* This characteristic measures how much influence a group has on the policymaking process. Often a good proxy for this measure is the group's size. The power

of a group in this context is its ability to bring an issue to a policymaker's attention and convince him or her to take a position on the issue to recognize and legitimize it. For example, homosexual men originally had little power to influence the course of AIDS policy because they were widely considered societal misfits.

3. *Commitment of the group.* The degree of the group's commitment to its position will influence an issue's expansion. As an example, the many AIDS pressure groups, through their commitment, have helped promote both research and spending on the disease.

Time considerations. The other three active factors (issue characteristics, issue catalysts, and characteristics of the interest groups) help determine the processing time and whether a long processing time will increase or decrease tension.

Time can influence issue development in a number of ways. A rapid transition from proto-issue stage to emergence can produce a crisis, leaving decision makers unprepared to face the problem. A volatile situation can also result from a long processing time that frustrates the interested parties. If the move from proto-issue to recognition is slow, more groups can begin to understand how the issue will affect their interests, inducing them to get involved and expand the issue and making it harder to control. Long processing time can make issues more controversial, but it can also reduce the interest in proto-issues. The public can be aroused by a proto-issue but may lose interest when tension does not continue increasing. As an example, the concern during the mid-1970s that chlorofluorocarbons were depleting the ozone layer abated with little effect on public policy (the chemicals were banned from spray cans in the United States, but no other restrictions were passed). This was primarily because a lack of research findings stymied efforts to develop the issue. When the famous "ozone hole" was discovered over Antarctica in the mid-1980s, the issue ripened, and international agreements were signed in 1987 and 1990 to reduce and then halt chlorofluorocarbon production.

Issue recognition stage

A proto-issue that passes through the emergence process reaches the issue recognition stage, in which a decision maker acknowledges that the problem is indeed an issue. The issue recognition stage exists when the intensity of group demands forces a decision maker to take a position on the issue, even if it is superficial or just an acknowledgment of the problem.

Perhaps the most significant change in the environment between awareness and recognition is the increase in tension caused by strengthening demands by interest groups. For example, the increasing interest in environmental issues shows how interest groups have increased the tension for change.

One implication of an issue moving to the awareness stage is that the policymaker must allocate organizational resources to analyze what steps are needed to approach it. This often causes a policy analysis unit to propose an in-depth analysis to raise the level of understanding on the issue. Typically a policymaker's environment is made up of several proto-issues, some of which will never emerge as issues. Usually, the resources for policy analysis are limited. The decision maker must choose which issues merit exploration from the total set of proto-issues.

Issues can be intrinsically important (on moral, ethical, or social grounds) and/or politically important (on the basis of political "ripeness"). The policymaker and policy team must monitor issues on the basis of ripeness and intrinsic importance in order to be prepared to address, in a timely fashion, issues that break into the political arena. See Chapter 13 for a model to quantify the constructs of intrinsic importance and ripeness.

Issue resolution stage

If a consensus emerges that some form of resolution is necessary, the issue moves from the legitimacy stage to the resolution stage. During this stage, policy options are identified, an option is selected to be the new policy, and procedures to implement it are developed.

A resolution will be shaped by debate and negotiation in the political arena if it is not under a policymaker's direct control. As the issue becomes more controversial and outside groups increase their involvement, resolutions reflect compromises among the positions of the interest groups. Political power and influence play major roles in determining the kind of compromise reached. The policymaker is forced to incorporate the expectations, values, and goals of others into the resolution, which may not be aligned with his or her best interests or judgment. In any case, the policymaker should use input from interest groups to broaden the appeal of the resolution.

If the process remains under the policymaker's control, the issue may go from resolution to dormancy. For example, suppose the legislature passed a mental health act containing a bill of rights for patients, but no money was allocated to implement the grievance procedure in the law. Nevertheless, the director of the bureau of mental health mandated the creation of a patient's rights advocate office, and the staff drafted grievance procedures for patients. In drafting these procedures, the staff used data from similar programs in other states, as well as input from influential and well-informed interest groups who were kept abreast of the process and the drafts. Groups that mildly opposed the procedures did not force the issue into the public arena because they thought it was not sufficiently important, their concerns were being heard, or they could not do any better. For whatever reason, the community, media, and legislature remained uninformed, and the issue was resolved wholly under the control of the bureau director.

Several factors can prevent an issue from reaching resolution. One is the delaying action of someone in the political system, which is particularly effective if an interest group or a coalition is weak, disorganized, or unstable. A policymaker can also prevent an issue from reaching resolution by refusing to acknowledge the issue and/or referring it to another policymaker or agency.

Issues may also enter dormancy if interest groups actively obstruct the resolution. For example, although a clear demand exists for national health insurance in the United States, the issue remains stalemated because of the multiplicity and power of opposing interests.

Issue realignment stage

When a resolution is implemented, an issue moves from the resolution stage to the realignment stage, wherein the compromises that led to the resolution are put into practice. To implement a response, the issue goes through two substages: (1) the political/managerial stage, when responsibilities and resources are assigned, and (2) the technical stage, when the decisions made at the political stage are translated into specific procedures and the policy change is implemented. Four phases can be identified in the political/managerial substage:

1. *Problem definition.* The issue is translated into a concrete series of goals.
2. *Information search.* Alternative solution strategies are studied.
3. *Adoption of the "best" alternative.* The policymaker decides on a particular strategy and attempts to convince all parties that the solution is indeed the best possible one.
4. *Resource development.* Funding and personnel are allocated under the constraints established by economic and political realities.

The technical substage can be further broken down into two parts. *Project administration* is the development of rules and procedures for the daily operation of the solution. *Implementation* is the actual enforcement of the rules and procedures. At this point, obstacles and flaws may become apparent. Change raises the level of tension whether the public perceives the change as positive or negative. Inertia and resistance to change can become serious roadblocks. If there is enough resistance to implementing the proposed resolution, the issue may either become dormant or remain in the realignment stage until the obstacles are overcome.

Issues rarely die. For example, the danger of rubella to pregnant women seemed resolved when a rubella vaccine was discovered, a plan for its distribution was carried out, and the issue entered into remission. In recent years, however, it has reappeared because of problems with implementing the distribution plan.

Another potential example is the patient's rights grievance procedure described above. The issue could easily spring back if the doctors and staff complained about excessive formality and paperwork and refused to follow the procedures.

If an interest group reacts negatively to a resolution, it might attempt to keep the issue in the active stage until a better (from its point of view) resolution can be achieved. This occurred when the U.S. Supreme Court ruled in favor of women's right to abortion. Interest groups that viewed this resolution and consequent realignment with extreme disfavor kept the issue alive by concentrating on subissues such as the legality of federal funding for abortions and parental notification or consent for delivering family planning services to minors. Until a resolution meets the minimal standards and expectations of the interest groups, those who evaluate the resolution unfavorably will try to keep the issue (or some subset of it) in the active stage.

Dormant stage

We can think of the dormant stage as a steady state in which all forces acting on an issue are at low intensity and in equilibrium. A dormant issue that was never in the public eye may not be recognized at all. Issues may transit between dormancy and any other stage. But just because issues are silent does not mean they can be forgotten. A smart policy analysis unit will constantly monitor the environment to see if it is likely to awaken dormant issues.

An issue may remain dormant until a change occurs in

1. The influence of a catalyst (e.g., the new governor has targeted the issue in the campaign)
2. The issue characteristics (e.g., the media picks up the issue and increases its visibility)
3. The alignment of interest groups (e.g., a national organization supports local proponents)
4. Time pressures (e.g., a deadline is set for action)

The Life Cycle as a Master Plan for a PIMS

We have offered a paradigm of the evolution of a policy issue. This section will examine how that life cycle can be used to define and specify the role and function of a policy information management system (PIMS) for issues in each phase of the life cycle. Here, PIMS is broadly defined as consisting of the policy analysis team, a set of decision-theoretic models, a series of internal and external data bases, and the computer technology to collect, retrieve, analyze, and communicate the necessary information.

The PIMS in the transition from awareness to recognition

Surprises are one of the hazards of life in the public arena, and they are especially problematic for issues in the awareness stage. If a PIMS is to be a credible tool in policy analysis, it must minimize surprises that can arise if it fails to

- Notice that a proto-issue will eventually turn into a major issue
- Predict problems that may arise during the implementation process
- Recognize that the right evaluation measures are not in place

During the transition between awareness and recognition, the key role for the PIMS may be identifying and monitoring issues. To prevent issues from emerging unexpectedly, the PIMS should help generate a list of "promising" proto-issues and keep the policymaker informed of their status.

At this early stage of the life cycle, position papers should describe the proto-issue and discuss its current magnitude and potential for development. Initial sets of options should also be prepared and an analysis suggested (but not carried out). In the early stages of designing a PIMS, it is unlikely that a policymaker will ask or expect a PIMS to carry out these functions. The policymaker may lack confidence in the PIMS or may not even believe it can play those roles, so this is a good time for the PIMS to establish its legitimacy on the issue.

The PIMS has five specific roles in the process of issue emergence:

1. *Identify and prioritize issues that are intrinsically important.* Issues become important either because the process of producing a service does not work properly or because a group of powerful opinion leaders decide that improvements must be made. Because dissatisfaction with one process may spill over into other processes, there is an advantage to identifying and acting on opportunities to improve processes before opinion leaders begin to complain. Therefore, some of the DSS's resources must be allocated to studies that reflect the concerns of opinion leaders in the field or that identify gaps in the operation of the organization.
2. *Monitor promising issues so the PIMS can anticipate when they will ripen.* This role is important because the PIMS must be ready to offer information in a timely fashion so it can prepare the policymaker for changes in issue status. Sources of information for this stage of monitoring include major events and policies being implemented in jurisdictions that are innovators and early adopters. Also valuable are comments from public officials, spokespersons for powerful interest groups,

academic experts, and publications. See Chapter 13 for a model to help quantify the degree of ripeness; a questionnaire to determine issue ripeness is presented in Appendix 13-A.

3. *Monitor the agenda of the policymaking elites to understand when information is needed.* Because policymakers are short of time, the PIMS must serve as a filter and provide information only when needed. Selectivity—what kind of information is presented and when and how it is presented—is vital.
4. *Document the character of issues that appear not only intrinsically important but likely to ripen soon.* This documentation should present the issue, background data on its magnitude, and preliminary suggestions about options to resolve it. Procedures should be in place to trigger in-depth analysis of a given issue on request (see Chapter 12).
5. *Help the policymaker decide when to take action on an issue.* Techniques such as decision trees and multiattribute value modeling can help crystallize options available and improve the policymaker's judgment on timing any action. These techniques can help in deciding whether and how to study an issue. They can also help in selecting and implementing a policy to deal with it.

The PIMS in the transition from recognition to resolution

Once the decision has been made to resolve an issue, the role and focus of a PIMS change. Now the PIMS must assist the policymaker in structuring the problem, integrating judgment and data, and conducting sensitivity analyses.

Help structure the problem. One of the most common difficulties in effectively resolving problems is not clearly understanding the nature of the problem. Policymakers may lack the time and inclination to ask reasonable questions: Why are people upset? What is wrong with our performance? Is the flaw permanent or temporary? Is it a problem of measurement, perception, or power? What are the consequences of not acting?

Help integrate judgment and data. In the transition from awareness to recognition, the PIMS must provide a description of the problem, its background, and its implications for the larger environment. This material, however, still leaves many questions unanswered. It does not generate options for policymakers, nor does it specify the outcomes, objectives, uncertainties, and values that help select the best solution.

The difficulties of addressing issues in a political environment may be compounded by the involvement of several policymakers or opinion leaders, all of whom may have different assumptions, perceptions, and values regarding the problem. A further complication may be perceptions that policymakers have about one another's values. In reality, the policymakers may structure problems similarly, but that is unimportant if the two perceive themselves as having substantially different approaches. To reduce such conflicts, the PIMS can clarify the values and preferences of the various policymakers and constituencies as discussed in Chapter 14.

The typical paucity of pertinent empirical data is another barrier to resolution. Unfortunately, as discussed in Chapter 12, prospective collection of data is expensive and time-consuming, so many policy analyses rely on data that were collected for other purposes. This forces policymakers to make decisions without appropriate information.

Experts are often asked to fill the information vacuum, but using experts can raise problems as well, as discussed in Chapter 5. For example, they may not represent the total spectrum of opinions on the subject. Or their "intellectual baggage" may prevent them from seeing the issue from the policymaker's point of view. The PIMS can foster the effective use of experts in filling the information vacuum by listing all recognized experts on specific topics and describing their belief systems. The PIMS can also provide a formal mechanism for integrating the input of experts with whatever empirical data are available. In so doing, the PIMS may help identify the best resolution to the issue.

Help in performing sensitivity analyses. Sensitivity analysis is the practice of varying the factors that influence a resolution to see which are most important. One difficulty with systematic and quantitative analysis is that it often produces a single answer to a complex problem. If a policymaker disagrees with some of the assumptions, data, or values used in the analysis, the answer is immediately discredited, along with the whole process. Sensitivity analysis prevents these problems because it allows assumptions on values, probabilities, and other data to be quickly and easily modified. To enrich the policymaker's understanding of the importance of various factors and how they interact, the PIMS must present results under a variety of scenarios. For example, a hospital considering expansion would want to project future usage patterns using a variety of local population estimates for the next 20 years.

The PIMS in the transition from resolution to realignment

A PIMS can play two important roles in the issue resolution stage: performing an implementation analysis and designing a process to predict the success of the proposed solution.

While issue resolution—supporting the choice of one alternative out of a set—has been a primary focus of PIMS in the past, policymakers may have greater concerns. Often, policymakers know exactly which solution they want to implement but wonder how to do it. In other words, these policymakers need support in playing the complex role of change agent.

Implementation analysis is the process of identifying the factors that could inhibit effective implementation, examining the status of the new policy in light of those factors, and identifying the characteristics that must be present in the implementation plan. The likelihood of implementation is an important factor in the desirability of options, so the PIMS should measure the implementation difficulty of each option with its implementation analysis tool (see Chapter 15).

Many factors can prevent the effective implementation of a technically reasonable policy. The policy could be perceived as being too complex or rigid. The groups most directly affected may not be ready for the change. The implementation process may lack the support of opinion leaders. The people responsible for implementing the policy may be technically weak or, more commonly, lack prestige and credibility in the eyes of those who must accept the change. Few of these obstacles are insurmountable if they are identified early and the proper strategies are used in carrying out the resolution.

By identifying these obstacles, the PIMS facilitates unfreezing, the first step described by change theory. (The other two steps are changing and refreezing.) As in previous phases of policy analysis, the PIMS provides a formal structure, or model, that increases the chance that all important factors will be considered adequately. The model also minimizes the impact of cognitive biases (see Chapter 2).

The second important role of the PIMS in the resolution stage is to help design a process to *predict the adequacy of proposed solutions.* Evaluation based on decision analysis can be used, for example, to choose whether to implement a policy statewide or to demonstrate it in a few counties. If a decision is made to start a demonstration project, the PIMS should be able to support an evaluation of the demonstration. Standards of performance (comparative or normative) must be established, methods of measuring actual performance developed, sample sizes determined to answer the basic hypotheses, and statistical analyses planned and executed. Data collection methods should be developed to permit measurement and explanation of whatever effect is found.

The PIMS in a dormant issue

Once a solution has been successfully implemented, the issue moves into a dormant stage. The issue can reemerge from this stage at any time. The likelihood of such reemergence depends on various factors, including the quality of the solution and the stability of the political realignments that were stimulated by implementation. This means that issues that are "solved" should still be monitored. Monitoring can be done in one of two modes: active or passive.

Active monitoring entails a formal study of the resolution. The study should assess not only the technical effectiveness of the solution but also its political acceptance by the various power groups.

Passive monitoring entails setting up a system that will react only when exceptional conditions arise. This "management-by-exception" process will provide early warning if something begins to go awry with the solution. Under passive monitoring, no specific studies are performed, but a few indicators are watched. If an indicator suggests dissatisfaction with some aspect of the solution is increasing, then a study can be started to determine whether the issue is about to reemerge.

Summary

As an issue moves through its life cycle, the role and focus of the PIMS change. Table 3-2 summarizes the type of policy information management required by a policymaker at different stages of the cycle.

Table 3-2 Roles of PIMS during the Stages of the Issue Life Cycle

Stages in Issue Life Cycle				
Awareness	*Recognition*	*Resolution*	*Realignment*	*Dormancy*
Issue identification	Issue structuring	Decision conferencing	Active monitoring	Active monitoring
Issue monitoring and anticipation	Clarification of values, uncertainty and assumption		Passive monitoring	Passive monitoring
	Conflict analysis		Decision-analytic evaluation	Active restructuring
			Implementation analyis	

Decision analysis and *sensitivity analysis* are performed in all stages of the life cycle.

Early in an issue's life, the PIMS should focus on predicting what proto-issues may enter the political arena by assessing issue ripeness. The PIMS should prepare background papers on the proto-issues and monitor the policymaker's agenda and priorities to ensure that position papers can be developed as needed. Decision-analytical support should be available to help the policymaker describe and quantify preferences at all times.

The focus of the PIMS changes as an issue moves to the active stages. The PIMS must help the policymaker understand the issue, options, possible outcomes, uncertain events, and positions of other policymakers and interest groups. Then the PIMS can elicit judgments about probabilities and utilities from experts and integrate those judgments with empirical data to analyze the desirability of each option.

Finally, sensitivity analyses can be conducted to help understand how variation in assumptions, data, values, and probabilities will affect the desirability of options.

Once the issue moves from resolution to realignment or implementation, the PIMS must focus on implementation and evaluation. Decision-analytical tools can be used to decide whether to implement the proposed solution on a full-scale or demonstration basis. The PIMS can design demonstrations to clearly determine whether the desired change actually occurs.

Finally, when an issue enters the dormant stage, the PIMS should monitor—either actively or passively—the status of the implemented solution. The objective is to identify weaknesses in the solution and detect when the issue is about to reenter an active phase.

Reference

Cobb, R. W., and C. D. Elder. 1972. *Participation in American Politics: The Dynamics of Agenda Building.* Boston: Allyn & Bacon.

4

Individual Differences

Previous chapters have introduced variations that place demands on policy information systems. Chapter 2 argued that different decision-making environments (rational, administrative, and political) require different types of information. For instance, an administrator operating in a rational environment is likely to be more interested in facts, while an administrator in a political environment will probably be more interested in the constituencies' positions. Chapter 3 showed how issues change over time. Each variation in a life cycle position of an issue places unique demands on a policy information system. Issues that have reached the resolution stage may require extensive analysis of alternative solutions. Issues entering the dormancy stage demand a mechanism to periodically monitor performance and satisfaction with the system. Issues in dormancy may call for information on current system performance.

These varying demands help define the role and resource requirements of a policy information system. Another influence on a policy information system is the policymakers who use it. Policymakers have different needs for information and support that may or may not be related to the issue they are addressing and the environment in which they are addressing it. Policymakers have various degrees of knowledge about issues. They also have various abilities to process information, make decisions, and communicate with customers and suppliers. This chapter will review the variations among policymakers and outline the features of a policy information system needed to respond to this variation.

Knowledge Variation

There are eight dimensions to knowledge variation. Effective resolution of issues within any system depends on a clear understanding of each of these sources of variation.

1. There can be wide variation in understanding about the *purpose of a system,* especially one that has existed for some time. One person may feel the purpose of a prison is primarily to protect the public from offenders, while another may believe its purpose is to rehabilitate. The relative importance of those two purposes should have substantial influence on the way resources are allocated to prisons. The purpose of some systems has long since passed away. For others the purpose is changing. Given the unstable nature of things, a policymaker's assumption about a system's purpose may be wrong or in conflict with the assumptions of other opinion leaders. These variations in understanding can lead to conflict in virtually every aspect of a policy resolution effort.

2. There can also be wide variation in the level of knowledge a policymaker has about the *needs of the system's customers.* Who are the customers of this system? What are their needs and expectations? How well are they being satisfied? Policymakers who deal with familiar systems are more likely to understand customers and their needs. But policymakers who are remote from their system and customers may be unable to understand these characteristics without study and support.

3. Policymakers can differ substantially in their understanding of what *products and services* are delivered by a system and the effectiveness with which those products and services meet the identified goals. In order to address an issue, then, some policymakers will need detailed information about the types, amounts, and adequacy of the products and services. Other policymakers may need just enough information to test their assumptions about how well those services meet customer needs.

4. Policymakers may vary widely in their understanding of the *production process.* It is essential to understand how products and services are produced, what steps are involved, and the possible sources of waste, rework, duplication, and needless complexity. Misunderstanding of the production process can lead to confusion. Some may hold a vague notion of what goes on. Some may need to test their assumption that the process they designed is being consistently fol-

lowed. Others may have recently tested the process and simply need improved documentation to use in describing the process to others.

5. Wide variation can also exist in the policymaker's understanding of the *employees* in the production process: the extent of their training, the effectiveness of their labor, and the problems keeping them from optimal performance. This is particularly important because employees who work in the system are often blamed for a system's failure when the system actually prevents them from doing their best. Yet they are often the ones who if properly trained and empowered can easily improve the system and reduce the need for complex studies and expensive changes. Some policymakers have no idea of a system's employee needs, expectations, and potential, while others need little more than summary statistics to present to others in order to resolve an issue.

6. Policymakers may vary in their understanding of environmental pressures. The *political and economic environment* can place constraints not only on how a system functions but also on the range of possible improvements. A state policymaker will find it difficult to reduce regulatory constraints on nursing homes at a time when abuses in the industry are receiving extensive press attention. When costs in nursing homes are increasing at 20 percent annually, it may not be appropriate to close down a hospital rate-setting commission, even if it has been ineffective. Most policymakers would quickly recognize the significance of these political and economic realities. But environmental pressures can be more subtle, in which case variations in individuals' sensitivity to them will be less apparent. A policymaker who is unaware of how governmental constraints influence how and what services can be delivered is at a clear disadvantage in attempting to improve the system realistically.

7. A policymaker must also understand what kind of *information* is being collected to monitor the system, what those data say about system performance, and what are the barriers to accessing the needed information. This understanding is necessary to defend the system's performance against critics, to prepare evidence supporting positions, and to provide clues about how and where the system can be improved. Yet some policymakers are much better informed in this area than others.

8. Finally, wide variations can exist in a policymaker's understanding of the *technical resources* needed to make the system function effectively. Policymakers must understand what resources are actu-

ally available, what barriers limit access to needed resources, and what resources might significantly improve the system performance.

These eight dimensions are important to understanding and resolving issues, and, as we have said, policymakers vary substantially in each of them. A properly structured policy information system can narrow these information gaps and improve policymakers' effectiveness but must not overload policymakers with information and support they don't need.

As the amount of information available has burgeoned, it has threatened to overwhelm even the best policymakers. Therefore, proper information management is becoming more and more important. When providing information, recall the physician's first rule: "Do no harm." Policy information systems should not contribute to gridlock. We can reach this goal by tuning these systems so they provide selective support where it's needed most, where that information can reduce variation and reduce uncertainty in the dimensions mentioned above.

Information Processing Variations

Policymakers also vary in their ability to process available information. Some focus very well on values, some deal well with uncertainty, and some readily absorb information about complex situations, even while others are overwhelmed in similar situations. As we all know, there are many ways to say the same thing. Some people are more sensitive to emphasis than others. A simple change of emphasis can change the impact of a sentence from rage to reverence. Some of this variation in processing ability results from the natural tendencies of an individual, but some of the variation is a reflection of environmental pressures facing the individual.

Healthy smokers may find very little emotional or psychological impact from information about their risk of dying of lung cancer; smokers with lung cancer may be fascinated by the same topic.

The variations in the extent to which the policymaker is in control of a situation also influence how to present information to him or her. Crises are characterized by rapid change that occurs regardless of the policymaker's actions, so the challenge is to decide which aspects can be controlled and which aspects must be tolerated and altered slightly if at all. Early in a crisis, policymakers may not even know what questions to ask about alternatives, and panic may limit their ability to process information. At this stage, they may prefer pictures or summary statistics to gain a fundamental understanding of the issue. As matters begin to calm down and policymakers can seek detail about specific subjects, it may become vital to perform detailed analysis of some microscopic topics. The choice between detailed information and summaries is one of several to be made in deciding how information should be presented to best promote processing by policymakers. Other variations include using pictures, words, or numbers; using concepts or examples; and using audio or visual presentations. Each choice can substantially influence the power of the information presented. A policy information system must be capable of responding to these varying presentation needs.

These variations in information processing mean that effective advisers must understand the policymaking customers very well and tailor the communication to allow these busy people to easily understand and incorporate the information into their deliberations. This ability to adapt to policymaker variations is rare even among the most skilled policy analysts. Computer-based policy support systems cannot be expected to automatically tailor their presentations to specific policymakers. But the system can make it easy for a policymaker to specify the depth and format for presenting information.

Communication Variations

Policymakers also vary in their ability to communicate with employees, constituencies, and colleagues. People frequently provide important policy-related information. In some cases geographic dispersion is a major source of variation in ability to communicate. Customers and suppliers, whether located in the next office or in the next country, must be reached in a timely fashion. Telephone and fax systems are powerful communication vehicles, but they may be inadequate if data must be communicated in a processable form. In those cases, access to user-friendly computer support is vital.

A second source of variation in communication ability is personal relationships. Certain personalities do not work well together. In extreme instances, just the knowledge that a certain person is participating in the same decision process can substantially impair another's ability to deal with the issue productively. In some cases, anonymous or confidential communication can facilitate problem solving, while in other cases it can inhibit progress.

Role of a Policy Information System

A policy information system must be able to respond to variations in knowledge, information processing, and communication needs of policymakers to effectively support information sharing, consensus management, and decision making. But we are hardly able to produce a policy information system that can decide which variations are most appropriate for a policymaker at a specific time and for a specific issue. The key is to have an adaptable information system that can provide support in the desired form.

We believe technology has reached the point where such an adaptive policy information system can be created. Computer-mediated communication can facilitate anonymous, immediate interaction among individuals and groups across virtually any distance. The processing and storage capabilities of computers allow immediate access to large data bases that can be tailored to the user's needs. Decision analysis and other modeling techniques allow us to capture and quantify subjective values and uncertainty. Existing interfaces allow us to turn over control of the system to users with minimal training and to communicate complex concepts in a variety of ways. With these five resources (interfaces, models, speed, storage capacity, and communication), the computer can respond to the individual differences of policymakers. In the next section, we present an example of one such system.

An Adaptive Policy Information System

By 1990, many in the health care field had become intrigued with the potential of Japanese management principles (Berwick 1989; Scholtes 1989) to improve the quality and reduce the cost of health care. These principles advocate much more attention to identifying customers, understanding their needs, committing to meeting those needs first and worrying about long-term rather than short-term bottom lines, improving process instead of blaming people, using statistics to improve management, educating all employees in process improvement tools, and treating suppliers as long-term partners. Hospitals and health care organizations across the United States sent top and middle managers to training programs on continuous quality improvement (CQI) and later on quality planning. Quality improvement (QI) structures were formed within organizations. Corporatewide in-house training was initiated, and quality improvement teams began improving processes for delivering health care. Consulting organizations began offering services to support this growing interest in quality improvement.

The concepts and tools seemed simple at first, but when implementation began, it became obvious that a true commitment to quality improvement would be a very difficult task. Training programs and consultants (at best sporadically available) could not reduce delays and promote continuous quality improvement unless they were available just-in-time.

In response to this demand, we began to develop a system called the Quality Improvement Support System (QISS). At this writing, QISS is operating in a network of 51 independent hospitals, clinics, and health maintenance organizations (HMOs) located across the United States. QISS uses the same basic interfaces as a related system called CHESS (Comprehensive Health Enhancement Support System), which is designed to help people on college campuses, in communities, and in employee assistance programs obtain access to health and human services. We will describe the full design of QISS, even though only 9 of its 12 components are currently operating. We present the full design because it can illustrate how a policy information system can meet the individual needs of policymakers.

QISS operates on an IBM-compatible personal computer. Local sites use a PC with a VGA monitor, hard disk, and modem connecting a host microcomputer in Madison, Wisconsin. QISS has four major categories of services: information, analysis, communication, and referral. After we describe each category, we will give an example of how QISS can be used.

Information

As we have stressed above, the type of information needed by policymakers varies depending on the stage of the issue and the state of the policymaker's knowledge. QISS responds to that variation in several ways, such as providing detailed or summarized information as needed.

Tailored information

Summaries can be used to introduce policymakers to important issues or to give an overview of options to resolve an issue. For example, involvement of physicians is an important issue in CQI. The user could be provided with a menu of choices about this issue, such as the purpose of such involvement, what physicians might select first, different strategies for involvement, and so on. A user could choose to be briefed on the most relevant topic or receive intensive coverage.

Instant library

The Instant Library stores, retrieves, and displays complete articles and annotated bibliographies about issues. The storage capacity of a compact disk allows enormous amounts of literature to be maintained on a personal computer. Another use of the Instant Library is to store fugitive literature—articles, reports, and analyses not available in many libraries (e.g., an executive summary of a quality improvement team report in which physicians played an important role).

The first screen of the library shows a shelf of books, each listing a subject. As the user selects a topic, he or she can narrow the search with a key word or see the entire list of relevant articles. If a key word such as *physician* is selected, the computer searches the topic area and lists all articles containing the key word, then the user chooses to see the full text or abstracts. The key word is highlighted throughout.

Project tracking

The Project Tracking component contains a data base of health issues being addressed by organizations in the QISS network. The data base is intended to promote collaboration and communication among organizations with similar interests; it also helps management follow the progress of issue resolution efforts inside their organizations.

The first screen allows the user to choose the type of organization to monitor (e.g., hospitals or HMOs). The computer displays a list of issue types that can be examined. QISS next lists those organizations addressing the selected issue, and the user can select one for more intensive analysis. Then a menu appears with information on the project, including timetable, objective, project contact, resources expended, tools employed, and results so far. The touch of a key allows the policymaker to instantaneously send a message via the computer to the contact in the other organization. Reports can be printed summarizing key points on the project.

Skills training

Individuals obviously have different levels of the skills required for making needed changes, and they also have varying degrees of confidence in their ability to implement changes. Recognizing that the most difficult part of a change is actually carrying it out, not deciding to make it, QISS provides Skills Training—a safe environment for people to gain experience and confidence in their ability to carry out a change.

By its nature, change destabilizes the status quo, creates resistance, and increases tension. A key issue in successful implementation of CQI in health care is that physicians feel comfortable with the tools of quality improvement. Skills Training allows users to practice flowcharting in a safe environment.

Feedback

QISS includes several mechanisms with which the policymaker can obtain and analyze feedback from the customers of a health care organization. A satisfaction survey can be input directly into the computer or as pencil-and-paper surveys. Policymakers can evaluate customer satisfaction with different aspects of the organization by monitoring satisfaction ratings over time or comparing satisfaction among different services of the same organization. They can also analyze the number and content of customer complaints and compliments. This system allows the user to identify upcoming issues and track the progress of efforts to improve performance.

Feedback starts with the user selecting which department or system to observe and deciding whether to examine comments or satisfaction scores. Comments must be further described as either complaints or compliments. After specifying the time period over which the analysis should be done and deciding the form of the data presentation (Pareto chart or control chart), the data are displayed on the screen. Specific satisfaction questionnaires can be examined and analyses quickly modified to address issues in greater detail.

Analyses

At times policymakers must examine values and feelings as well as empirical data. They may feel that the situation involves many factors with differing importance to the decision or project. Often the factors cannot be measured on a common numerical scale. Some policymakers seek help in separating a problem into manageable parts for individual treatment. The three analysis components of QISS help deal with values and uncertainties to make and implement decisions.

Decision analysis

QISS provides a program to help users think through important decisions. Within the CQI example, these could include whether to make the commitment to CQI, what quality improvement projects to chart, and whether to approve the recommendations of a quality improvement team.

The user is taken through five steps of analysis: (1) describing the problem, (2) generating options, (3) generating decision criteria, (4) rating the criteria's importance, and (5) rating the options against decision criteria. The directions at each stage of the process are tailored to the user's level of expertise with computers, and numbers are not used. In general, users are asked to respond graphically—for example, by creating bars to represent the relative importance of the decision criteria. After completing the analysis, the program displays the results graphically and helps the user understand how the results were reached. At any time users can return to rework the analysis as they gain understanding of the problem.

Decision Analysis can be used as a general tool for addressing problems and generating options and decision criteria. If a decision is frequently made (for instance, deciding what QI projects should be charted by an organization), a panel of experts can be convened to select decision criteria, decide how those criteria should be measured, and establish their relative importance. When similar decisions must be made, users can simply describe the options and decide how well they satisfy each criterion. The analysis model then uses the guidelines programmed into the computer to advise which project should be selected.

Conflict analysis

An important aspect of policymaking is understanding which conflicts are likely to erupt as an issue progresses. Who are the constituencies? What are their needs? How and why do these needs conflict with one another? What resolution of the issue would come closest to satisfying all constituencies? The Conflict Analysis component within QISS is designed to help policymakers gain a better understanding of the conflict.

The computer follows the same five-step process used in decision analysis. The difference is that the decision criteria, criteria weights, and ratings of how well each option satisfies each criterion are identified separately for each constituency. The computer compares and contrasts the models of each constituency, graphically pointing out sources of conflict and opportunities for compromise.

Implementation analysis

Implementation Analysis allows an organization to determine the likelihood that it will be able to undertake a major change, such as CQI. This component is based on a statistical model in which a panel of experts selects from change theory the factors likely to raise or lower an organization's chance of implementing a change. In devising the model, the experts estimate the relative importance of these factors and decide how various scores should be combined to yield a risk estimate.

In use, the computer asks questions about the organization's approach. How much planning went into this decision? Who are the major proponents and opponents? How much does each stand to gain or lose by implementation of this change? Then the computer displays a thermometer comparing this organization's chance of successful adoption to that of the average organization. Next the computer shows histograms pointing out the major barriers this organization would have to overcome to adopt the change. Finally, QISS explains why those barriers are important and suggests ways they can be overcome.

Communication

Policymakers differ in their need to understand constituent or adviser views on issues. Some hesitate to act at all without at least understanding these views in their deliberations. Constituents also vary substantially in their willingness to share views with policymakers—some do so freely, while others fear retribution or criticism. When policymakers want insights, a system such as QISS can be useful because it facilitates communication from many locations while protecting anonymity. Computer-mediated communication occurs in three formats, as follows.

Bulletin board

The Bulletin Board provides a forum for open communication on a variety of subjects. In some cases it may simply announce relevant events. It receives comments from anyone interested in a subject. Policymakers can use it to notify other users about events or interests; they can also observe the reaction of users to issues or possible resolutions. For example, a CEO who is implementing a CQI program could use the Bulletin Board to announce to the management team the times and locations of seminars on CQI.

Although many bulletin boards already exist, they are rarely used by policymakers because of complicated log-on procedures, poor displays, and constantly scrolling screens. As in other QISS components, it is vital that users need virtually no training to use the Bulletin Board.

Electronic mail

Electronic mail reduces barriers, offers anonymity, and simplifies access. Its primary goal is to increase users' willingness to ask difficult questions and share true feelings. E-mail allows users to exchange messages anonymously with a variety of peers and experts. Ideally such access will do two things: allow users to resolve problems with policies or implementation while they are still manageable, and reduce unnecessary face-to-face meetings with expensive staff.

For instance, an organization attempting to implement CQI could use E-mail to let people who are uncertain about the organization's commitment ask questions that could seem threatening or obtain information without embarrassment.

In QISS, experts are available at a central site linking 32 health care organizations. Upon selecting E-mail, the user sees a menu (in the form of an envelope) that lists the types of experts available to answer questions. The user sends a message to a specific expert, a colleague at another institution, or a subgroup of the network. When the message arrives at QISS Central, the triage officer will send an existing answer if the question has been answered previously for another writer. Otherwise, the officer transmits the message to an expert and ensures that the response is timely.

Discussion groups

QISS offers Discussion Groups for issues that interest several users and are too complex to be resolved in a single communication. An expert can lead a "seminar," which may last for several days. Members can debate subjects of mutual interest with other QISS users. Unlike telephone conferences, QISS interchanges can be stored for reference. Because comments are composed off-line, they can be carefully formulated. Anonymity increases the audience's willingness to participate.

For example, suppose a policymaker was considering policies on incest treatment. It might be very difficult to convene a panel of incest victims to discuss their problems or react to a proposed policy, but incest victims (possibly as part of their treatment) might be willing to interact anonymously with the policymaker on this delicate subject.

The functions of Discussion Groups are similar to those of the Bulletin Board. The component starts by showing a list of topics currently under discussion. If room is available, anyone can join a topic unless access is restricted (as it might be if it were discussing something like incest). Participants can raise issues, respond to comments, or simply observe the interaction.

Referral

Policymakers often need expert advice on issues but too frequently limit their search to trusted advisers who, whatever their degree of intelligence, may have similar values and thought processes. In the worst case, the adviser simply reinforces the policymaker's personal biases instead of adding fresh insights. Thus, a policymaker must seek ideas from other organizations and industries.

There are two challenges to seeking advice. One is to find the right person; the other is to overcome the economic and socio-emotional barriers to seeking help. QISS provides two services to help.

Matchmaker

Matchmaker contains a data base of experts organized according to the type of help they can offer and the issues in which they are interested. For instance, the data base for CQI classifies experts according to their theoretical and practical knowledge about statistics, group process, management, analytical services, and planning. The practical experts are further divided into administrators who have implemented CQI and consultants.

Upon entering the Matchmaker component, a user interested in physician involvement would select the topic, such as "planning," then narrow the search by choosing "medical staff" and other sorting criteria, such as a practical expert who has implemented CQI. QISS would then list experts meeting the specifications, complete with details on each.

Barrier reduction

A primary goal of our referral programs in the CHESS but not the QISS version is reducing the fear that often prevents people from reaching for help until their problem becomes a crisis. CHESS is designed for such users as rape victims, people with AIDS, or children of alcoholics. For these groups, searching for help can be threatening and embarrassing. The component allows users to learn what a service can do and how to contact the agency without embarrassment. Service locations are displayed on a community map along with bus routes and schedules. In some cases, a photo of the office is displayed to help the user find it.

Linkages

An advantage of QISS is its quick linkup among various components. If the Implementation Analysis concludes that the user has a high risk of failing to implement a new policy because the key actors had minimal involvement in promulgating it, QISS can offer to display an article on group process techniques from the Instant Library. A user of the Project Tracking component can immediately write an electronic mail message to a contact in an agency running a similar project. The Decision Analysis component can quickly link policymakers to tailored information to get brief explanations for why a certain selection criterion is part of project selection. These and other linkages transform QISS from a set of independent services to an integrated system that provides extremely powerful support.

Another form of integration occurs because each module of QISS can offer different information about the different aspects of an issue. In the physician involvement issue, the Instant Library allows them to review literature on the subject. Decision Analysis allows them to examine alternative forms of involvement, and Implementation Analysis allows them to develop and examine their plan for carrying out this policy. Project Tracking allows them to identify others who have attempted physician involvement. E-mail, Bulletin Board, and Discussion Groups facilitate communication with others who have dealt with this issue. So QISS supports individual differences regardless of where an issue is in its life cycle and where policymakers are in their understanding of it.

Example of QISS Use in a Hospital

The examples we have used so far were all related to policy. The following example shows how QISS might function in the operational context of analyzing and implementing CQI. It also shows how QISS can serve as an integrated system.

Susan Fleming, the quality coach of City Hospital, was asked by the hospital management team to prepare a list of suggested quality improvement projects for its consideration. In preparation, she turned on her microcomputer, opened the QISS program, selected the Customer Satisfaction module, and asked QISS to prepare a histogram showing (for the last six months) the number of complaints about various hospital functions. The histogram shows that, after Billing, Discharge Planning received the highest number of complaints. Fleming asked QISS to display the number of Discharge Planning complaints per month over the last year. The resulting chart showed that complaints had been rising steadily for the last five months. Next she asked to see all Discharge Planning complaints for the previous month and found patients complained most often that they and their families were not getting adequate training in "postdischarge responsibilities," such as medication instructions and activity limitations.

Fleming next turned to QISS's Implementation Analysis component to examine the organization's readiness to undertake a project in discharge planning. QISS helped her identify the strengths and weaknesses that would affect such a project, and the analysis brought to mind several steps that could improve prospects for success. After consulting several people involved in discharge planning, Fleming recommended formation of a quality improvement team on patient and family training.

Two weeks later, Fleming and the new team leader, Jim Konopacki, selected QISS's Project Tracking module. They entered information about their new project to make it available to other hospitals on the network and then asked QISS to list similar projects in those other hospitals. They found seven projects, two of which pertained to patient education. A review of the project summaries showed General Hospital's project was furthest developed. By pressing a function key, QISS automatically dialed the contact person at General Hospital and allowed Konopacki to send an E-mail message introducing City Hospital's project and asking if General had a flow diagram for its patient education process. The next day the diagram arrived on City's facsimile machine, and an E-mail message from General encouraged further contact about patient education.

The project tracking system suggested that Konopacki check the Instant Library to determine if anything was available on patient education. He found two relevant, unpublished reports of projects at other QISS hospitals. From the abstracts, he decided one was relevant to his project, so he called it up to the screen and read that University Hospital had developed videotapes that were given on discharge to certain patients and their families. The idea was intriguing.

Konopacki decided to see if other hospitals had been doing some creative thinking about the subject, so he posted a notice on the QISS Bulletin Board asking if anyone wanted to join an electronic Discussion Group on patient education at discharge. Nine hospitals offered to join, and the QISS central office formed a closed discussion group, one that would be open only to staff from those ten hospitals. For the next five weeks the members used the computer to discuss improvements in patient education and devised several ideas Konopacki thought worthy of pursuit. Two hospitals joined with City in a project to prepare material for discharged patients, allowing all three hospitals to learn from each other and cut costs.

While reviewing announcements on the Bulletin Board, Fleming noticed that Ricardo Gonzales would host this week's open electronic forum on "Quality Function Deployment—What Is It? How Does It Fit with Quality Improvement?" Gonzales had agreed that for the next week he would offer his thoughts and respond to comments and questions. At any time during the week, Fleming could review the ongoing discussion and make comments or inquiries. She found that the forum allowed her to discuss quality with important thinkers who otherwise would not be available.

One day Fleming and Konopacki were discussing a conflict brewing between two members of the quality improvement team. City hospital had a large transplant population, and a survey of patients had shown that they needed more information about prednisone therapy. There had been a growing debate among staffers who felt that videotapes were at least as good as patient and family classes and in the long run far less expensive. Konopacki was confused because the parties to the conflict were getting pretty emotional about something that didn't seem to justify it, so he decided to use the QISS Conflict Analysis module. This module (originally developed for international negotiations) has proven helpful in field tests of resolving conflict. Conflict Analysis helped Konopacki understand the sources of the conflict and led him to generate ideas for resolving it.

Over time the conflict was resolved, and the group decided to adopt videotapes as a new instructional medium. Konopacki wondered whether anyone on the system had begun working on such tapes, so he placed a notice on the QISS Bulletin Board. He also entered Matchmaker, where he asked for a list of consultants or organizations on patient education in the data base. Two organizations apparently had worked on this topic with other QISS hospitals, and the name, address, and telephone number of each were displayed, along with the name of someone at a QISS hospital who had been in contact with that organization. Konopacki called the two organizations and found that while neither was developing videotapes, one knew of a university that was producing and testing such tapes for transplant patients. Konopacki contacted the university and arranged for City to become a test site for the tape. He updated the Project Tracking program in QISS (as he had done periodically throughout the project) to allow others to share his experiences.

Summary

We have described our current efforts to construct a policy support system that responds to the individual differences of policymakers.

It employs a variety of the tools (e.g., Bayesian and multiattribute value models, conflict and implementation analysis, priority setting, and group process techniques) which will be discussed later in this book. The idea behind this system is to provide easy access to these resources for policymakers when they are ready to use them. In order to gain complete acceptance, this support system must be simple to use. We have not designed it for the analyst but for policymakers themselves. This is not to say that analysts should not use it—rather, we want it to be so easy to use that no training or technical sophistication is required. The system is designed so that people of all backgrounds can draw on the resources as they need them. We are convinced that such simplicity is needed for policymakers to actually use policy support systems.

References

Berwick, D. 1989. "Continuous Improvement as an Ideal in Health Care." *New England Journal of Medicine* 320 (January 5): 53–56.

Scholtes, P. 1989. *The Team Handbook.* Madison, WI: Joiner Associates.

Section II

Methodological Foundations

5

The Integrative Group Process

The integrative group process (IGP) is a method of facilitating meetings of experts. It is especially useful in efforts to model the experts' thought processes. Because the IGP borrows from several processes developed during 1960s and early 1970s, this chapter selectively reviews the literature on group processes from that period. Most group processes help members arrive at a judgment, but IGP goes a step further by recording details of their discussions, which enables their reasoning to be modeled, documented, and used by others.

Experts are a valuable resource for making a wide variety of decisions. While many analysts tend to rely on a single expert's knowledge, for several reasons we prefer using panels of experts. The biases and inconsistencies of individual experts working alone are difficult to detect. A group of experts is generally more accurate than an individual, especially if group members examine the merit of one another's ideas and are not unduly influenced by factors unrelated to the substance of the deliberations (such as personality conflicts).

Groups meet for different reasons, and the meeting procedure chosen should reflect those reasons (Wright and Ayton 1987). When the goal is to resolve a policy issue, the process should manage time, conflict, and member participation:

This chapter has benefited from comments by John Rohrbaugh, Judith Hall, and Armando Rotondi.

- Time management is important because issue resolution requires many more judgments than the average meeting. Without effective time management, a meeting can end before the experts' reasoning has been completely modeled.
- The process should use conflict among members in a productive fashion because group members with different backgrounds are likely to disagree. Some disagreements are necessary and increase the group's awareness of different perspectives, but others are dysfunctional. A group process should accommodate some degree of difference among members.
- The process should maximize participation by group members. Without effective participation and a sense of ownership, the model will lack advocates and in time will be abandoned, instead of being used and improved upon.

Successful meetings need structure and a great deal of preparation. Otherwise, meetings can frustrate participants and produce a misleading result.

This chapter will outline the integrative group process first used by Gustafson, Fryback, and Rose (1981). The name derives from the fact that the process integrates aspects of several existing techniques for conducting groups. We feel the success of IGP depends on its eclectic borrowing of techniques that have been proven to increase the effectiveness of meetings. In the ensuing sections, we refer to other approaches to managing groups to explain how IGP depends on research findings.

Despite the research showing the effectiveness of IGP's components, there are no systematic data on IGP as a whole and thus no proof that it is the best means of modeling experts' judgments. Our preference for using IGP to model group judgment stems from our experience with a variety of group processes. Specifically, IGP has these advantages:

- It reduces the likelihood that the facilitator will be surprised by events in the face-to-face meeting.
- It increases the commitment and participation of all members because it treats each person as a valuable asset.
- It produces high-quality results that are comparable across groups and facilitators (Gustafson et al. 1986) and does so within a guaranteed time frame.

Because the IGP is not linked to a particular mathematical model, it can be used with approaches as diverse as value models (Chapter 7), Bayesian models (Chapter 8), decision trees (Chapter 10), needs assessment and priority setting (Chapter 13), and other methods (such as rule-based systems) which are not presented here. In focusing on IGP, this chapter does not concentrate on the model type but rather on managing groups so that effective models can be constructed. (A more in-depth treatment of the IGP can be found in Gustafson 1990.)

Something New, Something Old

The literature on group processes is large and diverse. Here we selectively review processes from which IGP borrows key concepts.

Nominal group technique

Nominal group technique (Delbecq, Van de Ven, and Gustafson 1975) is a generic name for face-to-face group techniques that consist of these steps: silent idea generation, round-robin sharing of ideas, clarifying discussion and debate, individual reassessment, and mathematical aggregation of revised judgments. The primary appeal of this technique is that it produces a prioritized list of ideas in two hours or less.

While the nominal group technique is widely used and is good at fostering originality, it is not intended for all situations. In tasks that require ordering or evaluating several alternatives, the results are inferior to the judgment of the most knowledgeable participant (Holloman and Hendrick 1972; Nemiroff, Pasmore, and Ford 1976). In addition, the nominal group technique does poorly in terms of acceptance, which is a crucial determinant in whether a model is actually put to use. In these situations, processes with less structure tend to produce group decisions with a higher degree of acceptance (Stumpf 1978; Maier and Hoffman 1964; Miner 1976).

Extensive research on the nominal group technique (Campbell 1966; Dunnett Campbell, and Jaastad 1963; Vroom, Grant, and Cotton 1969; Gustafson et al. 1973; Van de Ven 1974; Stumpf 1978) suggests that three principles contribute to its success. First, ideas should not be evaluated one at a time—rather, the facilitator should collect many ideas before any is evaluated. Postponed evaluation promotes more creative solutions. Second, in assigning priorities, discussion and reestimation improves accuracy because a sort of bootstrapping occurs, in which members listen to fellow participants and revise their own opinions based on what they have learned. Third, ideas generated individually in silence are more numerous than those arrived at while listening to fellow participants, and they tend to be more creative as well. These data have induced us to borrow several methods from the nominal group technique: silent generation of ideas, postponement of evaluation, and revision of numerical estimates after group discussion.

Delphi technique

The Delphi technique (Dalkey and Helmer 1967) is a non–face-to-face procedure for aggregating group members' opinions. In this procedure, members answer questionnaires, and the facilitator summarizes the responses and mails the synthesis back to members for comment.

In some applications, for example in forecasting technological changes based on the insights of a large group of experts, the delphi method has proven useful (Campbell 1966; Linstone and Turoff 1975; Basu and Schroeder 1977). Delphi is also useful if conflict or status differences among group members are strong enough to threaten the group's functioning (Delbecq, Van de Ven, and Gustafson 1975). In other circumstances, face-to-face interaction is superior to Delphi's remote and private opinion gathering (Sackman 1974). Two studies comparing nominal group technique to Delphi found the techniques about equal (Seaver 1977; Miner 1976). In other studies, Delphi has been less accurate than nominal group technique. In one study, Delphi's remote feedback reduced the accuracy of the members' estimates (Gustafson et al. 1973). This research induced us to structure IGP so that individuals' ideas are assessed remotely and later subjected to face-to-face interaction and improvement.

Group communication strategy

The group communication strategy originates from a set of normative instructions proposed by Hall and Watson (1970). According to these instructions, conflict-reducing techniques such as changing opinions, majority voting, and bargaining should be avoided in group interactions, as should behavior such as arguing and making win-lose statements. In addition, the strategy teaches that differences of opinion are natural and initial agreements should be suspect.

Nemiroff, Pasmore, and Ford (1976) found that the group communication strategy leads to high-quality judgments and increases the group's later acceptance of the judgments. These studies used student subjects in situations where status differences may not have a been major factor. Holloman and Hendrick (1972) showed that members of groups give more weight to the opinions and suggestions of people with high status, so this process may not serve such groups. Furthermore, little is known about the success of normative instructions if group members are in conflict or have substantial stakes in the final judgment. The IGP facilitator uses the normative instructions of group communication strategy to maintain a productive discourse among the group members.

Social judgment analysis

Concern over the inadequacy of feedback in the Delphi process led to Rohrbaugh's (1979) study of two types of feedback. In one, the outcome of the group's deliberations is fed back to the group. In the other, named cognitive feedback, the logic behind the decisions, but not the outcome, is fed back. The social judgment analysis (Hammond et al. 1976) is a group process that also provides cognitive feedback. It asks each member to rate a series of scenarios and provides feedback using statistical analysis techniques like regression to identify the importance of the components in the scenarios in the ratings. An initial consensus model is built and used to predict a new series of scenarios that was also rated, noting instances in which the predictions of the statistical model seem inadequate. Members revise the statistical model until they reconcile the discrepancy between it and their intuition.

Rohrbaugh's study found this group process to be more accurate than the delphi method, but judgments were inferior to those of the best member. Unfortunately, no intervention has been shown to consistently raise group performance to the level of the best member. In contrast, about half of the groups in Rohrbaugh's study of social judgment analysis did achieve results better than their best member.

The effect of social judgment analysis on member acceptance

of the group results has not been studied. Excessive use of computers and statistical models may reduce both interaction among group members and acceptance. On the other hand, reasonable use of both tools could simulate and inform interaction, arouse interest and debate, and enhance members' acceptance of the final model.

The IGP borrows from the social judgment analysis the ideas of scenario analysis and the notion that feedback to the group should focus on the reasons behind an individual's judgment. This focus on reasons may be superior to focusing on conclusions because it gives individuals reasons to change their opinion. In contrast, stressing the consensus may increase peer pressure and herd mentality at the expense of correct judgment.

Cognitive mapping

At about the time that IGP was put together, Eden, Jones, and Sims (1983) developed a group process for modeling expert judgments called cognitive mapping. This process starts by constructing two parallel statements of the problem, one showing the factors creating the problem, the other showing the factors leading to a satisfactory solution. For example, if a cause of high labor costs is "shortage of qualified workers," a solution could be "increase availability of qualified workers." Through linguistic manipulation of the problem statement, cognitive mapping tries to spark new ideas. Clearly, causes of the problem and factors leading to its solution are related, and usually a minor change in wording is sufficient to convert one into the other.

When using cognitive mapping, Eden and colleagues often collect a member's ideas about the problem and solutions individually and then revise these ideas in a meeting. Occasionally they quantify the influence of causes and effects through a round-robin process, in which members write down their estimates and share them afterward. The facilitators then simulate how changing one factor would affect the problem and use these simulations to stimulate new insights into the problem.

Both cognitive mapping and IGP solicit opinions of group members before face-to-face interaction, but there are some differences. IGP does not insist on the creation of two opposite views of the same problem, a linguistic manipulation that Sims, Eden, and Jones (1981) report is helpful in problem solving. IGP is not tied to the use of any particular mathematical models and can be used with different modeling approaches.

Summary of what is borrowed

IGP and the nominal group technique both attempt to improve estimates through repetition. Both require each member to assess, discuss, and revise their numerical estimates, and both postpone the evaluation of ideas until the facilitator has collected everyone's ideas. Delphi and IGP share private, remote generation of opinions, but only in IGP are these remote interactions followed by in-person interaction. The IGP facilitator uses the group communication instructions as ground rules for group interaction. Both the IGP and the social judgment analysis ask group members to rate scenarios and focus discussion on a member's reasoning rather than the group's choice and consensus. Like cognitive mapping, the IGP facilitator privately solicits ideas and revises them in a group meeting.

Steps in the Integrative Group Process

The IGP has these steps:

1. *Choose the group's participants,* and secure their involvement.
2. Develop a *straw model* through telephone interviews or meetings with individual members. In an evaluation task, for example, the straw model would be a listing of the evaluation criteria or attributes.
3. Convene the group, and ask members to *revise the straw model* to reach a behavioral consensus on a set of attributes and measures.
4. *Design case scenarios* based on the group's model, and ask the group to judge them one by one.
5. Ask the members to individually *enumerate the model* by estimating the relative importance of each attribute.
6. *Identify sources of conflict,* and ask for clarification, discussion, and reassessment.
7. *Average the smaller differences,* using a weighting scheme if appropriate.
8. *Report* the group's judgment, *analyze* whether it agrees with the experts' in-person judgments, and *document* the findings.

Step 1. Composing the group

The composition of the group is an important and relatively controllable aspect of the process. The facilitator can choose members based on their expertise in the subject and whether they represent an interest group affected by the judgment. (Expertise is essential for its own sake, but it also helps if users of the model recognize the expertise that helped develop it.)

Some thought should be given to choosing organizational insiders or outsiders. Some authors believe that certain meetings should be staffed by people from outside, but others disagree.

> When there is no quality requirement for a decision, the use of experts would probably waste organizational resources. If the group's decision affects individuals or groups outside the organization, co-worker membership is inappropriate. Co-workers are less likely to develop a decision acceptable to outsiders because of their local orientation and because outsiders would have no influence in the decision. (Stumpf 1978)

We think that if the coworker is an expert on the subject and is well respected, there is no reason to ignore an expert at hand in favor of one from the outside.

It is best to use representatives of particular perspectives when acceptance of the decision is the prime criterion. For example, if we were developing a model to predict cocaine abuse among teenagers, we would want both theoretical and practical experts who approach cocaine abuse from psychological, sociological, and biological perspectives. Early involvement of representatives of perspectives that will eventually implement the proposed solution increases the chance that their interests and insights will be included.

The number of experts in the group should depend on its purpose. Experiments with group sizes have shown that if the quality of the solution is paramount, it is useful to include the diverse input of a large group (e.g., seven to nine people). If the degree of consensus is most important, it is useful to choose a smaller group (e.g., five to seven), so that all members can have their opinions considered and discussed (Cummings, Huber, and Arendt 1974; and Manners 1975). It is a general rule of thumb that the group size should not be smaller than five, because domination is likely in such groups; and it should not be larger than nine, because this size prevents some members from fully participating.

Heterogeneity of the group's background is closely related to the size of the group and is another important aspect of design. A necessary, though not sufficient, requirement for accurate group judgments is to have an appropriate knowledge pool present. Since no one is expert in all aspects of a problem, diverse backgrounds and expertise are imperative for this knowledge pool. The inclusion of theoretical and practical expertise from sociological, biological, and psychological perspectives is a good example of amassing sufficient expertise. Difference in background and knowledge could, however, accentuate the conflict among members, and if the purpose of the group is to create a strong consensus behind a decision, the facilitator should select members to minimize differences in backgrounds.

Getting people to devote their time to a meeting is often difficult because many people have wasted time and effort in meetings. The analyst can take a number of steps to increase participation. First, examine the purpose of the meeting. If it is difficult to obtain participation, perhaps the analyst is attempting to solve a trivial problem or proposing a solution that is so impractical as to be meaningless. Invited participants are likely to accept if the meeting addresses a problem they consider important and the process seems likely to result in action. One can approach the problem of action by asking, "What will be different after this meeting? What do you or others plan to do with your model of the group's reasoning?"

Several other points tend to increase participation:

- Remind participants of who else is being invited.
- Explain who nominated the person you are soliciting.
- Stress that very few people are being invited.
- Emphasize that each person's contribution is unique.
- Explain that the meeting is an ad hoc group, not an ongoing commitment.
- Clarify the logistics (when, where, and how)
- Partially reimburse invitees for their time (with an honorarium).

Step 2. Constructing the straw model

After being chosen, group members are interviewed individually, by phone if necessary, before the face-to-face meeting. During the interview, which will take roughly one hour, the facilitator reiterates the group's task and walks the expert through the steps in the IGP.

This prevents surprises at the meeting. The bulk of the interview, however, is devoted to listening to the member and trying to understand the model (the desired outcome, the predictive factors, and any measures of these factors) that underlies his or her opinions.

Suppose we were developing a model (or equation) to measure severity of illness for burn victims. The interviewers with panelists would tell them to assume they were asked to evaluate the severity of burn victims but could not see or touch the patients. All they could do is ask questions about the patients. Someone else would go get the answers for them. What questions would they ask? What answers would make them happy for the patient, and what answers would make them sad? These sample answers give clues to how a factor might be measured. For instance, a happy answer to the factor "body area covered by full thickness burn" might be "less than 5 percent with no burn on face." Such an answer suggests a measure that includes location and percentage of body area covered.

The following process might be used in constructing a model to evaluate alternative courses of action.

Early in the telephone conversation, the facilitator prompts the participant with open-ended questions and encourages reflection on personal experiences. The facilitator could ask the participant to recall an occasion where he or she attempted a similar evaluation. Later, the facilitator would ask questions with more focus, such as "What are the most important evaluation criteria?"

After interviewing each group member, the facilitator collates all responses into the straw model. This model is designed as a working framework for further revisions. Constructing the straw model before the meeting has three advantages. First, it increases the accuracy of the group's judgment by ensuring that the group process does not prevent an individual from presenting ideas. Second, the phone interviews collect ideas beforehand and save group time. Third, because the straw model serves as a blueprint for the discussion, it saves additional time by helping partition the group's task into smaller, more manageable segments.

In the next step, the face-to-face meeting starts with a presentation of the straw model and a request for improvements to it.

Step 3. Revising the straw model

During step 3, the group integrates the ideas of its members while revising the straw model. The facilitator opens the meeting by introducing himself or herself, explaining the task and restating its importance, and asking members to introduce themselves. Without these self-introductions, members are likely to waste time by implicitly introducing themselves throughout the deliberations. Instead, encourage members to describe their achievements and importance at the start so the group can move on to its task.

After the introductions, the facilitator presents the straw model, the purpose of the meeting, and the agenda, and the panelists are asked to suggest improvements to the straw model. Sometimes these will result in a whole new model, but most often there will be additions, deletions, or restatements of factors and/or the possible measures. In the burn example, gender had been suggested to be an important factor in the telephone interviews. But at the meeting it was decided the model could function without gender.

We find it very useful to have the facilitator record key elements of the discussion on flip charts displayed before the group. This lets panelists know their comments are being heard and helps to keep the focus of the discussion on the task and not the panelists. It also allows panelists to correct the recording if their ideas have not been properly represented.

The facilitator uses the instructions from group communication strategy as guidelines for interaction during the meeting. Active listening (such as nodding and asking for clarification) and recording ideas on the flip charts help the facilitator play the vital roles of directing the discussion, preventing premature closure of ideas (Van de Ven and Delbecq 1971), returning the group to pertinent activities and topics, distributing the group's time over different aspects of the task, and restraining criticism while ideas are being generated. During this process, the facilitator must continually address the group as a whole so no member begins to "own" the resulting ideas. The process continues until the group identifies, discusses, and revises all relevant factors.

Step 4. Designing case scenarios

Many people have difficulty describing how they think and reason about certain issues, and in particular they have trouble translating many years of experience into a few simple rules or mathematical formulas. In real life, many judgments are intuitive and difficult to express. To help members with this trouble, we ask them to make so many judgments that they begin to consciously derive a heuristic that simplifies the task. These heuristics are the first step toward making members aware of their unconscious reasoning.

Some investigators have questioned the reliability of experts' reports of their thought processes (Nisbett and Wilson 1977). The reliability of these reports, like any other data, depends on the mechanism by which they were generated. Experts' reports made shortly after the reasoning is done are more reliable than those made much later (See Ericsson and Simon 1980, for a more detailed discussion). Therefore, one purpose of this step of the IGP is to ensure that experts make many judgments before the facilitator asks them to weight the role of different factors in their reasoning. Furthermore, because different situations can highlight different issues, experts are encouraged to make many judgments and evaluate many different situations. This prevents any one issue from being artificially overemphasized.

Since the group members are removed from their usual work, it is useful to devise exercises that solicit their judgments and refresh their memory about how they approach complicated issues. At this point, the meeting may be interrupted to allow the facilitator to design these exercises. By scheduling step 3 of the model building from two to seven P.M., the evening and, if needed, the early morning are available to prepare materials for the following steps.

One can elicit experts' judgments with a case scenario—a description of a hypothetical situation or person. For example, if the panel suggests that three attributes (age, burn site, and degree of burn) determine the severity of a burn injury, then a case scenario might look like this:

Case 1

Age	55
Burn site	Head
Degree of burn	2nd

New case scenarios are defined by changing at least one attribute. A second scenario might look like this:

Case 2

Age	40
Burn site	Arms
Degree of burn	3rd

The patient described in case 1 is more severely burned and has a poorer prognosis than the patient described in case 2. By changing the levels of the attributes, different patients with different levels of burn severity can be described and assessed.

Some people generate scenarios by choosing levels of various attributes at random. Although such an approach is easy to do, it frequently creates unrealistic and/or extremely rare situations. Hammond (1955) encouraged facilitators to use scenarios that represent likely situations. Few studies, however, show how this can be done. These methods (none of them tested by research) are feasible:

- Ask members to compose case scenarios.
- Abstract cases from the actual population under study.
- Use correlation between attributes to generate scenarios.

We prefer to generate scenarios randomly and then ask members to review them and identify which are realistic and which are not. If cases are judged improbable, the reasons are recorded so future users of the model can determine the suitability of the model to their conditions. Alemi (1986) showed that such an approach can improve the accuracy of improperly specified models.

After relevant case scenarios are chosen, the experts evaluate them to help sort out their thinking about the problem. In the burn severity example, the experts rated the extent of a patient's severity on a scale from 0 to 100, where 100 was the most severe burn. Each scenario is written on a separate piece of paper. The group members rate the first five to ten scenarios, discuss their assumptions, and rate them again. The ratings for each problem are averaged, and consensus is not forced, because the discussion here aims to clarify assumptions and not to change others' ratings.

If members refuse to rate the first set of scenarios, the facilitator's inquiries may reveal that a crucial variable is missing. Because one purpose of the scenarios is to find what other information an expert would need to rate a case of this nature, the suggestions should be added to the scenarios and the process continued. However, revisions of the scenarios at this stage delay the group and signal that step 3 was not performed thoroughly. It is far better to do a careful job in step 3 and obviate the need for modifications in step 4.

Once the first set of scenarios has been rated, members are asked individually to rate the remaining scenarios (there may be as many as 100 more). As members do this, the facilitator lists the ratings in front of the group. When large discrepancies in ratings arise, the group pauses to discuss, using the estimate-talk-reestimate process (Gustafson et al. 1973).

If the goal of the meeting is to make members more aware of plausible situations (not to actually construct a model of the group's reasoning), the meeting may end after the ratings are done. If, for example, the purpose of the meeting is to make group members more sensitive toward the variations in burn severity, then rating several scenarios is sufficient, and there is no need to follow the remaining steps in IGP. Under the circumstances, the scenarios about different burn patients would be constructed, adorned with details to increase their credibility, written into narrative, and given to decision makers for discussion. Shoemaker (1989) provides examples of the use of scenarios in strategic planning. Chapter 15 provides additional examples of scenario-based planning.

While some group meetings end at this stage, others proceed beyond scenario development to quantification of the model in step 5. When the aim of the group meeting is to develop a quantitative model, repetitive judgment over a large number of scenarios makes members aware of their own thought processes and enables them to answer specific questions about how they weight various factors. This is the next step in the IGP.

Step 5. Quantifying the model

Now the facilitator chooses a model and designs a questionnaire to estimate its parameters. The IGP can be used to construct various mathematical models, but whatever model is being produced, you must always ask the group to estimate its parameters. For example, when group members construct a value model, they must estimate the relative importance of various attributes (see Chapter 7 for information on constructing such a model). If a Bayesian model is to be constructed, members must specify the likelihood ratios associated with different factors (see Chapter 8).

Each parameter of the model is estimated on a separate page in this questionnaire. A question about likelihood ratios associated with one factor would be placed on a different page from a question about likelihood ratios for another factor. Similarly, a question about the utility function in one attribute would be put on a different page from a question about the utility function in another attribute. Separating the questions in this manner has two purposes. First, it focuses the members' attention on one topic at a time. Second, it enables the facilitator to direct the discussion to a particular page when discrepancies arise among various estimates.

Group members work individually in front of the group. Seeing each other working encourages the members to exert more effort on the task. As the group proceeds, the facilitator collects the responses on a flip chart. When there are major differences among the ratings, the facilitator asks the members to explain their reasoning.

Step 6. Identifying sources of conflict

With a group of experts, conflict is inevitable. Several studies (Hammond et al. 1976; Rohrbaugh 1979) emphasize the role and usefulness of cognitive feedback in resolving conflicts over social issues. In IGP, such feedback is provided by asking group members to describe their logic.

The facilitator directs attention to the conflict, not to the proponents of the various positions, because disclosing the participants' identity could have the unwanted effect of polarizing the group. Castore and Muringham (1978) also showed that disclosing individuals' preferences lowers the degree of ultimate support for the group's product.

To save time, the facilitator should identify major differences among members and focus attention on them. Small differences are probably caused by errors in estimating numbers and not substantive issues, and therefore there is no need to discuss them. When the facilitator identifies an area of major disagreement, he or she asks for a discussion of the pertinent issues and a reestimation of the importance of the criterion. This process, which is used in rating scenarios, is used here to improve the accuracy of model parameters.

Disagreements in a group arise from many sources. Those resulting from an unclear specification of the problem may be reduced by clarifying the assumptions behind a member's perspectives. Disagreements resulting from differences in knowledge and experience may be reduced if members communicate the rationale behind their judgments. Disagreements resulting from value differences, however, are not always ameliorated by communication. Value differences caused by lack of clarity can be reversed by communication to reduce apparent disagreements, but those that are truly caused by underlying, irreconcilable value differences cannot. In these circumstances, discussion may clarify assumptions and reduce conflict.

Step 7. Averaging smaller differences

Although the facilitator has encouraged members to resolve their differences through discussion, at some point it is necessary to end the interaction and resolve minor group differences mathematically, perhaps by averaging estimates from individuals. An averaging scheme assumes that each group member has the same input in the final model. Several researchers have experimented with procedures for differential weighting of group members. Dalkey et al. (1969) asked members to rate themselves and used these ratings to differentially weight individual opinions. Stael Von Holstein (1970) found that differential weighting schemes (when one expert's views are given more weight than those of another expert) based on group members' past performance were more accurate than equal weighting schemes. Morris (1974) has formulated a scheme for differentially weighting members' opinions according to opinions of the model's users.

Rowse, Gustafson, and Ludke (1974) investigated five different methods of differential weighting of group members, none of which consistently improved the group's judgment. Despite a large number of studies on differential weighting (Seaver 1977; Sticha 1977; Hogarth 1977), we recommend using equal weights because this method seems to serve just as well.

Step 8. Evaluating

The final two steps in the meeting—evaluating it and writing a report of it—may occur days after it has ended.

This report contains topics such as:

- Why was the meeting convened?
- What methods were used to facilitate the meeting and analyze its findings?
- Did participants reach a consensus?
- If so, what was the model? (This section includes a detailed account of the model, including factors, their operational definition, and the scoring system for them.)

- Does the model simulate the opinions of the experts who developed it?
- How could the model be used in a realistic environment?

The evaluation could be broken into two sections: whether the group reached a consensus, and how well the model simulated this consensus.

We often rely on the scenario ratings to evaluate consensus among the experts. Recall step 4, when members rated 100 hypothetical cases.

If there is a consensus, the report also examines whether the model can simulate the opinions of its expert developers. The model score is calculated for each of the 100 scenarios and compared to the group's rating of each scenario. If the scores and the group's intuitive ratings agree, the facilitator has effectively modeled the panel's expertise. Otherwise, even though the group was in consensus, the members' expertise and knowledge are not simulated by the model, and a new model must be constructed. As before, correlations can be used to examine the similarity between model scores and the group's ratings of the scenarios. Correlations above 0.75 suggest the model is very predictive of the group's consensus; correlations below 0.60 suggest the model is marginally predictive at best.

Features of the IGP

The IGP has several advantages for distilling the expertise of a group into a mathematical model available to nonexperts. In every step, the IGP attempts to increase the availability of ideas and to evaluate those ideas on merit alone. For these reasons, we expect the process to lead to more accurate models. Commitment to the group's decision is also important, and the IGP encourages group members to believe in their model and thus increases their support for its use.

We feel that IGP's potential for greater accuracy and increased acceptance merits further experimentation.

Quality ideas

Groups are likely to arrive at "good" decisions if many quality ideas are available and they process these ideas expeditiously on the basis of merit. Some steps in the IGP increase the number and quality of ideas, while others increase the chance that ideas are processed promptly and based on merit. In step 1, the facilitator manipulates group composition and size to achieve a useful, heterogeneous background, which is intended to increase the availability and quality of ideas.

In step 2, the facilitator interviews each group member to construct a straw model. Because opinions are sought privately, dominant members cannot prevent less assertive ones from expressing ideas. This step allows the group more time to sort out the merits of various ideas. The straw model also disassociates the people who generated the ideas from the arguments about them. Preventing ownership of ideas encourages members to judge the merit of an idea rather than the person who presented it. Thus, creating the straw model may reduce distortions caused by excessively persuasive group members.

In step 3, the group meets to discuss the straw model. Seeing other members at work on the task encourages individuals to increase their efforts and search their memories more extensively (Hackman and Morris 1975), thus further enhancing the pool of available ideas. Because group interaction at this stage is focused on the reasons behind the judgments, members check one another's assumptions, thus further evaluating the merit of various ideas. Finally, because the facilitator records ideas in full view of the group, IGP reduces distractions during the processing of ideas.

Steps 4, 5, and 6 are concerned with quantification of ideas already presented, not with generating new ones. Nevertheless, in judging scenarios, new ideas emerge as group members discover that other pieces of information are needed before a judgment can be made.

The IGP is unlikely to sacrifice ideas to time pressures. Quantification helps the facilitator save time by focusing attention on major disagreements, not minor ones. The facilitator also directs attention away from small points which can soak up time at the expense of larger issues.

Promotion of commitment

Because many groups exist not only to take advantage of expertise of their members but also to gain a constituency for their products, a group process has a political dimension. Subsequent commitment to the group judgment appears to be a function of the amount of socialization that occurs during the process. If this is true, unstructured groups, with their unrestricted interaction, should produce higher commitments than structured groups. Miner (1976), for example, argues that unstructured groups also allow freer expression of feelings (as opposed to facts) and thus are more likely to lead to higher acceptance.

The formation of new social ties is another advantage of IGP. The two days usually needed for the process are ample for members to meet informally. Our experience suggests that eating together seems especially effective at increasing cohesiveness. Participants often use time before and after meetings to discuss matters of mutual interest. The group activity, by virtue of both its placement outside normal work and its length, provides new opportunities for communication among members. This communication can enhance social ties among members and in turn reinforce commitment to the group modeling effort.

Another way to increase commitment of members and outsiders is to make sure everyone understands the rationale for the decision. In IGP, this is done by documenting the discussions and revisions that produced the model. Burnstein et al. (1973) had individual subjects list the persuasiveness of arguments pro and con. The net balance of persuasiveness of the arguments correlated with attitude change produced by group discussion. But, more important, when individuals not present in the discussion were asked to read the same arguments, their attitudes changed similarly. Cinokur and Burnstein's work shows that the information content of group discussion is important for reproducing the group judgment. The IGP, by documenting the group's reasoning, increases commitment to the group's judgment.

IGP use in various settings

IGP is easy to learn. Gustafson, Fryback, and Rose (1983) trained five facilitators at different universities (some of whom had no background in modeling) to model the opinions of five panels of physicians. A comparison showed all models to be equally accurate, suggesting that IGP is easily learned.

The IGP is applicable to many problem-solving efforts, including mathematical or rule-based modeling as well as any other effort, so its application is not limited to assessment of utilities or probabilities. Most key elements of the process, including constructing and refining the straw model, stay the same. Investigators have used IGP to construct various types of mathematical and computing models, including probability models (Gustafson, Tianen, and Greist 1979; Alemi and Rice 1990), additive multiattribute utility models (Alemi et al. 1987; Alemi et al. 1990; Gustafson et al. 1990), and rule-based algorithms (Rotondi 1986).

References

Alemi, F. 1986. "Explicated Models Constructed under Time Pressure: Utility Modeling versus Process Tracing." *Organizational Behavior and Human Decision Processes* 38: 133–40.

Alemi, F., and J. Rice. 1990. "Acute Respiratory Track Infections in Children: Development of a Diagnostic Aid." In *Judgment and Decision Making,* edited by W. H. Loke. Singapore: Times.

Alemi, F., J. Stokes, III, J. Rice, E. Karim, R. Nau, W. Lacorte, and L. Saligman. 1987. "Appraisal of Modifiable Hospitalization Risks." *Medical Care* 25 (7): 582–91.

Alemi, F., B. Turner, L. Markson, and T. Maccaron. 1991. "Severity of the Course of AIDS." *Interfaces* (May).

Basu, S., and R. G. Schroeder. 1977. "Incorporating Judgments in Sales Forecasts: Application of the Delphi Method at American Hoist & Derrick." *Interfaces* 7: 18–27.

Burnstein, E. A. Vinokur, and Y. Trope. 1973. "Interpersonal Comparison versus Persuasive Argumentation." *Journal of Experimental Social Psychology* 9: 236–45.

Campbell, J. P. 1966. "Individual vs. Group Problem Solving in an Industrial Sample." *Journal of Applied Psychology* 52: 205–10.

Castore, C. H., and J. K. Muringham. 1978. "Determinants of Support for Group Decisions." *Organizational Behavior and Human Performance* 22: 75–92.

Cummings, L. L., G. P. Huber, and E. Arendt. 1974. "Effects of Size and Spatial Arrangements on Group Decision Making." *Academy of Management Journal* 17: 460–75.

Dalkey, N. C., and O. Helmer. 1967. "An Experimental Application of the Delphi Method to the Use of Experts." RAND Corporation, RM727PR.

Dalkey, N. C., B. Brown, and S. Cochran. 1969. "The Delphi Method." RAND Corporation, RM6115-PR.

Delbecq, A., A. Van de Ven, and D. Gustafson. 1975. *Group Techniques for Program Planning—A Guide to Nominal Group and Delphi Processes.* Chicago: Scott, Foresman.

Dunnett, M. D., J. Campbell, and K. Jaastad. 1963. "The Effect of Group Participation on Brainstorming Effectiveness for Two Industrial Samples." *Journal of Applied Psychology* 471: 30–37.

Eden, C., S. Jones, and D. Sims. 1983. *Messing About in Problems: An Informal Structured Approach to Their Identification and Management.* New York: Pergamon Press.

Ericsson, K. A., and H. A. Simon. 1980. "Verbal Reports as Data." *Psychological Review* 87: 215–51.

Gustafson, D. H., ed. 1990. *Modeling Risk.* Center for Health Systems Research and Analysis, University of Wisconsin-Madison.

Gustafson, D. H., D. G. Fryback, and J. H. Rose. 1981. Final Report of the Severity Index Methodology Development Research Project. Center for Health Systems Research and Analysis, University of Wisconsin-Madison.

Gustafson, D. H., D. G. Fryback, J. H. Rose, et al. 1983. "An Evaluation of Multiple Trauma Severity Indices Created by Different Index Development Strategies." *Medical Care* 21: 674–91.

Gustafson, D. H., D. Fryback, J. Rose, V. Yick, C. Prokop, D. Detmer, and J. Llewelyn. 1986. "A Decision Theoretic Methodology for Severity Index Development." *Medical Decision Making* 6 (1): 27–35.

Gustafson, D. H., F. C. Sainfort, R. Van Konigsveld, and D. R. Zimmerman. 1990. "The Quality Assessment Index (QAI) for Measuring Nursing Home Quality." *Health Services Research* 25 (1): 97–125.

Gustafson, D. H., R. K. Shukla, A. Delbecq, and G. W. Walster. 1973. "A Comparative Study of the Difference in Subjective Likelihood Estimates Made by Individuals, Groups, Delphi Groups, and Nominal Groups." *Organizational Behavior and Human Performance* 9: 280–91.

Gustafson, D. H., B. Tianen, and J. Greist. 1979. "Computer Based System for Identifying Suicide Attemptors." *Computers and Biomedical Research* 14 (December).

Hackman, R. J., and C. G. Morris. 1975. "Group Tasks, Group Interaction Process, and Group Performance Effectiveness: A Review and Proposed Integration." In *Advance in Experimental Social Psychology,* edited by Leonard Berkowitz. New York: Academic Press.

Hall, J., and W. Watson. 1970. "The Effects of a Normative Intervention on Group Decision Making Performances." *Human Relations* 23: 299.

Hammond, K. R. 1955. "Representative or Systematic Design in Clinical Psychology." *Psychological Bulletin* 51: 255.

Hammond, K. R., J. Rohrbaugh, J. Mumpower, and L. Adelman. 1976. "Social Judgment Theory: Applications in Policy Formation." University of Colorado Institute of Behavioral Science, Center for Research on Judgment Policy.

Hogarth, R. M. 1977. "Methods for Aggregating Opinions." *Decision Making and Change in Human Affairs,* 231–35.

Holloman, C. R., and H. W. Hendrick. 1972. "Effects of Status and Individual Ability on Group Problem Solving." *Decision Science* 3: 55–63.

Linstone, H., and M. Turoff. 1975. *The Delphi Method: Techniques and Applications.* Reading, MA: Addison Wesley.

Maier, N. R. F., and L. R. Hoffman. 1964. "Quality of First and Second Solution in Group Problem Solving." *Journal of Applied Psychology* 41: 320.

Manners, G. E. 1975. "Another Look at Group Size, Group Problem Solving, and Member Consensus." *Academy of Management Journal* 18: 715–24.

Miner, F. C. 1976. "The Effectiveness of Problem Centered Leadership,

Nominal Leadership, and the Delphi Process in a High Quality–High Acceptance Problem." Thesis, University of Minnesota.

Morris, P. A. 1974. "Decision Analysis of Expert Use." *Management Science* 20: 1233–41.

Nemiroff, P., W. Pasmore, D. Ford. 1976. "The Effects of Two Normative Structural Interactions on Established and Ad Hoc Groups: Implications for Improving Decision Making Effectiveness." *Decision Sciences* 7: 841.

Nisbett, R. E., and T. D. Wilson. 1977. "Telling More than We Can Know: Verbal Reports on Mental Processes." *Psychological Review* 84: 231–59.

Rohrbaugh, J. 1979. "Improving the Quality of Group Judgment: Social Judgment Analysis and the Delphi Technique." *Organizational Behavior and Human Performance* 24: 73–92.

Rotondi, A., C. Cleeland, et al. 1986. "A Model for Treatment of Cancer Pain." *Journal of Pain and Symptom Management* 1 (4): 209–15.

Rowse, G. L., D. H. Gustafson, and R. L. Ludke. 1974. "Comparison of Rules for Aggregating Subjective Likelihood Ratios." *Organizational Behavior and Human Performance* 12: 274–85.

Sackman, H. 1974. "Delphi Assessment: Expert Opinion, Forecasting and Group Process." RAND Corporation, Santa Monica, CA.

Seaver, D. A. 1977. "How Groups Can Assess Uncertainty: Human Interaction versus Mathematical Models." Paper presented at the IEEE 1977 International Conference on Cybernetics and Society, Washington, D.C.

Shoemaker, P. J. H. 1989. "Scenario Thinking." University of Chicago, Center for Decision Science.

Sims, D., C. Eden, and S. Jones. 1981. "Facilitating Problem Definitions in Teams." *European Journal of Operational Research* 6: 360–66.

Stael Von Holstein, C. A. S. 1970. *Assessment and Evaluation of Subjective Probability Distributions*. Stockholm: Economic Research Institute at the Stockholm School of Economics.

Sticha, P. J. 1977. "Coalition Formation and the Distribution of Influence in Decision Making Groups." Michigan Mathematical Psychology Program, technical report.

Stumpf, S. A. 1978. "Identifying Optimal Groups for Making Judgmental Decisions: An Experimental Study of Meta Decision Making." Dissertation, Graduate School of Administration, New York University.

Van de Ven, A. H. 1974. "Group Decision Making Effectiveness." Kent State University, Center for Business and Economic Research Press.

Van de Ven, A. H., and A. L. Delbecq. 1971. "Nominal versus Interacting Group Processes for Committee Decision-Making Effectiveness." *Academy of Management Journal* 14: 203–12.

Vroom, V. H., L. D. Grant, and T. S. Cotton. 1969. "The Consequences of Social Interaction in Group Problem Solving." *Organizational Behavior and Human Performance* 4: 77–95.

Wright, G., and P. Ayton. 1987. "Eliciting and Modeling Expert Knowledge." *Decision Support Systems* 3: 13–26.

6

Introduction to Decision Analysis

This chapter describes policy information management from the perspective of decision analysis. A decision model is presented first in a simplified form and later expanded to incorporate competing power groups (also called constituencies). A process for using decision analysis in policy analysis is introduced. Following chapters will elaborate on important components of this process, and later chapters will present case studies of how these components have been used in policy analyses. While the decision analysis process is presented here as a conceptual whole, we have often used parts of the process to address policy issues and have rarely used the whole process.

Uncertainty and values are both important in determining which action to select. The means by which the PIMS can help capture uncertainties and values is exemplified by a hypothetical situation faced by the head of the state agency responsible for evaluating nursing home quality: A nursing home has been overmedicating its residents in an effort to restrain them, and the state must take action to improve care at the home. The possible actions include fining the home, prohibiting admissions, and teaching the home personnel more appropriate use of psychotropic drugs.

Any real-world decision has many different effects. For instance, the state could institute a training program to help the home improve its use of psychotropics, but the state's action could have effects beyond changing this home's drug utilization practices. The nursing home could become more careful about other aspects of its care such as how it plans care for its patients. The nursing home industry could become convinced that the state is enforcing stricter regulations on administration of psychotropic drugs. All these effects are important dimensions that should be considered during the analysis and in any assessment performed afterward.

The problem becomes more complex because the policymaker must consider which constituencies' values should be taken into account and what their values are regarding the proposed actions. For example, the policymaker may want the state to portray a tougher image to the nursing home industry, but one constituent, the chairman of an important legislative committee, may object to this image. So the choice of action will depend on which constituencies' values are considered and how much importance each constituency is assigned.

The problem becomes even more complex because a particular outcome cannot be guaranteed. This is because three types of uncertainty typically confront the policymaker: diagnostic, future, and cause-and-effect.

The policymaker is unsure of the nature of the problem in *diagnostic uncertainty*. This causes difficulties because a successful solution must start with correct identification of the problem. For example, if the nursing home was simply ignorant of how to use psychotropic drugs appropriately, a decision to prohibit further admissions would just alienate the nursing home without solving the problem. In this case, training would not only solve the problem but also foster a positive relationship between the home and the state. If, however, the nursing home knew exactly what it was doing—cutting costs with chemical restraints—a consultation and training program could be fruitless. But while it is important to know as much as possible about the nursing home's reasons for its drug policy, one can never be certain about those reasons, and almost any decision includes an element of diagnostic uncertainty.

Future uncertainty reduces confidence that a particular action will lead to the predicted outcome. In characterizing a decision problem, the PIMS must identify which future events will affect the relationship between action and outcome, and analyze what the impact of those events will be. Suppose that during the time the policymaker is deciding to prohibit further admissions, the constitutionality of state regulation of admissions is being challenged in court. If the state loses the court test, then selecting that option (prohibiting admissions) could be futile.

Future events can be divided into two important groups: (1)

unlikely events that would have monumental consequences and (2) likely events that would have lesser but significant consequences. We believe the first category tends to receive undue importance during decision making.

Cause-and-effect uncertainty reflects the basic tenuousness in the relationship between action and outcome. In general, too little is known about most problems—even those that seem clearly understood—to be sure that a specific action will lead to a particular outcome. This is true even if no major events intervene. In the nursing home example, even if the inappropriate chemical restraint was clearly based on ignorance and the state had already won the court case, one cannot be sure training would correct the problem. Perhaps the home personnel did not have the education or motivation to use the drugs properly.

Organizing Thoughts

Some of the PIMS's greatest contributions should be in helping policymakers or problem-solving teams structure their thoughts. The basic model for doing this is portrayed in Figure 6–1. Specifically, a PIMS should help policymakers:

- Identify possible actions
- Identify possible outcomes
- Attach values to those outcomes
- Identify major uncertainties (events that, if they were to occur, would interfere with an action's leading to a certain outcome)
- Estimate the chance that these uncertain events will occur

Figure 6-1 is a simplified—perhaps oversimplified—representation of the real world. One step toward reality is to incorporate a mechanism for recognizing that certain constituencies (or power groups) will see the same issue differently. Various constituencies often attach different values to the same outcome and may even see events as carrying different levels of uncertainty. A PIMS (especially in the policy arena) must be capable of representing these variations in how different constituencies look at the same problem.

Figure 6-1 Conceptual Structure for a Decision Problem

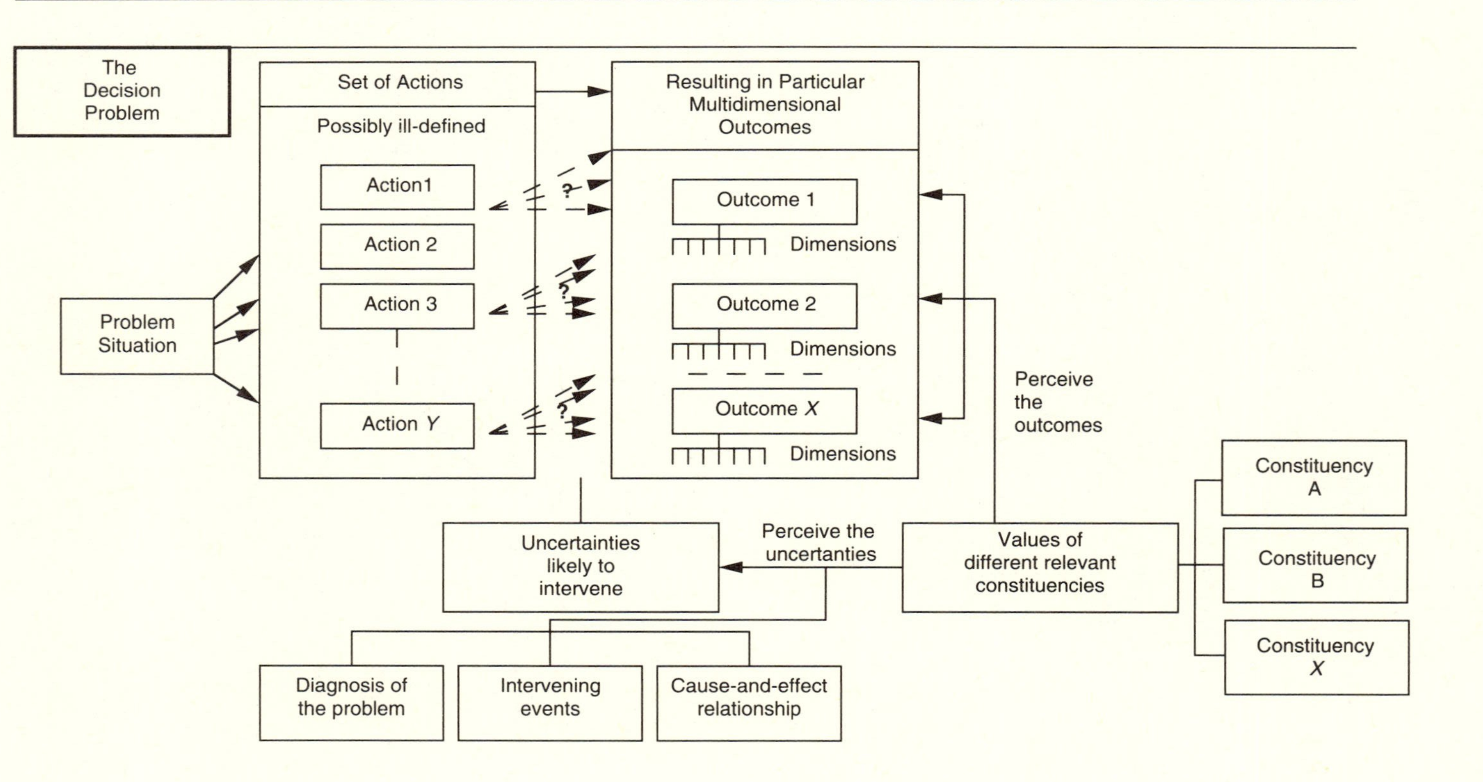

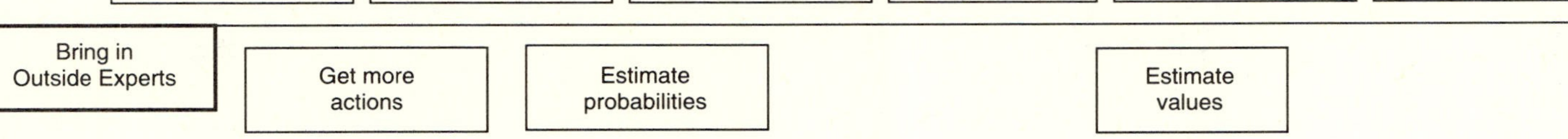

DSS Can Help
Generate a creative set of actions
Identify the principal uncertainties in the decision
Estimate the perceptions of the uncertainties
Identify the principal outcome dimensions
Identify the principal constituencies
Estimate the preferences about each outcome
Bring in Outside Experts
Get more actions
Estimate probabilities
Estimate values

Simplifying a Decision

While the above conceptual structure is present in almost every policy problem, the cost of fully considering each element of that structure may be prohibitive—and such consideration may be unnecessary. Often, in fact, some simplifications are appropriate. A useful simplification is to ignore some uncertainties, so the value of an action is assumed to be more "certain" than it really is—in other words, the chance of an event is either near zero or one. For instance, in deciding which counties need federal assistance to improve health services, the policymaker might choose to assess current levels of medical services and ignore the uncertainty about future service levels. Of course, such simplifications are only appropriate when using them will make little difference to the results of the analysis.

Alternatively, we may assume that uncertainty is the only issue and that the other values and actions can be addressed without the help of the PIMS. For example, the principal challenge in dealing with thyroid disease may be diagnosing whether a patient is hyperthyroid, enthyroid, or hypothyroid. Presumably, after diagnosis, the treatment (or action) selection would be relatively clear. The analyses could be more complex, but as stated in earlier chapters our goal should be not to overwhelm the policymaker with information but rather to provide just enough to meet his or her needs. An important challenge then is to determine how to simplify an analysis without diminishing its usefulness.

The following sections introduce a variety of ways that decision analysis can assist policymakers. These are summarized in Figure 6-2.

Problem Structuring

Problem structuring is useful when policymakers do not truly understand the problem they are addressing. This lack of understanding can be manifested in disagreements about the proper course of action. Each member of a policy team may prefer a reasonable action based on his or her limited perspective of the issue, but this is not enough to make a proper decision. We believe a PIMS can promote better understanding of the issue by forcing policymakers to explicitly identify:

- Individual assumptions about the problem and its causes

- Objectives being pursued by each policymaker
- Constituencies having different perceptions and values
- Options available
- Factors that influence the desirability of various outcomes
- Principal uncertainties that complicate the problem

One way to structure the problem is for the PIMS to conduct an independent analysis and present the results to the policymaker in a brief paper. Another method is to conduct a decision conference in which a policy information management staff member facilitates the policy team's efforts to identify the issues. A decision conference is a day-long retreat during which the policy team agrees upon the conceptual structure of the problem. Ideally, when the conference is over, the group will agree on objectives, possible actions, uncertainties, outcomes, values, probabilities, and perhaps other topics. If this information is fed into a computer and manipulated, participants can see the implications of alternative formulations of the problem during the conference. Most computerized analysis requires that the policymakers quantify their values and uncertainties. This quantification should prod policymakers to think carefully about the issues. In any case, the real contribution of the decision conference is to structure the problem. Values and uncertainties are quantified at this stage only to help promote understanding and agreement about the structure of the problem, not to reach agreement about what action is preferred.

Analyzing Uncertainty

We use the term *uncertainty* to mean the situation when a policymaker is not sure what will happen if a certain action is taken or not sure what state the environment is really in. What is the chance that a fine will really change the way the nursing home uses psychotropic drugs? What is the change that a patient is suffering from hyperthyroidism? Through the process of analyzing uncertainty, the decision's options and their relative desirability can be clarified. What causes system malfunctions? What are the chances a similar event will recur? How likely is action A to lead to outcome B? Often the answers to such questions are vague at best. Returning to our example, although you probably have some clues about whether the nursing home's overmedication was caused by ignorance or greed, usu-

Figure 6-2 Different Types of Problems

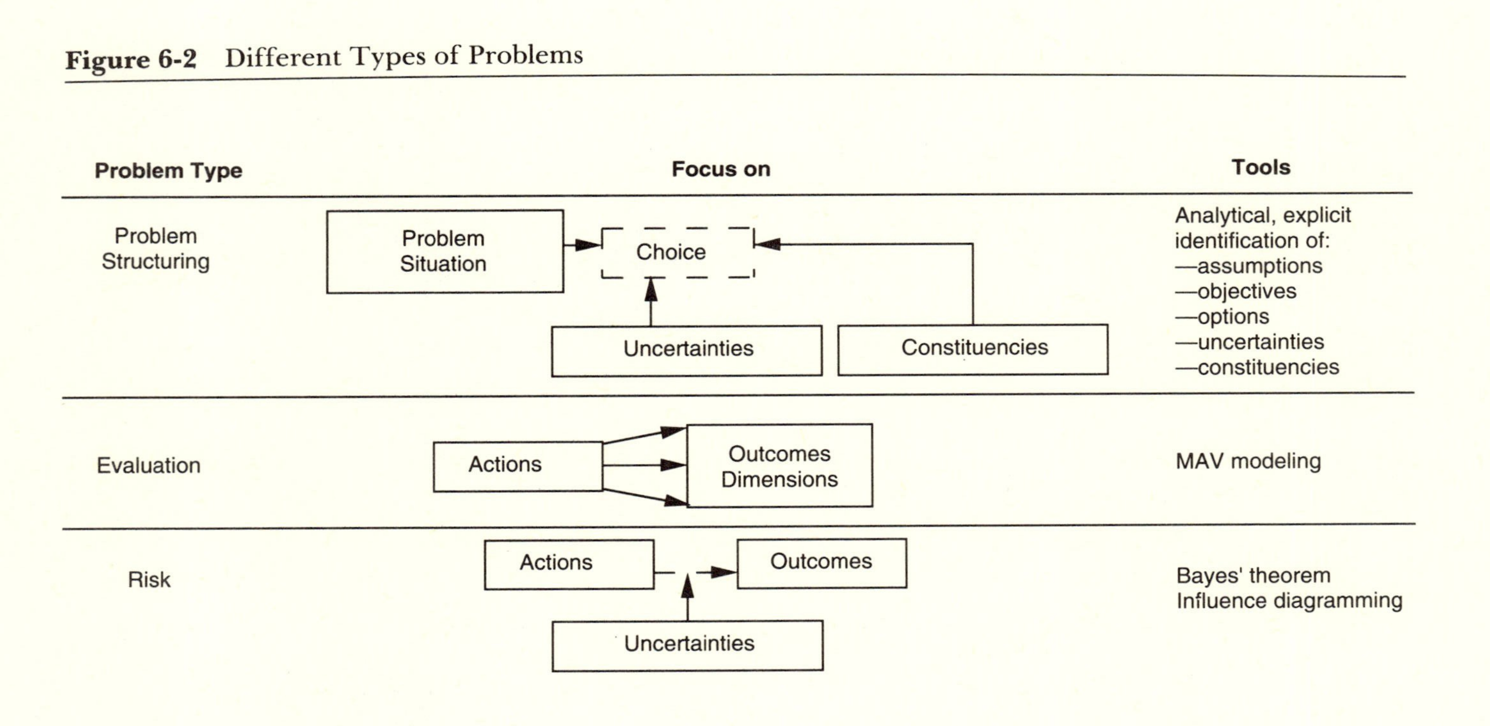

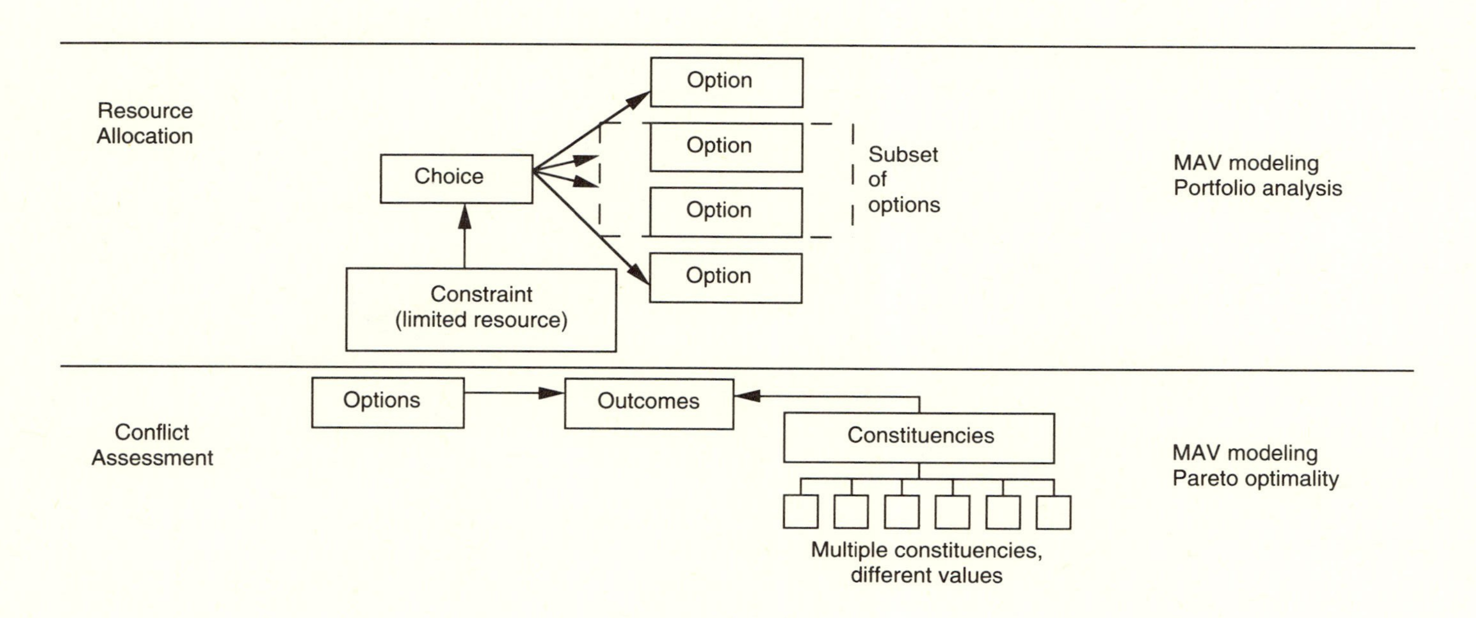
Resource
Allocation
Choice
Option
Option
Option
Option
Subset
of
options
Constraint
(limited resource)
MAV modeling
Portfolio analysis
Conflict
Assessment
Options
Outcomes
Constituencies
Multiple constituencies,
different values
MAV modeling
Pareto optimality

Figure 6-3 Pareto Optimality

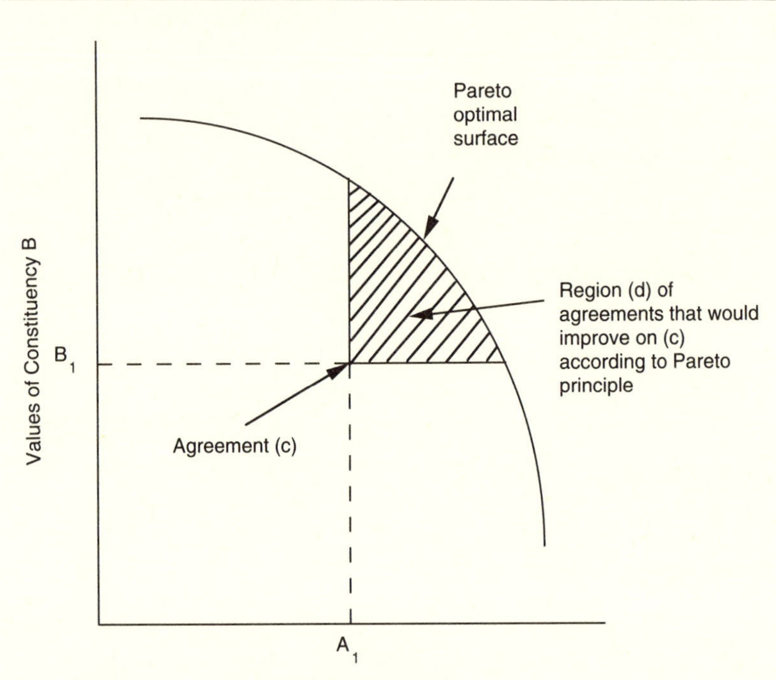

ally the clues are neither equally important nor measured on a common scale. The PIMS helps to compress the evaluations to a single scale for comparison.

Deciding on the nature and relative importance of these clues is difficult because people tend to assess complex uncertainties poorly unless they can divide them into manageable components. Decision analysis can help make this division by using probability models that combine components after their individual contributions have been determined. We address such a probability model—Bayes' theorem—in Chapter 8.

Analyzing Values

If the policy options are clearly identified and uncertainty plays a minor role, the primary concern is to examine the options in terms of a complex set of value dimensions that have differing levels of importance and are measured on different scales. One option may be preferable on one dimension but unacceptable on another. In traditional attempts to debate policy, advocates of one option focus on the dimensions that show it having a favorable outcome, while opponents attack it on dimensions on which it performs poorly. Optimally, a decision analysis provides a mechanism to force consideration of all important dimensions—a task that requires answers to these questions:

- Which objectives are paramount?
- How can an option's performance on a wide range of measuring scales be collapsed into an overall measure of relative value?

The decision analysis approach to these questions uses a process called multiattribute value (MAV) modeling, which is introduced in Chapter 7.

Allocating Resources

If uncertainty and evaluating options are not major factors, the problem is often how to allocate limited resources to various options. The British National Health Service, with a fixed budget, deals with this issue quite directly. Some money is allocated to hip replacement, some to community health services, and some to long-term institutional care for the elderly. Many people who request a service after the money has run out must wait until the next year.

Decision analysis can address resource allocation problems with MAV modeling, which helps policymakers determine their objectives and allocate funds accordingly.

Conflict Assessment

Common sense tells us that people with different values tend to choose different options. The principal challenge facing a policymaking team may be understanding how different constituencies view and value a problem and determining what trade-offs will lead to a win-win, instead of a win-lose, solution. Decision analysis addresses situations like this by developing a MAV model for each constituency and using the concept of Pareto optimality to address the resulting conflict assessment. Pareto optimality starts when an agreement (c) between two constituencies is being considered. Suppose the utility of (c) to constituency A is A_1 and to constituency B is B_1. An agreement is Pareto optimal if no alternative agreement would improve one constituency's utility without reducing that of the other constituency (gains on either or both sides are acceptable or desirable, of course). Because Figure 6-3 shows region (d), where improvements for one constituency's utility would not damage the other's utility, agreement (c) is not Pareto optimal. Chapter 14 describes in detail how this use of decision analysis can help policymakers examine conflicts.

Steps in a Decision Analysis

Figure 6-1 presented the basic decision analysis model before it is expanded to include multiple constituencies. This section introduces a process by which that model can be formulated and used to help policymakers. The decision analysis process can be divided into the seven steps shown and analyzed below.

Step 1. Problem exploration and goal clarification

Problem exploration is the process of understanding the problem and what its resolution will achieve. This understanding is crucial because it helps identify creative options for action and sets some criteria for evaluating the decision.

Let's return to the head of the Division of Nursing Home Administration who was trying to decide what to do about the nursing home that was restraining its residents with excessive medication. The problem exploration might begin by understanding the problem statement: "Excessive use of drugs to restrain residents." Although this type of statement is often taken at face value, several questions could be asked. How should nursing home residents behave? What does *restraint* mean? Why must residents be restrained? Why are drugs used at all? When are drugs appropriate, and when not? What other alternatives does a nursing home have to deal with problem behavior?

The questions at this stage are directed at (1) helping to understand the objective of an organization, (2) defining frequently misunderstood terms, (3) clarifying the practices causing the problem, (4) understanding the reasons for the practice, and (5) separating desirable from undesirable aspects of the practice.

During this step, the decision analyst must determine which ends, or objectives, will be achieved by solving the problem. In the example, the policymaker must determine whether the goal is primarily to

- Protect an individual patient without changing overall methods in the nursing home
- Correct a problem facing several patients—in other words, change the home's general practices
- Correct a problem that appears to be industrywide

Once these questions have been answered, the decision analyst and policymaker will have a much better grasp on the problem. The selected objective will significantly affect both the type of actions considered and the particular action selected.

Step 2. Problem classification

During step 2, the analyst will decide which aspects of the decision model should be emphasized, which constituencies should be included, and where information for the analysis will be obtained.

The analysis could emphasize

- Uncertainty analysis (diagnosis or prediction)
- Value analysis (evaluation)
- Both uncertainty and value analysis
- Single or multiple constituencies

Deciding which constituencies must be considered in the analysis is critical. A decision analysis can always assume that only one constituency exists and that disagreements arise primarily from misunderstandings of the problem, not from different value systems among the various constituencies. But when several constituencies with different assumptions and values approaches are involved, the PIMS must examine the problem from the perspective of each constituency.

In step 2, a choice must also be made about who will provide input into the decision analysis. Who will specify the options, outcomes, and uncertainties? Who will estimate values and probabilities? Will outside experts be called in? Which constituencies will be involved? Will members of the policymaking team provide judgments independently, or will they work as a team to identify and explore differences of opinion?

Step 3. Problem structuring

Problem structuring adds conceptual detail to the general structure provided by step 2. The goals of structuring the problem are to clearly articulate

- What the problem is about, why it exists, and whom it affects
- The assumptions and objectives of each affected constituency
- A creative set of options for the policymaker
- Outcomes to be sought or avoided
- The uncertainties (diagnostic, future events, cause-and-effect relationships) that affect the choice of action

Structuring is the stage in which the specific set of decision options is identified. Although generating options is critical, it is often overlooked by decision makers—a pitfall that can easily promote conflict in cases where diametrically opposed options falsely appear to be the only possible alternatives. Often, creative solutions can be identified that better meet the needs of all constituencies.

To generate better options, one must (as in step 1) understand the purpose and operation of the system under investigation. One must also understand how and why the system has failed thus far. The process of identifying new options relies heavily on reaching outside the organization for theoretical and practical experts, but the process should also encourage insiders to see the problem in new ways.

Other stages in problem structuring are also important. One is the process of explicitly identifying the objectives and assumptions of the primary constituencies. Objectives are important because they lead to the preference of one option over the other. If the policymaking team can understand what each constituency is trying to achieve, the team can analyze and understand its preferences more easily. The same argument holds for assumptions. Two people with similar objectives but different assumptions about how the world operates can examine the same evidence and reach widely divergent conclusions.

Take, for example, the issue of whether a policy should be

Figure 6-4 A Typical Decision Matrix

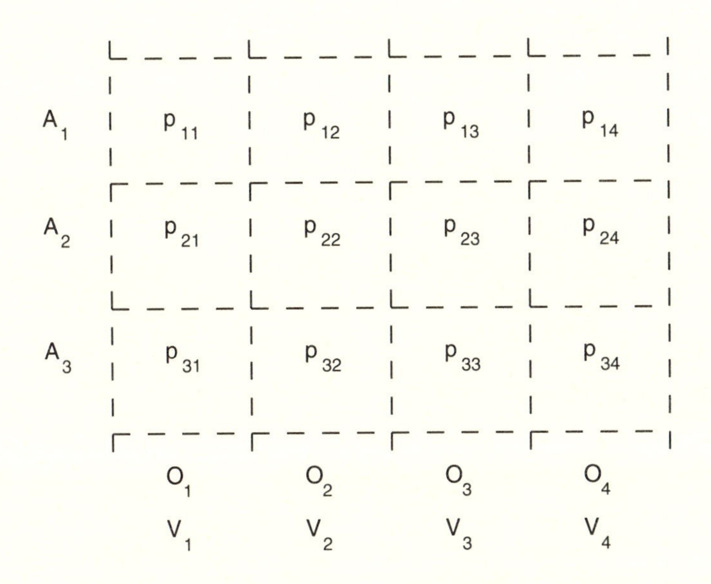

established requiring parental notification when a teenager uses family planning services. Assume that both constituencies—those favoring and those opposing such notification—want teenagers to reach their potential and that neither side wants a teen to have an unwanted pregnancy. One side believes that the child's moral development is paramount for his or her long-term well-being and that this development will be impeded by "condoning" or "assisting" premarital intercourse. The other side believes that whatever the moral issue, many teens are having premarital intercourse and confidential birth control services are essential to preventing pregnancy. In each case, the assumptions (and their relative importance) influence the choice of objectives and action, and that is why they should be identified and examined during problem structuring.

The next phase of problem structuring is to explicitly identify desirable and undesirable outcomes. Here the constituency objectives should be examined. In the example of the nursing home, the policymaker may want to improve the quality of care in the home and portray a certain image to the industry. At the same time, the

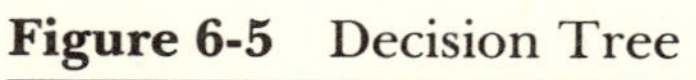

Figure 6-5 Decision Tree

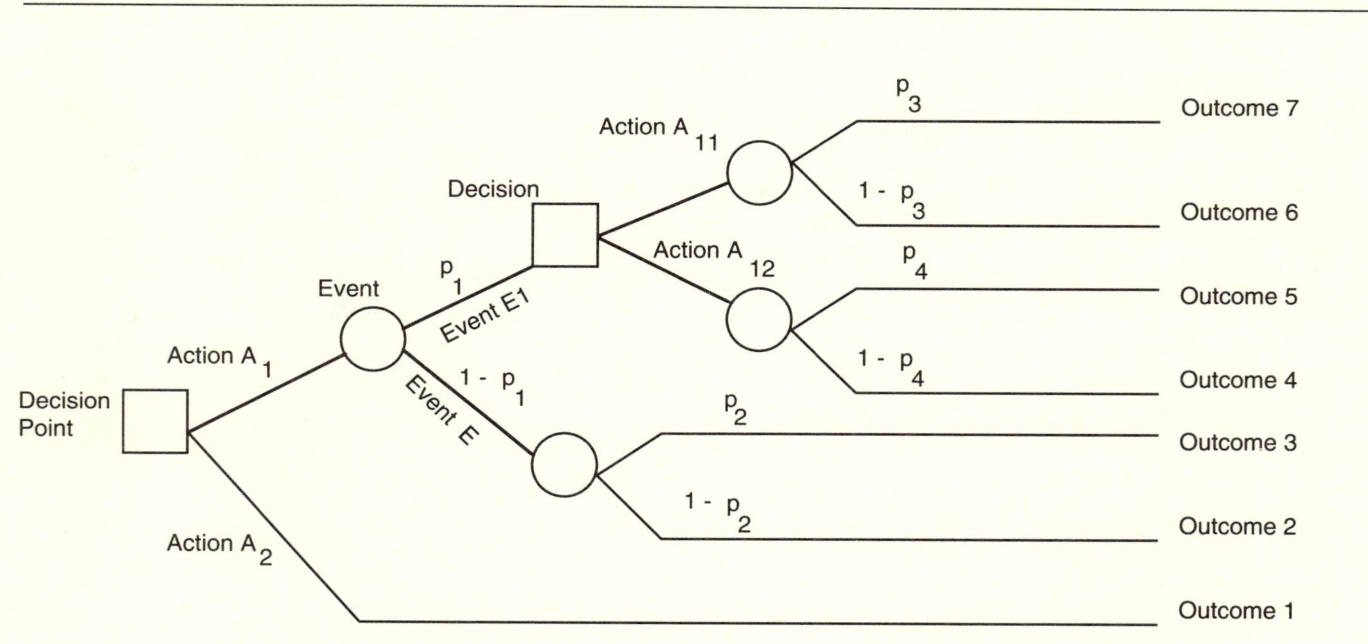

policymaker may wish to avoid other outcomes, such as increasing the amount of paperwork required of the industry.

Two methodologies can assist in outcome identification: (1) the integrative group process (see Chapter 5) which provides a means for helping groups identify outcomes, and (2) multiattribute utility modeling (see Chapter 7), a system that recognizes that policymakers often try to reach several goals while solving one problem.

The final phase of problem structuring is identifying uncertainties. Earlier, three types of uncertainty were identified: (1) uncertainty about the source or diagnosis of system malfunction, (2) uncertainty about whether particular events will occur and affect the success of an action, and (3) uncertainty about the basic cause-and-effect relationship between actions and outcomes. The PIMS must include in the analysis those uncertainties that are likely to have an important influence on which option is superior. Once selected, the level of each uncertainty must be measured.

Two useful ways of representing the structure of a decision problem are the decision matrix (Figure 6-4) and the decision tree (Figure 6-5). In the decision matrix:

A_1–A_3	Represent actions
O_1–O_4	Represent outcomes
V_1–V_4	Represent the values of each outcome
p_n	Represent probability (p_{34} is the probability that action 3 will lead to outcome 4.)

Actions (A_1–A_3) are listed as rows in the matrix. Outcomes (O_1–O_4) are listed in columns. Values (V_1–V_4) are assigned to each of the outcomes to reflect the relative desirability of each. Probabilities (p_{11}–P_{34}) are shown in the cells to reflect the chance (p_{ij}) that an action will lead to an outcome.

The decision matrix representation is quite helpful for simple decision problems, but it cannot portray the situation adequately when values and uncertainties are multidimensional. In this case, if uncertainties are the principal focus of the analysis, a decision tree (Chapter 9) can be more helpful. Figure 6-5 shows a sample decision tree.

This decision tree suggests that two actions (A_1 and A_2) are available to the policymaker. One action (A_2) leads with certainty to outcome O_1. (Each decision point is signified by a square; options arise from decision points.) If action A_1 is taken, event 1 (signfied by an O node) will occur with probability p_1. Event 1 will occur with probability (1 – p), outcome O_3 will occur with probability p_2, and outcome O_2 will occur with probability $(1 - p_2)$. If event 1 does occur, the policymaker faces another decision (A_{11} vs. A_{12}). A_{11} leads to outcome O_7 with probability p_3 and to outcome O_6 with probability $1 - p_3$. A_{12} leads to outcome O_5 with probability p_4 and outcome O_4 with probability $(1 - p_4)$. Each outcome can be multidimensional in nature, but this is not always obvious from the decision tree. The tree's greatest value is to portray the uncertainties in a decision, and it is less useful in portraying the complexity of the multidimensional values or outcomes.

We wish to emphasize that problem structuring is a cyclical process—the structure may change once the policymaking team has estimated values and uncertainties and determined how the structure, values, and uncertainties apply to the action selected. The cyclical nature of the structuring process is desirable, not something to be avoided.

Step 4. Quantifying values

In step 4, complex outcomes are broken into their components and weighted in terms of relative value. The components can be measured on the same scale, called a utility scale, and an equation can be constructed to permit the calculation of the weighted average of the scores. The concept of weighted utility scores is very simple, but the process of properly implementing the concept is not. Chapter 7 presents procedures for developing high-quality models.

The concept of value has two sides: the concept of cost and the concept of benefit. Cost is typically measured in dollars and often appears straightforward. Benefit is measured in terms of values and is often considered unquantifiable. Both cost and benefit impressions are misleading and dysfunctional. Two problems result from viewing cost as a monetary consideration. First, certain costs, such as loss of goodwill, must be ignored. Second, it may be impossible to set dollar figures on costs at the time the decision is made. Fortunately, absolute measures are not needed for decision analysis, just relative measures. For example, we may not know the cost of alternative A_1, but we can agree that it would cost twice as much as A_2.

Assuming that values are unquantifiable can be a major pitfall because it can place them subservient to costs in a formal analysis of policy, even though values often drive the actual decision. By assuming values cannot be quantified, the analysis team may put itself in a position of ignoring the concerns most likely to influence the policymaker.

Another factor in quantifying values is severity. Severity measures, for example, the difference in difficulty of successfully treating various customers of a health or social service system. When the Health Care Financing Administration began publishing mortality rates for hospitals in the United States, hospitals with high rates commonly responded, "Our patients are different because their conditions are more severe." They may have been right, and they correctly noted that severity is a valid factor when comparing performance of health care providers.

An early use of severity indexes resulted from a congressional directive to allocate HMO startup funds on the basis of degree of "medical underservice," or severity of the need to upgrade health services. As a result, more than 3,100 geographical areas were scored on their degree of "medical underservedness," an undefined but multidimensional concept. Decision analysis models were used to establish the Index of Medical Underservice (Health Services Research Group 1975). Since that time, severity indexes using decision analysis have been developed for a variety of health problems, including burns (Gustafson and Holloway 1975), heart disease (Gustafson et al. 1986), trauma (Gustafson et al. 1983), and psychiatric emergencies (Johnson and Gustafson 1989). A severity index to evaluate burn care is the basis for Chapter 15.

There is an important conceptual distinction between using decision analysis to combine several dimensions of an outcome (such as the benefits of changing the chemical restraint practice of a nursing home) and the use of decision models to combine different dimensions of severity (such as the index of underservice). Procedurally, however, there is no distinction. In both cases, the factors are identified and weighted, methods are developed to measure components on a common scale, and a procedure is selected to combine individual subscores into a composite score. The details are discussed in Chapter 15.

Step 5. Quantifying uncertainties

In step 5, the concepts of probability are used to translate three types of uncertainty (diagnostic, future event, and cause-and-effect) into scores that can be compared or added. Scores can range from 0 (no

Figure 6-6 Expected Utility Matrix

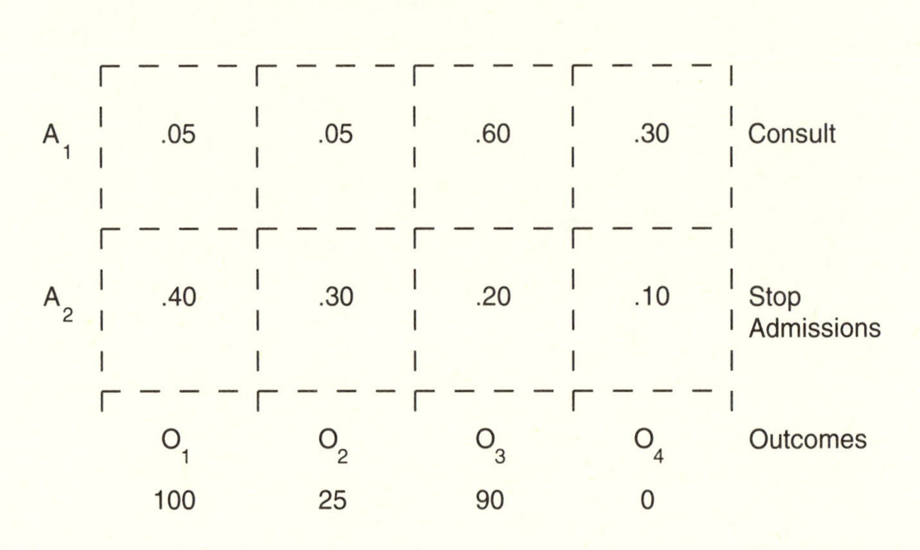

possibility) to 1.0 (absolute certainty). When all possible outcomes are identified, the sum of their probability scores must equal 1.0.

The process of assigning probabilities may be straightforward. If the nursing home inspectors were asked to estimate the chance that the home's chemical restraint practice resulted from ignorance or knowing intention to save money, they might agree that the chances were 90 percent ignorance and 10 percent intent. In some cases, additional data are needed to assess the probabilities. In other cases, there is too much data for the inspectors to process effectively. In both cases, the probability assessment must be divided into manageable components. Bayes' theorem (see Chapter 8) provides one means for disaggregation and reaggregation.

The previous discussion was conducted in the context of a one-time decision, but a policymaker may require a uniform procedure for estimating uncertainty for a series of similar situations. For instance, a computer-based interview might be developed to estimate the suicide potential of people who complain of suicidal thoughts. The procedures that would be used are virtually identical with the value models described in step 4.

Step 6. Expected utility and information processing

Once values and uncertainties are quantified, these data must be combined to rate the relative desirability of each possible action. This can be done using the *expected utility,* which is essentially the weighted average of consequences that could follow each action. The weights are the probabilities that the action will lead to each outcome. Suppose, in the nursing home example, two actions are possible:

A_1 = Consult

A_2 = Stop admission

The possible outcomes are:

O_1 = Chemical restraint is corrected and industry "gets the message that the state intends tougher regulation"

O_2 = No change in practice but industry "gets the message"

O_3 = Practice is corrected but has no impact on industry's view of the state

O_4 = No change in practice and no impact on industry's view of the state

Suppose the relative desirability of each outcome is $O_1 = 100$, $O_2 = 25$, $O_3 = 90$, $O_4 = 0$. The probability that each action will lead to each outcome is shown in the eight cells of the matrix in Figure 6-6.

The expected utility principle says the desirability of action A_1 is:

$$p_{11}U(O_1) + p_{12}U(O_3) + p_{13}U(O_3) + p_{14}U(O_4) =$$

$$.05(100) + .05(25) + .60(90) + .30(0) = 60.25$$

Similarly, the desirability of action A_2 is:

$$.40(100) + .30(25) + .20(90) + .10(0) = 65.5$$

where U_n = relative desirability of outcome *n,* and p_{11} = probability of O_1 given A_1

This analysis suggests that the best action would be to stop admissions because the expected utility of action $A_2 = 65.5$, greater than the utility of action A_1 (60.25).

Step 7. Sensitivity analysis

The previous analysis suggests that action A_1 is inferior to A_2. But this should not be taken at face value because the utility and probability estimates might not be accurate. Perhaps the source of those estimates was guesses, or the estimates were average scores from a group, some of whose members had little faith in the estimates. In these cases, it would be valuable to know whether the choice of highest expected utility would be affected by using a different set of estimates. Stated another way, it might make sense to determine how much an estimate would have to change to alter the choice of "preferred" action.

In the example, how large must p_{22} be (holding all other p_{2j}'s in constant proportion to one another) for A_1 to become the preferred action? It turns out that if p_{22} = .40 and retaining p_{21}, p_{23}, and p_{24} in basically a 4–2–1 relationship so that p_{21} = .34, p_{23} = .17, and p_{24} = .09, the utility of A_2 = 59.3. This suggests that a 40 percent chance that action A_2 would lead to outcome O_2 would be needed before A_1 would have a higher expected value than A_2, if all other estimates were held constant.

Of course, several estimates can be modified at once, especially using computers. Sensitivity analysis can be vital not only to examining the impact of errors in estimation but also to determining which variables need the most attention (e.g., reduction in disagreement and/or increase in confidence).

At each stage in the decision analysis process, it is possible and often essential to return to an earlier stage to

- Add a new action or outcome
- Add new uncertainties
- Refine probability estimates
- Refine utility estimates

This cyclical approach offers a better understanding of the decision problem and fosters greater confidence in the analysis. Often the decision recommended by the analysis is not the one implemented, but the value of the analysis remains because it increases the policymaking team's understanding of the issues. Phillips refers to this as the theory of requisite decisions—once all parties agree that the problem representation is adequate for reaching the decision, the model is "requisite":

From this point of view, decision analysis is more an aid to problem solving than a mathematical technique. Considered in this light, decision analysis provides the decision maker with a practical means for maintaining control of complex decision problems that involve risk, uncertainty, and multiple objectives. (Phillips 1984, p. 26)

References

Gustafson, D., and D. Holloway. 1975. "A Decision Theory Approach to Measuring Severity of Illness." *Health Services Research* 10: 97–196.

Gustafson, D., D. Fryback, and J. Rose. 1983. "An Evaluation of Multiple Trauma Severity Indices Created by Different Index Development Strategies." *Medical Care* 21: 674–91.

Gustafson, D., D. Fryback, J. Rose, et al. 1986. "A Decision Theoretic Model for Severity Index Development." *Medical Decision Making* 6 (1): 27–35.

Health Services Research Group. 1975. "Development of the Index of Medical Underservice." *Health Services Research* (Summer).

Johnson, S., and D. Gustafson. 1989. Final Report of the Psychiatric Emergency Severity Index Project, Maine Health Information Center, Augusta.

Phillips, L. D. 1984. "A Theory of Requisite Decision Models." *Acta Psychologica* 56: 29–48.

7

Value Models

This chapter introduces a flexible method for modeling policymakers' values and several means of validating such models. Our examples assume that a model is developed jointly by one policy analyst and one policymaker, but if this development process involves a group, we suggest using the integrative group process (Chapter 5) as the basis for the modeling effort.

Value models are based on Bernoulli's (1938) recognition that money's value did not always equal its amount. He postulated that increasing amount of income had decreasing value to the wage earner. Value models have received considerable attention from economists and psychologists (Savage 1972; Edwards 1974). A comprehensive and rather mathematical introduction to constructing value models is found in von Winterfeldt and Edwards (1986). Here we focus on detailed instructions for making the models and ignore their mathematical foundations.

Value models help us quantify treacherous concepts like a person's preferences. These models assume that the decision makers must select from several options and that the selection should depend on grading the preferences for the options. These preferences are quantified by examining the various *attributes* (characteristics, dimensions, or features) of the options. For example, if policymakers were examining whether to fund several social programs, the value of the programs could be scored by examining such attributes as impact on access to care and on quality and cost of care. First, the expected impact of each program on each attribute would be scored. Second, scores would be weighted by the relative importance of each attribute. Third, the scores for all attributes would be aggregated. Fourth, the option with the highest weighted score would be chosen.

If each program were described in terms of n attributes A_1, A_2, . . . , A_n, the programs would be assigned a score on each attribute, $V(A_1)$, $V(A_2)$, . . . , $V(A_n)$. The overall value of a program equals:

$$\text{Value} = \text{Function}\,[V(C_1), V(C_2), \ldots, V(C_n)]$$

What Good Are Value Models?

The function of value models is to assign scores so the higher numbers match the policymaker's more preferred options. Value models have two basic purposes: to communicate a policymaker's preferences and to measure hard-to-quantify concepts.

For several reasons, values and attitudes, which we can call *preferences,* play major roles in setting policy and management decisions. In organizations, decision making is often very complex and a product of collective action. Frequently, decisions must be made concerning issues on which few data exist, forcing policymakers to make decisions on the basis of opinion, not fact. Often there is no correct resolution to a problem—with all options having equally legitimate perspectives and values playing a major role in the final judgment. Value perspectives are often inadequately acknowledged or ineffectively communicated. As a result, judgments are misunderstood, and policymakers are unable to defend their final position. Value models can help in the recognition and quantification of the values behind the decision.

Modeling values helps policymakers communicate their policy positions by explicitly showing the importance of each priority. Value models divide multiattribute problems into components that are systematically analyzed. A specific formula then quantifies the outcome of the analysis. These models clarify the basis of decisions so others can see the logic behind the decision and—ideally—agree with it. For example, we constructed a value model to help the Wisconsin Division of Health determine the eligibility of nursing home residents for a higher level of reimbursement (the "super-skilled" level of nursing care). This model showed which attributes of an applicant affected eligibility and how much weight each attribute deserved. As a consequence of this effort, the regulator, the industry, and the patients became more aware of how eligibility decisions were made (Cline et al. 1982).

Value models have been used to model policymakers' priorities for evaluating standards for offshore oil discharges (von Winterfeldt

1980), energy alternatives (Keeney 1976), drug therapy options (Aschenbrenner and Kaubeck 1978), and family planning options (Beach et al. 1979). In Chapter 13, we report on using value models to analyze conflicts among constituencies.

As discussed so far, one purpose of value models is to model preferences. The second major use of value models is to quantify other hard-to-measure concepts like the severity of trauma (Detmer et al. 1977) the degree to which an area is medically underserved (Health Services Research Group 1975), or the quality of the remaining years of life (Pliskin et al. 1980). These hard-to-measure concepts are similar to preferences because they are subjective and open to disagreement. Evaluating preferences or severity requires a measurement of subjective attributes and characteristics of the problem.

Measuring a subjective construct may seem a contradiction in terms. After all, rational people look at the same things and evaluate them very differently. For some broad distinctions, such as whether there is more of one thing than another the task is easy. Thus, we might say that one patient is more ill than another, or that an area is more medically underserved than another. However, it is more difficult to say precisely how much more. Value models allow us to quantify estimations of such hard-to-measure concepts.

Numbers Can Be Misleading

Though value models allow us to quantify subjective concepts, the resulting numbers are rough estimates that should not be mistaken for precise measurements. It is important that managers and policymakers using these models do not read more into the numbers than they mean. Analysts must stress that the numbers in value models are intended to offer a consistent method of tracking, comparing, and communicating rough, subjective concepts.

An important distinction is whether the model is to be used for rank ordering (*ordinal*) or rating the extent of presence of a hard-to-measure concept (*interval*). Some value models produce numbers that are only useful for rank ordering options. Thus, some severity indexes indicate whether one patient is sicker than another, not how much sicker. In these circumstances, a patient with a severity score of 4 may not be twice as ill as a patient with a severity of 2. Averaging such ordinal scores is meaningless. In contrast, value models that score on an interval scale show how much more preferable one option is than another. For example, a severity index can be created to show how much more severe one patient's condition is than another's. A patient scoring 4 *can* be considered twice as ill as one scoring 2. Further, averaging interval scores is meaningful.

Interval scales are more difficult to construct than ordinal scales. In the remainder of this chapter, we will show how both ordinal and interval value models can be constructed. First we wish to introduce an example for reference throughout the chapter.

Example: Severity of AIDS Cases

A value model was used to create a severity index for acquired immune deficiency syndrome (AIDS) (Alemi et al. 1990). The U.S. Centers for Disease Control defines an AIDS case as an HIV-positive patient suffering from any of a list of opportunistic infections. These infections are referred to as defining diseases. AIDS patients typically suffer several illnesses besides defining diseases before passing away. Because AIDS patients often suffer a complex set of diseases, the cost of treatment is heavily dependent on the course of the illness. Patients with a defining disease of the skin cancer Kaposi's sarcoma have half the first-year costs of AIDS patients with defining disease of pneumocystic pneumonia, a lung infection (Pascal et al. 1989). Thus, we cannot accurately project costs until we can describe the severity of the course of AIDS.

In their search for an accurate measure of AIDS severity, the policymakers faced a dilemma because objective measures were unavailable or inadequate for at least three reasons. First, objective measures were not available on many aspects of the course of AIDS. Second, the available objective data described prognoses given a mix of aggressive and less aggressive care—the data did not describe prognoses under ideal care. Third, available objective measures were also outdated because of rapid advances in treatment which had prolonged life expectancies (Scitovsky 1989). The drug that had improved survival, AZT, had not been available just a few years before, nor was its use uniformly coded in the existing data bases. Thus, the existing objective data would have misled a policymaker about prognoses.

Of course, it is possible to collect new objective data that would control for quality and aggressiveness of care and the emergence of new treatments, but this is always expensive and usually slow. With money desperately needed for treatment, it was difficult to justify a large, expensive data collection for planning. Furthermore, the time needed for such data collection would have prevented findings from being available within the policymaker's time frame for action. Thus, the policymaker asked us to develop a severity index based on the input of clinicians.

Constructing Value Models

Using the example of the AIDS severity index, we will show how to examine the need for a value model and how to create such a model.

Step 1. Would it help?

The first, and most obvious, question is whether constructing a value model would help resolve the problem faced by the manager or policymaker. Defining the problem is the most significant step of the analysis, yet surprisingly little literature is available for guidance (Volkema 1981). Chapter 3 and Chapter 4 provide clues on how problems emerge, solutions are sought, and change is effectuated.

To define a problem, we must answer several related questions: Who is the decision maker? What objectives does this person wish to achieve? What role do subjective judgments play in these goals? Would it help to model the judgment? Should a value model be used? Finally, once a model is developed, how might it be used?

Who decides? In health and social service policy, there are often many decision makers. No single person's viewpoint is sufficient, and we need a multidisciplinary consensus instead. If, as is often the case, the input comes from several constituencies with different value systems, the analyst should think through the needs of these diverse decision makers. Chapter 5 discusses how a facilitator can guide a group to reach decisions or make judgments. For simplicity, the following discussion assumes that only one person is involved in the decision-making process.

The core of the problem here was that AIDS patients need different amounts of resources depending on the severity of their illness. The federal administrators of the Medicaid program wanted to measure severity of AIDS patients because the federal government paid for part of their care. The state administrators were likewise interested because state funds paid for another portion of their care.

For the study, we assembled six experts known for clinical work with AIDS patients or for research on the survival of AIDS patients. Physicians came from several states, including New York and California (which have the highest number of AIDS patients in the United States). California had more homosexual AIDS patients, while intravenous drug users were more prevalent in New York. We used the integrative group process to construct the severity index.

What must be done? Problem solving starts by recognizing a gap between the present situation and the desired outcome. Typically, at least one decision maker has noticed a difference between what is and what should be and begins to share this awareness with the relevant levels of the organization. Gradually a motivation is created to change, informal social ties are established to promote the change, and an individual or group receives a mandate to find a solution.

Often, a perceived problem must be reformulated to address the real issues. For example, a stated problem of developing a model of AIDS severity may indicate that the real issue is insufficient funds to provide for all AIDS treatment programs. When solutions are proposed prematurely, it is important to sit back and gain greater perspective on the problem. Occasionally, the decision makers have a solution in mind before fully understanding the problem, which shows the need for examining the decision makers' circumstances in greater depth. In these situations, it is the analyst's responsibility to redefine the problem to make it relevant to the real issues. VanGundy (1981) reviews 70 techniques used by analysts to redefine problems and creatively search for solutions, including structured techniques like brainstorming and less structured techniques like using analogies. Analysts can refer to VanGundy for guidance in restructuring problems. In addition, as mentioned earlier, Chapters 3 and 4 provide guidelines on recognizing problems and effectuating change.

What judgments must be made? After the problem has been defined, we must examine the role subjective judgments can play in its resolution. One can do this by asking several "what if" questions: What plans would change if the judgment were different? What is being done now? If no one makes a judgment about the underlying concept, would it really matter, and who would complain? Would it be useful to tell how the judgment was made, or is it better to leave matters rather ambiguous? Must we choose among options, or should we let things unfold on their own? Is a subjective component critical to the judgment, or can it be based on objective standards?

In the example, the administrators needed to budget for the coming years, and they knew judgments of severity would help them anticipate utilization rates and overall costs. Agencies caring for low-severity patients would receive a smaller allocation than agencies caring for high-severity patients. But there were no objective measures of severity available, so we opted for using clinician judgments instead.

How can the judgment be used? In understanding what judgments must be made, it was crucial to attend to the limitations of circumstances in which these judgments are going to be made. The use of existing AIDS severity indexes was limited because they relied on physiological variables that were unavailable in automated data bases (Redfield and Burke 1988). Policymakers asked us to predict prognoses from existing data. The only data base widely available on AIDS patients was Medicaid administrative data, which are routinely collected after every Medicaid encounter. Because these data did not include any known physiological predictors of survival (such as number of T4 cells), we had to find alternative ways to predict survival.

We thought Medicaid's data on diagnosis could be used to predict prognosis because the timing and nature of opportunistic infections suggests the amount of deterioration in the immune system and thus survival time. Therefore, we set out to create an index that relied on the timing and nature of infections rather than on physiological variables. This limited the accuracy of the index, as physiological data were perhaps the most accurate measure of the immune system, but the system could easily be applied to existing data. Our conversations with policymakers helped us define the problem in a way that allowed us to trade off accuracy against ease of application.

Would a model help? Another task in defining the problem is understanding whether a model would help, and what type of model would be best for the task. Although it was theoretically possible to have an expert panel review each case and estimate severity, from the outset it was clear that a model was needed because case-by-case review is extremely expensive. Moreover, the large number of cases would require the use of several expert panels, each judging a subset of cases, and the panels might disagree. Further, judgments within a panel can be quite inconsistent over time. In contrast, the model provided a quick way of rating the severity of a group of patients. It also explained the rationale behind the ratings, which allowed skeptics to examine the fairness of judgments, thus increasing the acceptance of those judgments.

Now that it was clear that a model of severity was needed, the natural next question was which of many possible mathematical models would be the best tool. The idea of transition from one state to another is a common allegory for thinking about many other diseases (Kao 1972; Humphrey and Humphrey 1975), but AIDS patients have so many different conditions that it seemed impractical to estimate the transition probabilities. Instead, we selected a value model because it did not require so many estimates of transition probabilities.

Step 2. Soliciting the attributes

After defining the problem, we solicited the attributes needed to make the judgment from the expert. There are three steps to interviewing an expert. First, the analyst introduces himself or herself and describes the purpose of the meeting. Second, the analyst asks about the expert's relevant experience. Third, using the expert's terminology, the analyst asks about the attributes.

Introduce yourself and your purpose. Briefly say who you are, why you are talking to the expert, the model's purpose, and how it will be developed. Be as brief as possible. An interview is going well if the analyst is listening and the expert is talking. If it takes you five minutes just to describe your purpose, then something is amiss. Probably you haven't understood the problem well, or possibly the expert is not familiar with the problem.

Be assertive in setting the interview's pace and agenda. Because you are likely to receive a comment whenever you pause, be judicious about pausing. Thus, if you stop after saying, "The purpose of modeling is to construct a severity index to work with existing data bases," your expert will likely discuss your purpose. But if you immediately ask about the expert's experience in assessing severity, the expert is more likely to begin describing his or her background. The point is that, as the analyst, you set the agenda, and you should construct your questions to resolve your uncertainties, not to suit the expert's agenda.

Introduce the expert. It is important to establish that you are a recordkeeper and that the expert will provide the content. A good way of doing this is to ask the expert to introduce himself or herself by describing relevant experiences. This approach also stresses your respect for the expert's expertise.

Start with tangible examples. Concrete examples help you understand which patient attributes should be used to predict severity and how they can be measured. Ask the expert to recall an actual situation and contrast it with other occasions to discern the key discriminators. For example, you might ask the expert to describe a severely ill patient in detail (to ensure that the expert is referring to a particular patient). Then ask for a description of a patient who was not severely ill and elicit the key differences between the two patients. These differences are attributes you can use to judge severity. Continue asking the expert to think about specific patients until you have identified a number of attributes. Here is a sample dialogue.

Analyst: Can you recall a specific patient with a very poor prognosis?

Expert: I work in a referral center, and we see a lot of severely ill patients. They seem to have many illness and are unable to recover completely, so they continue to worsen.

Analyst: Tell me about a recent patient who was severely ill.

Expert: A 28-year-old homosexual male patient deteriorated rapidly. He kept fighting recurrent influenza and died from gastrointestinal cancer. The real problem was that he couldn't tolerate AZT, so we couldn't help him much. Once a person has cancer, we can do little to maintain him.

Analyst: Tell me about a patient with a good prognosis, say close to five years.

Expert: Well, let me think. A year ago we had a 32-year-old male patient diagnosed with AIDS who has not had serious disease since—a few skin infections but nothing serious. His spirit is up, he continues working, and we have every reason to expect he will survive four or five years.

Analyst: What key difference between the two patients made you realize that the first patient had a poorer prognosis than the second?

Expert: That's a difficult question—patients are so different from each other that it's tough to point at one characteristic. But if you really push me, I would say two characteristics: the history of illness and the ability to tolerate AZT.

Analyst: What about the medical history is relevant?

Expert: If I must predict a prognosis, I want to know whether he has had serious illness in vital organs.

Analyst: Which organs?

Expert: Brain, heart, and lungs are more important than, say, skin.

The primary feature of this approach is its reliance on concrete examples for directing the expert to specific patients' attributes. First of all, this seems to help the expert recall the details; second, soliciting attributes by contrasting patients helps single out those attributes that truly affect prognosis. A further advantage is that this method does not produce a wish list of information that is loosely tied to survival—an extravagance we cannot afford.

After you have identified some attributes, you can ask directly for additional attributes that indicate prognosis.

Analyst: Are there other markers for poor prognosis?

Expert: Comorbidities are important. Perhaps advanced age suggests poorer prognosis. Sex may matter.

Analyst: Does the age or sex really matter in predicting prognosis?

Expert: Sex does not matter, but age does. But there are many exceptions. You cannot predict the prognosis of a patient based on age alone.

Analyst: What are some other markers of poor prognosis?

Arrange the attributes. Now you can create a hierarchy from broad to specific attributes (Keeney and Raiffa 1976). Some analysts suggest using a hierarchy to solicit and structure the attributes. For example, an expert may suggest that a patient's prognosis depends on medical history and demographics (see Figure 7-1). Demo-

Figure 7-1 An Example of Hierarchy of Attributes

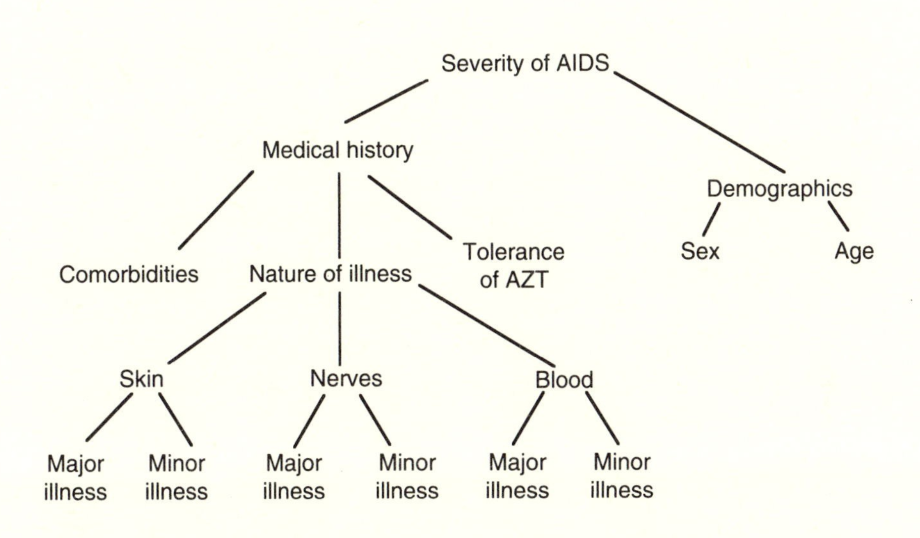

graphics include age and sex. Medical history involves the nature of the illness, comorbidities, and tolerance of AZT. The nature of illness breaks down into body systems involved (skin, nerves, blood, etc.). Within each body system, some diagnoses are minor and other diagnoses are more threatening. The expert then lists, within each system, a range of diseases. The hierarchical structure promotes completeness and simplifies tracking many attributes.

Be careful about terminology. Always use the expert's terminology, even if you think a reformulation would help. Thus, if the expert refers to "sex," do not substitute "gender." Such new terminology may confuse the conversation and create an environment where the analyst acts more like an expert, which can undermine the expert's confidence that he or she is being heard. It is reasonable, however, to ask for clarification—"sex" could refer to gender or to sex practices, and you must understand which meaning is intended.

If you must introduce your own terminology, for example in stating your purpose, make sure it does not alter the expert's responses. Psychological research suggests that changing the framing of a question alters the response. Consider these two questions:

- "What are the markers for survival?"
- "What are the markers for poor prognosis?"

One question emphasizes survival, the other mortality. We would expect that patient attributes indicating survival would also indicate mortality, but researchers have found this to be untrue (for a review, see Nisbett and Ross 1980). Experts may identify entirely different attributes for survival and mortality. This research suggests that value-laden prompts tap different parts of the memory and can evoke recall of different pieces of information.

Evidence about the impact of questions on recall and judgment is substantial (Hogarth 1975; Ericsson and Simon 1980). For example, in one study subjects used surprisingly different sets of attributes to judge whether a person was an introvert or an extrovert (Snyder and Cantor 1979). Studies like this suggest that analysts should ask their questions in two ways, once in positive terms and again in negative terms.

In general, the less esoteric prompts are more likely to produce the best responses, so formulate a few prompts and use the ones that feel most natural for your task.

Take notes, and do not interrupt. Have paper and pencil available, and write down the important points. Not only does this help the expert's recall, but it also helps you review matters while the expert is still available. Experts tend to list a few attributes, then focus attention on one or two. Actively listen to these areas of focus. When the expert is finished, review your notes for items that need elaboration. If you don't understand certain points, ask for examples, which are an excellent means of clarification.

For instance, after the expert has described attributes of vital organ involvement, you may ask the expert to elaborate on something mentioned earlier, such as "acceptance of AZT." If the expert mentions other topics in the process, return to them after completing the discussion of AZT acceptance. This ensures that no loose ends are left when the interview is finished and reassures the expert that you are indeed listening.

Other approaches. Other, more statistical approaches to soliciting attributes are available, such as multidimensional scaling and factor analysis. However, we prefer the behavioral approach to soliciting attributes because it involves the expert more in the process and leads to greater acceptance of the model.

Step 3. Do it again

After soliciting a set of attributes, it is important to examine and, if necessary, revise them. Several tests should be conducted to ensure that the solicitation process succeeded.

The first test ensures that the listed attributes are exhaustive by using them to describe several hypothetical patients and asking the expert to rate their prognosis. If the expert needs additional information for a judgment, solicit new attributes until you have sufficient information to judge severity.

A second test checks that the attributes are not redundant by examining whether knowledge of one attribute implies knowledge of another. For example, the expert may consider "inability to administer AZT" and "cancer of GI tract" redundant if no patient with GI cancer can accept AZT. In such cases, either the two attributes should be collapsed into one, or one must be dropped from the analysis.

A third test ensures that each attribute is important to the decision maker's judgment. You can test this by asking the decision makers to judge two hypothetical situations: one with the attribute at its lowest level and another with the attribute at peak level. If the judgments are similar, the attribute may be ignored. For example, gender may be unimportant if male and female AIDS patients with the same history of illness have identical prognoses.

Fourth, a series of tests examines whether the attributes are *related* or *dependent* (Keeney and Raiffa 1976; Keeney 1977). These words are much abused and variously defined. By independence we mean that in judging two different patients, the shared feature among these patients does not affect how other features are judged. We distinguish this from other senses of independence by calling it *preferential independence*. There are many situations in which preferential independence does not hold. In predicting three-year risks of hospitalization, age and lifestyle may be dependent (Alemi et al. 1987). Among young adults, drinking may be a more important concern than cholesterol risks, while among older adults, cholesterol is the more important risk factor. Thus, the relative importance of cholesterol and drinking risks depends on the age of the patients being compared.

In many circumstances, preferential independence holds. It often holds even when experts complain that the attributes are dependent in other senses. When preferential independence holds, it is reasonable to break a complex judgment into components. Or, to say it differently, with preferential independence, it is possible to find a formula that translates scores on several attributes into an overall severity score in such a manner as to resemble the expert's intuitive rating of severity. When preferential independence does not hold, it is often a sign that some underlying issue is poorly understood. In these circumstances, the analyst should query the expert further and revise the attributes to eliminate dependencies (Keeney 1980).

In the AIDS severity study, discussions with the expert and later revisions led to the following set of 18 patient attributes for judging severity of AIDS:

1. Age
2. Race
3. Transmission mode
4. Defining diagnosis
5. Time since defining diagnosis
6. Diseases of nervous system
7. Disseminated diseases
8. Gastrointestinal diseases
9. Skin diseases
10. Lung diseases
11. Heart diseases
12. Recurrence of a disease
13. Functioning of the organs
14. Comorbidity
15. Psychiatric comorbidity
16. Nutritional status
17. Drug markers
18. Functional impairment

As the number of attributes in a model increases, the chances for preferential dependence also increase. Our rule of thumb is that preferential dependencies are much more likely in value models with more than nine attributes.

Step 4. Determine the possible levels of each attribute

Now it is time to identify the possible levels of each attribute. We start by deciding if the attributes are discrete or continuous. Attributes such as age are continuous; attributes such as diseases of the nervous system are discrete. However, continuous attributes may be expressed in terms of a few discrete levels, so that age can be described in decades, not individual years. The four steps in identifying the levels of an attribute are to define the range, define the best and worst levels, define some intermediate levels, and fill in the other possible levels so that the listing of the levels is exhaustive (capable of covering all possible situations).

To define the range, we must select a target population and ask the expert to describe the possible range of the variable in it. Thus, for the AIDS severity index, we ask our expert to focus on adult AIDS patients and, for each attribute, suggest the possible ranges. To assess the range of nervous system diseases, we asked:

> Analyst: In adult AIDS patients, what is a disease that suggests the most extensive involvement of the nervous system?

Next we ask the expert to *specify the best and the worst possible level* of each attribute. In the AIDS index, we could easily identify the level with the best possible prognosis: the normal finding within each attribute—in common language, the healthy condition. We accomplished the more difficult task of identifying the level with the worst possible prognosis by asking the expert:

> Analyst: What would be the gravest disease of the central nervous system, in terms of prognosis?

A typical error in obtaining the best and the worst levels is failing to describe these levels in detail. For example, in assessing the value of nutritional status, it is not helpful to define the levels as:

- Best nutritional status
- Worst nutritional status

Nor does it help to define the worst level as "severely nutritionally deficient" because it is best to avoid using adjectives in describing levels, as experts perceive words like *severely,* or *best* in different ways. The levels must be defined in terms of the underlying physical process measured in each attribute, and the descriptions must be connected to the nature of the attribute. Thus, a good level for the worst nutritional status might be "patients on total parenteral treatment," and the best status might be "nutritional treatment not needed."

Next, ask the expert to *define intermediate levels.* These levels are often defined by asking for a level between the best and worst levels. In the example, this dialogue might occur:

Analyst: I understand that patients on total parenteral treatment have the worst prognosis. Can you think of other relatively common conditions with a slightly better prognosis?

Expert: Well, a host of things can happen. Pick up any book on nutritional diseases and you find all kinds of things.

Analyst: Right, but can you give me three or four examples?

Expert: Sure. The patient may be on anti-emetics or nutritional supplements.

Analyst: Do these levels include a level with a moderately poor prognosis and one with a relatively good prognosis?

Expert: Not really. If you want a level indicative of moderately poor prognosis, then you should include whether the patient is receiving Lomotil or Imodium.

(Appendix 7-D shows the levels that the AIDS expert specified for each attribute.)

It is not always possible to solicit all possible levels of an attribute from the expert interviews. In these circumstances, we can *fill in the gaps* afterward by reading the literature or interviewing other experts. The levels specified by the first expert are used as markers for placing the remaining levels, so that the levels range from best to worst. In the example, a clinician on the project team reviewed the expert's suggestions and filled in a long list of intermediate levels.

Step 5. Assess the value of a single attribute

Now we will evaluate a patient on a single attribute as a preparation for aggregating the value of all attributes to get the overall score. We prefer using the double-anchored estimation method for evaluating a single attribute (Kneppreth et al. 1974). This approach gets its name from the practice of selecting the best and worst levels first and rating the remaining levels according to these two "anchors." In this method, first the attribute levels are ranked, or, if the attribute is continuous, the most and least preferred levels are specified. Then the best and the worst levels are used as anchors for assessing the other levels.

For example, skin infections have the following levels:

- No skin disorder
- Kaposi's sarcoma
- Shingles
- Herpes complex
- Candida or mucus
- Thrush

The following interaction typifies the questioning for the double-anchored estimation method:

Analyst: Which among the skin disorders has the worst prognosis?

Expert: None is really that serious.

Analyst: Yes, I understand that, but which is the most serious?

Expert: Patients with thrush perhaps have a worse prognosis than patients with other skin infections.

Analyst: Let's rate the severity of thrush at 100 and place the severity of no skin disorder at 0. How would you rate shingles?

Expert: Shingles is almost as serious as thrush.

Analyst: This tells me that you might rate the severity of shingles nearer 100 than 0. Where exactly would you rate it?

Expert: Maybe 90.

Analyst: Can you now rate the remaining levels?

Recently, several psychologists have questioned whether experts are systematically biased in assessing value. Yates and Jagacinski (1979) showed that using different anchors produced different value functions. For example, in assessing the value of money,

Kahneman and Tversky (1979) showed that values associated with gains or losses are different from values related to the amount of monetary return. They argued that the value of money is judged according to the decision maker's current assets. Because value may depend on the anchors used, it is important to use different anchors from just the best or worst levels. Thus, if the value of skin infections is assessed by anchoring on "shingles" and "no skin infections," then it is important to verify the ratings relative to other levels. If the expert rated skin infections as follows:

0 No skin disorder
10 Kaposi's sarcoma
90 Shingles
95 Herpes complex
100 Candida or mucus
100 Thrush

We might ask:

Analyst: You have rated herpes complex halfway between shingles and candida. Is this OK?

Expert: Not really. Prognosis of patients with herpes is closer to patients with candida.

Analyst: How would you change the ratings?

Expert: Maybe we should rate herpes 98.

It is occasionally useful to change not only the anchors but also the assessment method. Appendix 7-A describes several alternative methods of assessing single-attribute value functions. When a value is measured by two different methods, you can expect inadvertent discrepancies, which you must ask the expert to resolve.

Step 6. Choose an aggregation rule

Now we must find a way to aggregate the scores across all attributes. Note that our conventions have produced a situation in which the value of each attribute is somewhere between 0 and 100. Thus, the prognosis of patients with skin infection and the prognosis of patients with various GI disease have the same range. Adding these scores will be misleading because skin infections are less serious than GI problems, so we must find an aggregation rule that differentially weights the various attributes.

The most obvious rule is the *additive model.* Assume that S represents the severity of AIDS. If a patient is described by a series of n attributes of $\{A_1, A_2, \ldots, A_i, \ldots, A_n\}$, then using the additive rule, the overall severity equals:

$$S = \Sigma_i W_i V(A_{ij})$$

where $V(A_{ij})$ is the value of the jth level in the ith patient attribute, W_i is the weight associated with the ith attribute in predicting prognosis, and $\Sigma_i W_i = 1$.

Several other models are possible in addition to the additive model. The multiplicative model is described in the appendix. The following discussion focuses on the additive model.

Step 7. Estimate weights

We can estimate the weights for an additive value model in a number of ways. In this section we present the method of rating the ratio of importance of the attributes. Appendix 7-B presents other methods. As we mention below, we often find it useful to mix several approaches.

Some analysts estimate weights by *assessing the ratio of the importance* of two attributes (Edwards 1977). The attributes are rank ordered, and the least important is assigned 10 points. Then the expert is asked to estimate the relative importance of the other attributes. There is no upper limit to the number of points other attributes can be assigned. For example, in estimating the weights for the three attributes (skin infections, lung infections, and GI diseases), the analyst and the expert might have the following discussion:

Analyst: Which of the three attributes is most important?

Expert: Well, they are all important, but patients with either lung infections or GI diseases have worse prognoses than patients with skin infections.

Analyst: Do lung infections have a worse prognosis than GI diseases?

Expert: That's more difficult to answer. No. I would say that for all practical purposes, they have the same prognosis. Well, now that I think about it, perhaps patients with GI diseases have a slightly worse prognosis.

Having obtained the rank ordering of the attributes, we must estimate the importance weights.

Analyst: Let's say that we arbitrarily rate the importance of skin infection in determining prognosis at 10 points. GI diseases are how many times more important than skin infections?

Expert: Quite a bit. Maybe three times.

Analyst: That is, if we assign 10 points to skin infections, we should assign 30 points to the importance of GI diseases?

Expert: Yes, that sounds right.

Analyst: How about lung infections? How many more times important are they than GI diseases?

Expert: I would say about the same.

Analyst: (Checking for consistency in the subjective judgments.) Would you consider lung infections three times more serious than skin infections?

Expert: Yes, I think that should be about right.

In the dialogue above, the analyst first found the order of the attributes, then asked for the ratio of the weights of the attributes. Knowing the ratio of attributes allows us to estimate the attribute weights. If the model has only three attributes, the weights for the attributes can be obtained by solving the following three equations:

$$W(\text{GI diseases}) / W(\text{skin infection}) = 3$$

$$W(\text{lung diseases}) / W(\text{skin infection}) = 3$$

$$W(\text{lung diseases}) + W(\text{skin infection}) + W(\text{GI diseases}) = 1$$

One characteristic of this estimation method is that its emphasis on the ratio of the importance of the attributes leads to relatively extreme weighting compared to other approaches. Thus, some attributes may be judged critical, and others rather trivial. Other approaches, especially the direct magnitude process, may judge all attributes as almost equally important.

In choosing a method to estimate weights, you should consider several trade-offs. You can introduce errors by asking experts awkward and partially understood questions, but you can also cause error with an easier, but formally less justified, method. Our preference is to estimate weights in several ways and use the resulting differences to help experts think more carefully about their real beliefs. In doing so, we usually start with a rank order technique, then move on to assess ratios, obtain a direct magnitude estimate, identify discrepancies, and finally ask the expert to resolve them.

A final but vital question is whether we can really expect experts to be able to describe how they weight attributes in their own thinking. Nisbett and Wilson (1977) argued that directly assessed weight may not reflect an expert's true beliefs, but other investigators have found to the contrary (John and Edwards 1978).

The only way to decide if the directly assessed weights reflect the expert's opinions is to look at how the resulting models perform. In a number of applications, value models based on directly assessed weights correlated quite well with the subject's judgments (Fischer 1979). The typical correlation is actually in the upper 80s, which is high in comparison to most social science correlations. This success confirms our belief in the accuracy (perhaps we should say adequacy) of the subjective assessment techniques for constructing value models.

Step 8. Evaluate the model

While researchers know the importance of carefully evaluating value models, policy analysts often lack the time and resources to do this. Because of the importance of having confidence in the models and being able to defend our analytical methodology, we will present several ways of testing the adequacy of value models (Gustafson et al. 1980).

The first concern is *accuracy*. Strictly speaking, some value models can be evaluated by comparing their predictions against observable behavior. If a model is used to measure a subjective concept, its accuracy can be evaluated by comparing predictions to an observed and objective standard, which is often called the gold standard, to emphasize its status as being beyond debate. In practice, gold standards are rarely available for judging the accuracy of subjective concepts (otherwise, we would not need the models in the first place). For example, the accuracy of a severity index can be examined by comparing it to observed outcomes of patients' care. If the severity index accurately predicts outcomes, we have evidence favoring the model, but an outcome prediction alone is rarely adequate because a patient's attitudes toward aggressive intervention and the quality of the intervention both influence it. Nevertheless, if we expect poor care and patient preferences to have relatively little impact on outcome, we can evaluate the accuracy of a severity index against observed outcomes. We can do this by randomly selecting a group of patients, using the value model to predict survival, and observing the accuracy of its predictions.

A model is considered *valid* if several different ways of measuring it lead to the same finding. For example, the AIDS severity model should be correlated with other measures of AIDS severity. If the analyst has access to other severity indexes (such as physiologically based indexes), the predictions of the different approaches can be compared on a sample of patients. We can also explore construct validity by comparing the model with a surrogate measure of severity. Because severely ill patients stay longer in the hospital, we could correlate length of stay to severity scores. The point is that convergence among several measures of severity increases our confidence in the model.

A less attractive, but often practical, construct validity measure is to correlate the model scores and the global judgment of the expert who developed the index (Fryback 1976). The expert is asked to score several (perhaps 100) hypothetical case profiles described only by variables included in the model. If the model accurately predicts the expert's judgments, our confidence in the model increases, but this measure has the drawback of producing optimistic results. After all, if the expert who developed the model can't get the model to predict his or her judgments, who can? It is far better to ask a separate panel of experts to rate the patient profiles.

In the AIDS severity project, we collected the expert's estimate of survival time for 97 hypothetical patients and examined whether the value model could predict these ratings. The correlation between the additive model and the rating of survival was –0.53. (The negative correlation means that high severity scores indicate shorter survival, and the magnitude of the correlation ranges between 0 and –1.0.) The –0.53 correlation suggests low to moderate agreement between the model and the expert's intuitions; correlations closer to –1.0 imply greater agreement.

We can also judge correlations by comparing them with agreement among the experts. The correlation between several pairs of experts rating the same 97 hypothetical patients was also similar. The value model agreed with our expert as much as the experts agreed with each other. Thus, the value model may be a reasonable approach to measuring severity of AIDS. (Appendix 7-C discusses an alternative value model with slightly better correlation.)

The value model should also seem reasonable to experts, something we term *face validity*. Thus, in our example, the severity index should seem reasonable to clinicians and policymakers. Otherwise, even if it accurately predicts the outcome of care and other measures of severity, we may experience problems with its acceptance. Clinicians who are unfamiliar with statistics will likely rely on their experience to judge the index, meaning that the variables, weights, and value scores must seem reasonable and practical to them. Face validity is tested by showing the model to a new set of experts and asking if they understand it and whether it is conceptually reasonable.

When different people apply the value model to the same situation, they must arrive at the same scores, which we term *inter-rater reliability*. In our example, different registered record abstractors who use the index to rate the severity of a patient should produce the same score. If a model relies on hard-to-observe patient attributes, the abstractors will disagree about the condition of patients. If reasonable people using a value model reach different conclusions, then we lose confidence in the model's utility as a systematic method of evaluation. Inter-rater reliability is tested by having different record abstractors rate the severity of randomly selected patients.

The value model should require only *available data* for input. Relying on obscure data may increase the model's accuracy at the expense of practicality. Thus, the severity index should rely on reasonable sources of data, usually from existing data bases. A physiologically based data base, for instance, would predict prognosis of AIDS patients quite accurately. However, such an index would be useless if physiological information is generally unavailable and routine collection of this information would take considerable time and money.

While the issue of data availability may seem obvious, it is a very common error in the development of value models. Experts used to working in organizations with superlative data systems may want data that are unavailable at average institutions, and they may produce a value model with limited utility. If we do not care about comparing scores across organizations, we can tailor indexes to each institution's capabilities and allow each institution to decide whether the cost of collecting new data are justified by the expected increase in accuracy. However, if scores will be used to compare institutions or allocate resources among institutions, then we need a single value model that must be based on data available to all organizations.

The index should be *simple* to use. The index of medical underservice (Health Services Research Group 1975) is a good example of the importance of simplicity. This index, developed to help the federal government set priorities for funding HMOs, community health centers, and health facility development programs, originally had nine variables, but the director of the sponsoring federal agency rejected it because of the number of variables. Because he wanted to be able to "calculate the score on the back of an envelope," the index was reduced to four variables. This version performed as well as one with nine variables; it was used for eight years to help set nationwide funding priorities. This example shows that simplicity does not always equal incompetence. Simplicity nearly always makes an index easy to understand and use.

Most value models are devised to apply to a particular context, and they are not portable to other settings or uses. We view such *context dependence,* in general, as a liability, but this is not always the case. For example, the AIDS severity index was intended for evaluating treatment programs and prognosis of groups of patients. Its use for predicting prognosis of individual patients is inappropriate and possibly misleading.

Appendix 7-A. Alternative Methods of Assessing Single-Attribute Values

We can assess value functions in a number of different ways aside from the double-anchored method (Johnson and Huber 1979).

The *mid-value splitting technique* sets the best and worst levels of the attribute at 100 and 0. Then we find a level of the attribute that psychologically seems halfway between the best and the worst level and set its value at 50. Using the best, worst, and mid-value points, we continue finding points that psychologically seem halfway between any two points. After several points are identified, the values of other points are assessed by linear extrapolation from existing points. The following conversation illustrates how the mid-value splitting technique could be used to assess the value of age in AIDS severity.

Analyst: What is the age with the best prognosis?

Expert: A 20-year-old has the best chance of survival.

Analyst: What is the age with the worst prognosis?

Expert: AIDS patients over 70 are more susceptible to opportunistic infections and have the worst prognosis. Of course, infants with AIDS have an even worse prognosis, but I understand we are focusing on adults.

Analyst: Which age has a prognosis half as bad as a 70-year-old?

Expert: I am going to say about 40, though I am not really sure.

Analyst: I understand. We do not need exact answers. Perhaps it may help to ask the question differently. Do you think an increase in age from 40 to 70 causes as much of a deterioration in prognosis as an increase from 20 to 40 years?

Expert: If you are asking roughly, I agree.

Analyst: If 20 years is rated as 0, 70 years as 100, do you think it would be reasonable to rate 40 years as 50?

Expert: I suppose my previous answers imply that I should say "yes."

Analyst: Yes, but this is not binding—you can revise your answers.

Expert: A rating of 50 for the age of 40 seems fine as a first approximation.

Analyst: Can you tell me what age would have a prognosis halfway between 20 and 40 years old?

Using the mid-value splitting technique, the analyst chooses a value score, and the expert specifies the particular attribute level that matches it. This is opposite to the double-anchored estimation, in which the analyst specifies an attribute level and asks for its value. The choice between the two methods should depend on whether the attribute is discrete or continuous. Often, with discrete attributes, there are no levels to correspond to a particular value scores, leading us to select the double-anchored method.

A third method for assessing a value function is to draw a curve in the following fashion. The *x* axis is the various attribute levels. The *y* axis is the value associated with each attribute level. The best attribute level is assigned 100 and drawn on the curve. The worst attribute level is assigned 0. The expert is asked to draw a curve between these two points showing the value of remaining attribute levels (Pai et al. 1971). Once the graph is drawn, the analyst and the expert review its implications. For example, a graph can be constructed with age (20 to 70 years) on the *x* axis and value (0 to 100) on the *y* axis. Two points are marked on the graph (age 20 at 0 value and age 70 at 100 value). The analyst asks the expert to draw a line between these two points showing the prognosis for intermediate ages.

Finally, an extremely easy method, which requires no numerical assessment at all, is to assume a linear value function over the attribute. This arbitrary assumption introduces some errors, but they will be small if you are constructing an ordinal value scale, and the single-attribute value function is monotonic (meaning that an increase in the attribute level will cause either no change or an increase in value) (Warner 1971).

For example, we cannot assume that increasing age will cause a proportionate decline in prognosis. In other words, the relationship between the variables is not monotonic: the prognosis for infants is especially poor, while 20-year-old patients have the best prognosis and 70-year-olds have a poor outlook. Because increasing age does not consistently lead to increasing severity—and in fact it can also reduce severity—an assumption of linear value is misleading.

Appendix 7-B. Alternative Methods for Estimating Attribute Weights

In the direct magnitude estimate, the expert is asked to rank order the attributes and then to rate their importance by assigning each a number between 0 and 100. Once the ratings are obtained, they are scaled to range between 0 and 1 by dividing each weight by the sum of the ratings. Subjects rarely rate the importance of an attribute near zero, so the direct magnitude estimation has the characteristic of producing weights that are close together, but the process has the advantage of simplicity and comprehensibility.

Weights can be estimated by having the expert distribute a fixed number of points, typically 100, among the attributes (Torgerson 1958). The main advantage of this method is simplicity, since it is only slightly more difficult than the ranking method. But if we have a large number of attributes, experts will have difficulty assigning numbers that total 100.

One approach to estimating weights is to ask the expert to rate "corner" cases. A corner case is a description of a patient with one attribute at its most extreme level and the remainder at normal levels. The expert's score for the corner case shows its relative importance. In multiplicative models (described in Appendix 7-C), the analyst can estimate other parameters by presenting corner cases with two or more attributes at peak levels. After the expert rates several cases, a set of parameters is estimated that optimizes the fit between model predictions and expert's ratings.

Another approach is to mix and match methods. Several empirical comparisons of assessment methods have shown that different weight estimation methods lead to similar assessments. A study that compared seven methods for obtaining subjective weights, including 100 point distribution, ranking, and ratio methods, found no differences in their results (Cook and Stewart 1975). Such insensitivity to assessment procedures is encouraging because it shows that the estimates are not by-products of the method and thus are more likely to reflect the expert's true opinions. This allows us to substitute one method for another.

Appendix 7-C. Alternative Aggregation Rules

The additive value model assumes that single-attribute value scores are weighted for importance and then added together. In essence, it calculates a weighted average of single-attribute value functions.

The multiplicative model is another common aggregation rule. In the AIDS severity study, discussions with physicians suggested that a high score in any single attribute value function was sufficient ground for judging the patient severely ill. Using a multiplicative model, overall severity would be calculated thusly:

$$S = \{-1 + \Pi_i [1 + kk_i U(A_{ij})]\}/k$$

where K_i and K are constants chosen so that $k = -1 + \Pi_i (1 + kk_i)$

In a multiplicative model, when the constant K is close to –1, high scores in one category are sufficient to produce an overall severe score even if other categories are normal. This model better resembled the expert's intuitions, so we investigated its accuracy for judging AIDS severity. To construct the multiplicative value model, the expert must estimate "n + 1" parameters: the *n* constants K_i and one additional parameter, the constant K.

In the AIDS severity project, we constructed a multiplicative value model. On 97 hypothetical patients, we separately compared the severity ratings of the multiplicative and the additive models to the expert's intuitive ratings and found correlations of –0.53 and –0.60, showing that the multiplicative model was slightly more accurate. The difference in the accuracy of the two models was statistically significant. Therefore, we chose the multiplicative severity model, details of which are presented in Appendix 7-D.

In situations where preferential independence does not hold, neither additive nor multiplicative models are appropriate. Instead, value assessments are needed for each combination of attributes. We faced this problem when creating an index to estimate an employee's risks of hospitalization from various lifestyles. The relative importance of various lifestyles was dependent on age. For young employees, the most important concern was drinking and driving. For older ones, high cholesterol and high blood pressure were more important. Additive and multiplicative models could not reflect the effect of age on lifestyles, so we solved our dilemma by constructing separate additive models for the younger and older employees (Alemi et al. 1987).

Appendix 7-D. Severity of the Course of AIDS

Step 1: In each category choose the lowest score that applies to patient's characteristic. If you cannot find an exact match, approximate the score by using the two markers most similar to it. Please indicate score in the box provided.

Age

Less than 18 years, do not use this index
18 to 40 years, 1.000
40 to 60 years, 0.9774
Over 60 years, 0.9436

Race

White, 1.000
Black, 0.9525
Hispanic, 0.95252
Other, 1.000

Defininіng AIDS diagnosis

Keposi's, 1.000
Candida Esophagitis, 0.8093
PCP, 0.8014
Toxoplasmosis, 0.7537
Cryptococcosis, 0.7338
CMV Retinitis, 0.7259
Cryptosporidiosis, 0.7179
Dementia, 0.7140
CMV colitis, 0.6981
Lymphoma, 0.6981
PML, 0.6941

Mode of transmission

Nontrauma blood transfusion, 0.9316
Drug abuse, 0.8792
Other, 1.000

Time since AIDS

Less than 3 months, 1.000
More than 3 months, 0.9841
More than 6 months, 0.9682
More than 9 months, 0.9563
More than 12 months, 0.9404
More than 15 months, 0.9245
More than 18 months, 0.9086
More than 21 months, 0.8927
More than 24 months, 0.8768

Skin disorders

No skin disorder, 1.000
KS, 1.000
Shingles, 0.9036
Herpes complex, 0.8735
Candida or mucus, 0.8555
Thrush, 0.8555

Lung disorders

No lung disorder, 1.000
Pneumonia unspecified, 0.9208
Bacterial pneumonia, 0.8960
TB, 0.8911
Mild PCP, 0.8664
Cryptococcosis, 0.8161
Herpes simplex, 0.8115
Histoplasmosis, 0.8135
PCP with respiratory failure, 0.8100
MAI, 0.8020
KS, 0.7772

Nervous system diseases

No nervous system involvement, 1.000
Neurosyphilis, 0.9975
TB meningitis, 0.7776
Crypto, 0.7616
Seizure, 0.7611
Myelopathy, 0.7511
CMV retinitis, 0.7454
Nocardia, 0.7454
Meningitis encephalitis, 0.7368
Histo, 0.7264
PML, 0.7213
Encephalopathy/HIV dementia, 0.7213
Coccidiomycosis, 0.7189
Lymphoma, 0.7139

GI diseases

No GI disease, 1.000

More than 36 months, 0.8172
More than 48 months, 0.7537
More than 60 months, 0.6941

Heart disorders

No heart disorder, 1.000
HIV cardiomyopathy, 0.7337

Disseminated diseases

No disseminated illness, 1.000
Thrombocytopenia ITP, 0.9237
KS, 0.9067
Nonsalmonella sepsis, 0.8163
Salmonella sepsis, 0.8043
Other drug induced anemia, 0.7918
Varicella zoster, 0.7912
TB, 0.7910
Nocardia, 0.7842
Non-TB myco bacterial disease, 0.7705
Transfusion, 0.7611
Toxo, 0.7591
AZT drug induced anemia, 0.7576
Crypto, 0.7555
Histoplasmosis, 0.7405
Hodgkin's, 0.7340
Coccidiodomycosis, 0.7310
CMV, 0.7239
Non-Hodgkin's lymphoma, 0.7164
Thrombocytopenia TTP, 0.7139

Reoccurring acute illness

No, 1.000
Yes, 0.8357

Drug markers

None, 1.000
Lack of prohilaxis, 0.8756
Starting AZT on 1 gram, 0.7954
Starting and stopping of AZT, 0.7963
Drop of AZT 1 gram, 0.7673
Incomplete treatment in HSV, VZV, MAI, or CM retinitis, 0.7593
Prescribed oral narcotics, 0.7512
Prescribed parenteral narcotics, 0.7192
Incomplete treatment of PCP, 0.7111
Incomplete treatment of Toxo, 0.7031

Isosporidium, 0.8091
Candidal esophagitis, 0.8058
Salmonella, 0.7905
TB, 0.7897
Nonspecific diarrhea, 0.7803
Herpes esophagitis, 0.7536
MAI, 0.7494
Cryptosporidium, 0.7369
KS, 0.7324
CMV colitis, 0.7086
GI cander, 0.7060

Organ involvement

None, 1.000

	Failure	*Insufficiency*	*Dysfunction*
Cerebral	0.7000	0.7240	0.7480
Liver	0.7040	0.7600	0.8720
Heart	0.7080	0.7320	0.7560
Lung	0.7120	0.7520	0.8000
Renal	0.7180	0.7910	0.8840
Adrenal	0.7640	0.8240	0.7960

Comorbidity

None, 1.000
Hypertension, 1.000
Influenza, 0.9203
Legionella, 0.9402
Alcoholism, 0.8406

Functional impairment

No marker, 1.000
Boarding home care, 0.7933
Home health care, 0.7655
Nursing home care, 0.7535
Hospice care, 0.7416

Psychiatric comorbidity

None, 1.000
Psychiatric problem in psych. hosp., 0.8872
Psychiatric problem in med. setting, 0.8268
Severe depression, 0.8268

Nutritional status

No markers, 1.000
Antiemetic, 0.9282
Nutritional supplement, 0.7687
Payment for nutritionist, 0.7607
Lomotil/Imodium, 0.7447
Total parenteral treatment, 0.7248

Incomplete treatment in Crypto,
0.6951

Step 2: Multiply all scores and enter here
Step 3: Subtract one from above entry and enter here
Step 4: Divide by –0.99 and enter here

This entry indicates the severity of the course of illness. The higher the score the worse the prognosis. Maximum score is 1, minimum score is 0. This score is a rough estimate of prognosis and should not be used to guide treatment of individual patients.

References

Alemi, F., J. Stokes, III, J. Rice, E. Karim, W. Lacorte, L. Saligman, and R. Nau. 1987. "Appraisal of Modifiable Hospitalization Risk." *Medical Care* 25 (7): 582–91.

Alemi, F., B. Turner, L. Markson, and T. Maccaron. 1990. "Severity of the Course of AIDS." *Interfaces* 21: 105–6.

Aschenbrenner, K., and T. Kaubeck. 1978. "Challenging the Cushing Syndrome: Multi-Attribute Evaluation of Cortisone Drugs." *Organizational Behavior and Human Performance* 22 (2): 215–34.

Beach, L. R., F. L. Campbell, and B. D. Townes. 1979. "Subjective Expected Utility and the Prediction of Birth-Planning Decisions." *Organizational Behavior and Human Performance* 24: 18–28.

Bernoulli, D. 1938. "Spearman theoria novai de mensura sortus." *Comettariii Academiae Saentiarum Imperialses Petropolitica* 5: 175–92. Translated by L. Somner in *Econometrica* 22 (1954): 23–36.

Cline, B., F. Alemi, and K. Bosworth 1982. "Intensive Skilled Nursing Care: A Multi-Attribute Utility Model for Level of Care Decision Making." *Journal of American Health Care Association.*

Cook, R. L., and T. R. Stewart. 1975. "A Comparison of Seven Methods for Obtaining Subjective Description of Judgmental Policy." *Organizational Behavior and Human Performance* 12: 31–45.

Detmer, D. E., J. A. Moylan, J. Rose, R. Shultz, R. Wallace, and R. A. Daly. 1977. "Regional Categorization and Quality of Care in Major Trauma." *Journal of Trauma* 17 (8): 592.

Edwards, W. 1974. "The Theory of Decision Making." *Psychology Bulletin* 51 (4): 320.

Edwards, W. 1977. "How to Use Multi-Attribute Utility Measurement for Social Decision-Making." *IEEE Transactions on Systems, Man and Cybernetics* SMC7: 326-40.

Ericsson, K. A., and H. A. Simon. 1980. "Verbal Reports as Data." *Psychological Review* 87: 215–51.

Fischer, G. W. 1979. "Utility Models for Multiple Objective Decisions: Do They Accurately Represent Human Preferences?" *Decision Science* 10: 451.

Fryback, D. G. 1976. "Comparison of Four Types of Linear Models of Subjective Evaluations: Cross Validation as a Function of Sample Dimensionality of Stimuli." Center for Health System Research and Analysis, University of Wisconsin-Madison.

Gustafson, D. H., M. E. Hiles, and C. Taylor. 1980. "Report on the Trauma Severity Index Conference." Center for Health Systems Research and Analysis, University of Wisconsin–Madison.

Health Services Research Group, Center for Health Systems Research and Analysis, University of Wisconsin. 1975. "Development of the Index for Medical Under-service." *Health Services Research* 10 (2): 168–80.

Hogarth, R. M. 1975. "Cognitive Processes and the Assessment of Subjective Probabilities." *Journal of American Statistical Association* 70: 271–94.

Humphrey, S. P., and A. Humphrey. 1975. "An Investigation of Subjective Preference Orderings for Multi-Attribute Alternatives." In *Utility, Probability and Human Decision-Making,* edited by D. Wendt and C. Vlek, pp. 119–23. The Netherlands: D. Reidel Publishing Co.

John, R. S., and W. Edwards. 1978. "Subjective versus Statistical Importance Weights: a Criterion Validation." Research report, Social Science Research Institute, SSRI 78–7, University of Southern California.

Johnson, E. M., and G. P. Huber. 1979. "The Technology of Utility Assessments: Issues and Problems." *IEEE Transactions in Systems, Man and Cybernetics.* Kahneman, D., and A. Tversky. 1979. "Prospect Theory: An Analysis of Decisions Under Risk." *Econometrica* 47: 263–91.

Kahneman, D., and A. Tversky. 1979. "Prospect Theory: An Analysis of Decisions Under Risk." *Econometrica* 47: 263–91.

Kao, E. P. C. 1972. "A Semi Markov Model to Predict Recovery Progress of Coronary Patients." *Health Services Research* 7: 191–208.

Keeney, R. 1980. "Analysis of Preference Dependencies among Objectives." Woodward and Clyde Consultants (July).

———. 1977. "The Art of Assessing Multi-Attribute Utility Functions." *Organizational Behavior and Human Performance* 19: 267–310.

Keeney, R. L., and H. Raiffa. 1976. *Decisions and Multiple Objectives: Preferences and Value Tradeoffs.* New York: John Wiley.

Kneppreth, N. P., D. H. Gustafson, and R. P. Leifer. 1974. "Techniques for Assessment of Worth." Technical Paper 254, U.S. Army Research Institute for the Behavioral and Social Sciences, Arlington, MD.

Nisbett, R., and L. Ross. 1980. *Human Inferences.* Englewood Cliffs, NJ: Prentice-Hall.

Nisbett, R. E., and T. D. Wilson. 1977. "Telling More Than We Can Know: Verbal Reports on Mental Processes." *Psychological Review* 84 (3): 231–59.

Pai, G. K., D. H. Gustafson, and G. W. Kiner. 1971. "Comparison of Three Non-Risk Methods for Determining a Preference Function." Center for Health Systems Research and Analysis, University of Wisconsin–Madison.

Pascal, A., C. L. Bennett, and M. C. Bennett. 1989. "The Cost of Financing Care for AIDS Patients: Results of a Cohort Study in Los Angeles." National Center for Health Services Research Conference Proceedings on New Perspectives on HIV-Related Illness, NCHSR publications, Rockville, MD.

Pliskin, J. S., D. S. Shepard, and M. C. Weinstein. 1980. "Utility Functions for Life Years and Health Status." *Operation Research* 28 (1): 206.

Redfield, R. R., and D. S. Burke. 1988. "HIV Infection: The Clinical Picture." *Scientific American* 259 (4): 90–98.

Savage, L. D. 1972. *Foundations of Measurement.* New York: John Wiley.

Scitovsky, A. A. 1989. "Past Lessons and Future Directions: The Economics of Health Services Delivery for HIV-Related Illness." National Center for Health Services Research Conference Proceedings on New Perspectives on HIV-Related Illness, NCHSR publications, Rockville, MD.

Snyder, M., and N. Cantor. 1979. "Testing Theories about Other People: Remembering All the History that Fits." Unpublished manuscript, University of Minnesota.

Torgerson, W. S. 1958. *Theory and Methods of Scaling*. New York: John Wiley.

VanGundy, A. B. 1981. *Techniques of Structured Problem Solving*. New York: Van Nostrand Reinhold.

Volkema, R. 1981. "Different Approaches to Problem Solving." Ph.D. thesis, University of Wisconsin.

Von Winterfeldt, D., and W. Edwards. 1986. *Decision Analysis and Behavioral Research*. New York: Cambridge University Press.

Warner, H. 1971. "Estimating Coefficients in Linear Models: It Don't Make No Never Mind." *Psychological Bulletin* 83 (2): 213–14.

Yates, J. F., and C. M. Jagacinski. 1979. "Reference Effects in Multi-Attribute Evaluations." *Organizational Behavior and Human Performance* 24 (3): 400–410.

8

Forecasting without Real Data: Bayesian Probability Models

While most forecasts rely on the past, we often want to predict events that have no antecedents at all, or whose antecedents are cloudy, ambiguous, or insufficiently documented. Statisticians have developed several tools to discern historical trends and forecast the future. These tools are quite useful for planning purposes and should be used when relevant historical data are available. The traditional approach might be used to predict demand for a product by projecting historical sales figures into the future. If sales have been increasing consistently for several years, some may be willing to assume they will continue to do so.

Often, however, analysts must forecast unique events that lack good antecedents. Sometimes the environment has changed so radically that trends are irrelevant. Even if we could gather historical data, we may lack the necessary time or money. In these circumstances, forecasting can be done with a Bayesian subjective probability model. This chapter introduces the concepts of subjective probability and the Bayesian model.

This discussion refers to the task of predicting demand for a proposal to establish a special type of HMO. HMOs are group health insurance packages sold through employers to employees. HMOs require employees to consult a primary care physician before visiting a specialist; the primary physician has financial incentives to reduce inappropriate use of services. Experience with HMOs shows they can cut costs by reducing unnecessary hospitalization.

At first, predicting demand for the proposal seems relatively easy, because there is a great deal of national experience with HMOs. But the proposed HMO uses technology to set it apart from the crowd. Members will be encouraged to save time by using computer terminals connected to their television sets. The member will initiate contact with the HMO through the computer, which will interview the member and send a summary to the primary care doctor, who would consult the patient's record and decide whether the patient should:

- Wait and see what happens
- Have specific tests done before visiting
- Take a prescription (phoned to a nearby pharmacy) and wait to see if the symptoms disappear
- Visit a physician's assistant, primary care physician, or specialist

With the decision made, the computer will inform the patient which action the primary physician recommends. If the doctor does not recommend a visit, the computer will automatically call a few days later to see if the symptoms have diminished. All care will be supervised by the patient's physician, and all communication will be in plain English. When the computer does not understand the patient, an operator will intercede to translate the complaint into language comprehensible to the computer.

Clearly, this is not the kind of HMO with which we have much experience, but let's blend a few more uncertainties into the brew. Assume that the local insurance market has changed radically in recent years—competition has increased, and businesses have organized powerful coalitions to control health care costs. At the federal level, national health insurance is again under discussion. With such radical changes on the horizon, data as young as two years old may be irrelevant. As if these constraints were not enough, we need to produce the forecast in a hurry. What can we do? How can we predict an unprecedented event?

Different Sources of Data

Uncertainty about the success of the proposed HMO can be expressed in two ways, one based on expected frequency of success and the other based on strength of belief in the truth of a statement about success or failure. A statement based on frequency of success might sound like this: "Experience with new health care technologies shows that x times out of n, such innovations succeed." A statement of strength of belief might sound like this: "Having read the business plan, I find it reasonable. I think the proposed HMO has a good chance of succeeding. In fact, I am so confident of this fact that I would bet my career on it."

Both statements measure the degree of uncertainty about the success of the HMO, but there is a major difference between them. Measurement of a frequency is based on observation and history, while measurement of strength of belief is based on an individual's opinion. And that simple difference allows us to forecast without real data, using the principles of subjective probability. Savage (1954) and DeFinetti (1964) argued that the rules of probabilities can work with uncertainties expressed as strength of opinion. Savage termed the strength of a decision maker's convictions "subjective probability" and used the calculus of probability to analyze them.

Subjective probability can be measured along two different concepts: (1) intensity of feelings and (2) hypothetical action (Ramsey 1950). We measure subjective probability on the basis of intensity of feelings by asking an expert to mark a scale between 0 and 1. We measure subjective probability on the basis of hypothetical actions by asking the expert about the hypothetical frequency that the event will occur.

Suppose we want to measure the probability that an employee will join the HMO. Using the first method, we would ask an expert on the local health care market about intensity of feeling:

> Q: Do you think employees will join the plan? On a scale from 0 to 1, with 1 being certain, how strongly do you feel you are right?

When measuring according to hypothetical frequencies, we ask the expert to imagine what frequency he or she expects. While the event has not occurred repeatedly, we can ask the expert to imagine that it has.

> Q: Out of 100 employees, how many do you think will join the plan?

Obviously, both methods try to tap the strength of the expert's belief, but the second method does so by suggesting a hypothetical situation. It assumes that the expert can imagine how 100 employees would choose, even though none has yet faced the choice.

Can we apply the methods used to analyze frequency counts to analyze subjective probabilities? The obvious differences between measurements of subjective probabilities and observational data do not require us to use different methods of analysis. Both approaches produce probabilities, and thus these probabilities, however measured, should follow the same set of rules. When strength of belief is measured as a hypothetical frequency, we can easily show that beliefs can be treated as probability functions. But how can we argue that subjective probabilities measured as intensity of feelings should be treated as probability functions? To answer this, we must examine the definition of a probability measure. A probability function is defined by the following characteristics:

1. The probability of an event is a positive number between 0 and 1.
2. One event certainly will happen, so the sum of the probabilities of the events is 1.
3. The probability of any two exclusive events occurring equals the sum of the probability of each occurring.

These assumptions are at the root of all mathematical work in probability, so any beliefs expressed as probability must follow them. The first assumption is always true, because we can assign numbers to beliefs so they are always positive. But the second and third assumptions are not always true, and people do hold beliefs that violate them. We can, however, take steps to ensure that these two assumptions are also met. For example, when the estimates of probabilities of two mutually exclusive and exhaustive events (e.g., probability of success and failure) do not total 1, we can standardize the estimates to do so. When the estimated probabilities of two mutually exclusive events do not equal the sum of their probabilities—$p(A) + p(B) = p(A \text{ or } B)$—we can ask whether they should, and adjust as necessary. In their thinking, decision makers may or may not follow the calculus of probability. But what people do and how they should do it are two different issues. Decision makers may wish to follow the rules of probability, even though they have not always done so.

Our argument is not that probabilities and beliefs are the identical constructs, but rather that probabilities provide a context in which beliefs can be studied. That is, if beliefs are expressed as probabilities, the rules of probability provide a systematic and orderly method of examining the implications of these beliefs. If it seems correct that strengths of beliefs should follow the calculus of probability, then we can standardize our information and adjust experts' estimates to do so.

In the remainder of this chapter, we show how experts' opinions and estimates are assessed and analyzed with the rules of probability, how to test a model's validity, and how to use a model to make a forecast. The first step, however, is to clarify exactly what we shall forecast.

Step 1. Keep it simple

To forecast a number of events, we first make sure that they are mutually exclusive (the events cannot occur simultaneously) and constitute an exhaustive set (one event in the set must happen). Thus, in the proposed HMO, we might decide to predict the following exhaustive list of mutually exclusive events:

- More than 75 percent of employees will join
- Between 50 percent and 75 percent will join
- Between 25 percent and 50 percent will join
- Fewer than 25 percent will join

The event being forecast should be expressed in terms of the experts' daily experiences and in terms they are familiar with. If we plan to tap the intuitions of benefit managers about the proposed HMO, we should realize they might have difficulty with our event types, which are described in terms of the entire employee population. If benefit managers are more comfortable thinking about individuals, we can calculate the four events from the probability that one employee will join. It makes no difference for the analysis how one defines the events of interest. It may make a big difference to the experts, however, so be sure to define the event of interest in terms familiar to them.

Expertise is a funny thing. If we ask experts about situations slightly outside their specific area or frame of reference, we often get erroneous responses. For example, some weather forecasters might predict rain more accurately than air pollution because they have more experience with rain. Therefore, it would be more reasonable to ask benefit managers about the probability of events, but focus on the individual, not the group:

- The employee will join the proposed HMO.
- The employee will not join the proposed HMO.

Many analysts and decision makers, recognizing that real situations are complex and have tangled interrelationships, tend to work with a great deal of complexity. We prefer to forecast as few events as possible. In our example, we might try to predict the following events:

- The employee will never join the proposed HMO.
- The employee will not join the HMO in the first year but will in the second.
- The employee will join the HMO in the first year but withdraw in the second.
- The employee will join the HMO in the first year to stay.

Again the events are mutually exclusive and exhaustive, but now they are more complex. The forecasts deal not only with applicants' decisions but also with the stability of those decisions. People may join when they are sick and withdraw when they are well. Turnover rates affect administration and utilization costs, so information about stability of the risk pool is important. In spite of the utility of such a categorization, we think it is difficult to combine two predictions and prefer to design a separate model for each—for reasons of simplicity and accuracy. As will become clear shortly, we can use simpler methods of forecast with two events than when more events are possible.

The events must be chosen carefully because a failure to minimize their number may indicate you have not captured the essence of the uncertainty. One way of ensuring that the underlying uncertainty is being addressed is to examine the link between the forecast event and the actions the decision maker is contemplating. Unless these actions differ radically from one another, some of the events should be combined. A model of uncertainty needs no more than two events unless there is clear proof to the contrary. Even then, it is often best to build more than one model to forecast more than two events.

For our purposes, we are interested in predicting how many employees will join the HMO, because this is the key uncertainty that investors need to judge the proposal. To predict the number who will join, we can calculate p(J), the probability that an individual employee will join. If the total number of employees is *n*, then the number who will join is *n*p(J). Having made these assumptions, let's return to the question of assessing the probability of joining.

Step 2. Is unaided intuition enough?

We suggested that demand for the proposed HMO can be assessed by asking experts, "Out of 100 employees, how many will join?" The suggestion was somewhat rhetorical, and an expert might well answer, "Who knows? Some people will join the proposed HMO, others will not—it all depends on many factors." Clearly, if posed in these terms, the question is too general to answer. When the task is complex, meaning that many contradictory clues must be evaluated, experts' predictions can be way off the mark.

Errors in judgments may be reduced if we can break complex judgments into several components and ask the experts to estimate their probabilities. Then we can use the rules of probability to assess the contribution these estimates will have on the overall, complex judgment (Beach 1975). In analyzing opinions about future uncertainties, we often find that forecasts depend on a host of factors. Because a direct estimation of what might happen can be too difficult for most experts, we need an indirect approach in which the forecast is decomposed into predictions about a number of smaller events. In talking with experts, the first task is to understand whether they can make the desired forecast with confidence and without reservation. If they can, then we rely on their forecast and save everybody's time. When they cannot, we can use the process discussed below to disassemble the problem and make an accurate forecast.

Nothing is totally new, and the most radical plan has components that resemble aspects of established plans. Though the proposed HMO is novel, experience offers clues to help us predict the reaction to it. The success of the HMO will depend on factors that have influenced demand for services in other circumstances. Experience shows that the plan's success depends on the composition of the potential enrollees. In other words, some people have characteristics that dispose them toward or against joining the HMO. As a first approximation, the plan might be more attractive to young employees who are familiar with computers, to older high-level employees who want to save time, to employees comfortable with delayed communications on telephone answering machines, and to patients who want more control over their care. If most employees are members of these groups, we might reasonably project good demand.

But a case is rarely that simple, so we usually must compile the impact of various clues before reaching our conclusion. Almost every employee will have a characteristic suggestive of a tendency to join or decline. We suggest summarizing the impact of diverse characteristics with the Bayes' probability theory.

Bayes' theorem is a formally optimal model for revising existing opinion (sometimes called prior opinion) in the light of new evidence. The theorem states:

$$p(J|C_1, \ldots, C_n) = (C_1, \ldots, C_n|J) \times p(J)/p(C_1, \ldots, C_n)$$

Where p designates the subjective probability of an event.

J identifies that an individual will join the new plan; thus, $p(J)$ is the probability that the individual will join the HMO.

C_i identifies the individual's *i*th characteristic.

$|$ designates "given"; thus, $J|C_1$ says "individual joining the plan given he or she has characteristic C_1."

$p(J|C_1, \ldots, C_n)$ measures our belief that the plan will succeed in attracting an individual with characteristics C_1 through C_n. This variable shows the probability of the event we wish to predict after examining clues C_1 through C_n.

$p(C_1, \ldots, C_n|J)$ is the probability of finding characteristics C_1 through C_n among joiners. This variable is called the likelihood associated with the characteristics and is a measure of the impact of the characteristics on our forecasts.

$p(J)$ is the probability of an employee joining the proposed plan if we are totally ignorant of whether the employee has characteristics C_1 through C_n. Thus, this variable measures our guess that an employee will join the plan if we know nothing about him or her. It is called the prior probability to highlight the fact that this is the probability of the event before information is available.

$p(C_1, \ldots, C_n)$ is the probability that characteristics C_1 through C_n will be found among the employees.

The difference between $p(J|C_1, \ldots, C_n)$ and $p(J)$ is the knowledge of characteristics C_1 through C_n. Thus, the theorem shows how our opinion about an employee's reaction to the plan will be modified by our knowledge of his or her characteristics. Because Bayes' theorem prescribes how opinions should be revised to reflect new data, it is a tool for consistent and systematic processing of opinions.

Bayes' theorem can be simplified considerably if it is written for different and mutually exclusive events. Suppose J marks the event that the individual will join the proposed plan, and N designates that the individual will decline. If Bayes' formula is written for both events, the formulas are:

$$p(J|C_1, \ldots, C_n) = p(C_1, \ldots, C_n|J) \times p(J)/p(C_1, \ldots, C_n)$$

$$p(N|C_1, \ldots, C_n) = p(C_1, \ldots, C_n|N) \times p(N)/p(C_1, \ldots, C_n)$$

We obtain a modified form of Bayes' formula by dividing the two equations by each other, giving this result:

$$p(J|C_1, \ldots, C_n)/p(N|C_1, \ldots, C_n) =$$

$$[p(C_1, \ldots, C_n|J)/p(C_1, \ldots, C_n|N)] \times [p(J)/p(N)]$$

In this form of Bayes' theorem, often called the odds form, the unconditional probability of the employee characteristics, $p(C_1, \ldots, C_n)$, does not enter the equation. As we need not measure this factor, we simplify the measurement task. Consequently, this form of the Bayes' model is used more widely than the original form. Each term in the equation has been named and given intuitive explanation.

The odds $p(J)/p(N)$ are called prior odds since they are the odds for the two hypotheses before information is available. The impact of the datum, $p(C_1, \ldots, C_n|J)/p(C_1, \ldots, C_n|N)$, is called the likelihood ratio. The forecast, $p(J|C_1, \ldots, C_n)/p(N|C_1, \ldots, C_n)$, is called the posterior odds because it shows the odds for the event after various information has been considered. Thus, the odds form of the Bayes' model can be written with the terms used throughout this chapter:

$$\text{Posterior odds} = \text{Likelihood ratio} \times \text{Prior odds}$$

Before proceeding, we must examine the logic behind the formula, not in terms of mathematical proof but in terms of how the formula relates to a set of intuitions about probabilities.

Bayes' theorem was first presented by Thomas Bayes, an English mathematician, although why he never submitted his paper for publication remains a mystery. Nevertheless, using Bayes' notes, Price presented a proof of Bayes' theorem. The following presentation of Bayes' argument differs slightly from the original (Bayes 1963).

We will consider for the sake of argument that there is only one characteristic of interest, say familiarity with computers. If we knew who will join and who will not, we could establish four groups:

- A group of size a joins the program and is computer literate.
- A group of size b joins the program and is not computer literate.
- A group of size c does not join and is computer literate.
- A group of size d does not join and is not computer literate.

The total number of employees then is a + b + c + d; among them, a + b join the program. If probability of a particular event is defined as the number of ways it can occur divided by the number of possible events, then the probability of joining the program, p(J), is:

$$p(J) = (a + b)/(a + b + c + d)$$

This says the probability of joining is the proportion of total employees who join. Similarly, the chance that an employee is computer literate, $p(C_1)$, is the total number of computer literates, a + c, divided by the total number of employees:

$$p(C_1) = (a + c)/(a + b + c + d)$$

But if we deal only with computer-literate employees, the total number of employees with whom we are concerned changes. Thus, the probability of joining or of having characteristic C_1 changes as well. Among all computer-literate employees (a + c), a group of size a joins the HMO. Therefore, the chances that a computer literate joins are:

$$p(J|C_1) = a/(a + c)$$

And similarly, the chances that we will find computer literates among joiners are:

$$p(C_1|J) = a/(a + b)$$

The Bayes' formula suggests that $p(J|C_1)$ can be calculated as:

$$p(J|C_1) = p(C_1|J) \times p(J)/p(C_1)$$

By inserting values calculated for the terms on the right-hand side, we observe:

$$p(J|C_1) = [a/(a + b)] \times [(a + b)/(a + b + c + d)]/[(a + c)/(a + b + c + d)]$$

$$p(J|C_1) = a/(a + c)$$

This value is identical to the value we intuitively calculated for $p(J|C_1)$—showing that the Bayes' formula agrees with our intuitions of how probability of joining should be calculated. The point is that if we partition employees into the four groups, count the number in each group, and define probability as described, then Bayes' formula holds and is a logically consistent approach.

Step 3. Identify clues

Usually, many clues are available for each forecast, and we must select a few under our usual time and data collection constraints. This requires some familiarity with existing data. For example, a benefit manager would need just a phone call to find the gender distribution of employees but might need a survey to find how many employee spouses are covered by outside health plans. Both are possible, but before ordering a survey, consider how much accuracy would be added. If it is minimal, then avoid the survey and use more readily available clues.

The identification of clues starts with the published literature. Even when we think our task is unique, it is always surprising how much has been published about related topics. To our surprise, there was a great deal of literature on predicting decisions to join an HMO, and even though these studies don't concern HMOs with our unique characteristics, reading them can help us think more carefully about clues. It is our experience that one seldom finds exactly what is needed in the literature.

Once the literature search is completed, we strongly advocate using experts to identify clues for a forecast. Even if there is extensive literature on a subject, we cannot expect to select the most important variables or to discern all important clues. In a few telephone interviews with experts, we can find the key variables, get suggestions on measuring each one, and identify two or three superior journal articles.

Experts should be chosen on the basis of accessibility and expertise. To forecast HMO enrollment, appropriate experts might be people with firsthand knowledge of the employees, such as benefit managers, actuaries in other insurance companies, and local planning agency personnel.

It's useful to start talking with experts by asking broad questions designed to help the experts talk about themselves. A good opening query might be:

> Q: Would you tell me a little about your experience with employee choice of health plans?

The expert might respond with an anecdote about irrational choices by employees, implying that a rational system cannot predict everyone's behavior. Equally, the expert might mention how difficult it is to forecast, or how many years he or she has spent studying these phenomena. The analyst should understand what is occurring here. In these early responses, the expert is expressing a sense of the importance and the value of his or her experience and input. It is vital to acknowledge this hidden message and allow ample time for the expert to describe historic situations.

After the expert has been primed by recalling these experiences, the analyst asks about characteristics that might suggest an employee's decision to join or not join the plan. An opening inquiry could be:

> Q: Suppose you were to decide whether an employee is likely to join but you could not contact the employee. I was chosen to be your eyes and your ears. What should I look for?

After a few queries of this type, ask more focused questions:

> Q: What is an example of a characteristic that would increase the chance of joining the HMO?

We refer to the second question as a positive prompt because it elicits factors that would increase the chances of joining. Negative prompts seek factors that decrease the probability. An example of a negative prompt is:

Q: Describe an employee who is unlikely to join the proposed HMO.

This distinction is important because research shows that positive and negative prompts yield different sets of factors. When Snyder and Swann (1978) asked subjects to identify clues for introversion and extroversion, they got differing responses. Though introversion and extroversion are opposite concepts and clues identifying one yield information about the other, the subjects identified two unrelated sets of clues. Thus, forecasting should start with clues that support the forecast, then explore clues that oppose it. The responses can later be combined so the model contains both sets.

It is important to get several opinions on what clues are needed. Each expert has access to a unique set of information; using more than one expert enables us to pool information and improve the accuracy of the recall of clues. Our experience suggests that at least three experts should be interviewed for about one hour each. After a preliminary list of factors is collected during that interview, the experts should have a chance to revise the list, either by telephone, by mail, or in a meeting. If time and resources allow, we prefer the integrative group process (see Chapter 5) for identifying the clues.

Let us suppose that our experts identified the following clues for predicting an employee's decision to join:

- Age
- Income and value of time to the employee
- Gender
- Computer literacy
- Current membership in an HMO

Some of these clues are more informative than others. In the next step, these clues are further fleshed out.

Step 4. Describe levels of each clue

A level of a clue measures the extent to which it is present. At the simplest, there are two levels, presence or absence, but sometimes there are more. Thus, age of employees may be described in terms of six discrete levels, each corresponding to a decade: younger than 21, 21–30, 31–40, 41–50, 51–60, older than 60. Occasionally we have continuous clues with many more levels. For example, when any year between 18 and 65 is considered, we have at least 47 levels.

In principle, it is possible to accommodate both discrete and continuous variables in a Bayesian model. In practice, discrete clues are used more frequently for two reasons: (1) experts seem to have more difficulty estimating likelihood ratios associated with continuous clues, and (2) in the health and social service areas, most clues tend to be discrete and virtually all other types of clue can be transformed to discrete clues.

As with defining the forecast event, the primary rule for creating discrete levels is to minimize the number of categories. Rarely are more than five or six categories required, and frequently two or three suffice.

We prefer to identify levels for various clues by asking the experts to describe a level at which the clue will increase the probability of the forecast event. Thus, we may have the following conversation:

Q: What would be an example of an age that would favor joining the HMO?

A: Young people are more likely to join than older people.

Q: How do you define young employees?

A: Who knows? It all depends. But if you really push me, I would say below 30 is different from above 30. This is probably why the young used to say "never trust anybody over 30." This age marks real differences in life outlook.

Q: What age reduces chances of joining the HMO?

A: Employees over 40 are different, too—pretty much settled in their ways. Of course, you understand we are just talking in general—there are many exceptions.

Q: Sure. I understand, but we are trying to model these general trends.

In all cases, each category or division should represent a different chance of joining the HMO. One way to check this would be to ask:

Q: Do you think a 50-year-old employee is substantially less likely to join than a 40-year-old?

A: Yes, but the difference is not great.

After much interaction with the experts, we might devise the following levels for each of the clues identified earlier:

- Age (younger than 30, 31–40, older than 41)
- Value of time to the employee (income over $50,000, income between $30,000 and $50,000, income less than $30,000)
- Gender (male, female)
- Computer literacy (programs computers, frequently uses a computer, routinely uses output of a computer, has no interaction with a computer)
- Tendency to join existing HMOs (enrolled in an HMO, not enrolled in an HMO)

In describing the levels of each clue, we also think through some measurement issues. How do we determine the value of time? We use income, hence hourly wage, as a surrogate, even though it would be more accurate to survey the group. This decision rests on the fact that income data are accessible while a survey would be slow and expensive. But such decisions may mask a major pitfall. If income is not a good surrogate for value of time, we have wrecked our effort by taking the easy way out. Remember the story about the man who lost his keys in the street but was searching for them in his house. Asked why he was looking there, he responded with a certain pinched logic: "The street is dark—the light's better in the house." The lesson is that surrogate measures must be chosen carefully to preserve the value of the clue.

Step 5. Test for conditional independence

Conditional independence means that the presence of one clue does not change the value of any other clue. Conditional independence is an important criterion that can streamline a long list of clues (Schum 1965). Conditional independence means that for a specific population, such as employees who join the HMO, presence of one characteristic does not change the value of another. Let us examine whether age and sex are conditionally independent in predicting the probability of joining the HMO.

Mathematically, if two clues, age and sex, are conditionally independent, then among the population of employees who join, J, we should have:

$$p(\text{age}|\text{sex}, J) = p(\text{age}|J)$$

This formula says that the likelihood for age remains the same whether it is conditioned on the presence of sex or not. Thus, the impact of age on the forecast does not depend on sex, and vice versa.

Conditional independence simplifies the forecasting task. The impact of a piece of information on the forecast, we noted earlier, is its likelihood ratio. For characteristic C_1 the likelihood ratio for joining, J, or not joining, N, is written as:

$$\text{Likelihood ratio for } C_1 = \frac{p(C_1|J)}{p(C_1|N)}$$

Similarly, the likelihood ratio for a different characteristic, say C_2, is:

$$\text{Likelihood ratio for } C_2 = \frac{p(C_2|J),}{p(C_2|N)}$$

And the likelihood ratio for the joint occurrence of C_1 and C_2 is:

$$\text{Likelihood ratio for } C_1 \text{ and } C_2 = \frac{p(C_1, C_2|J),}{p(C_1, C_2|N))}$$

Conditional independence implies that the joint likelihood ratio of C_1 and C_2 is:

$$\text{Likelihood ratio for } C_1 \text{ and } C_2 = \frac{p(C_1|J)}{p(C_1|N)} \times \frac{p(C_2|J)}{p(C_2|N)}$$

In English, assuming conditional independence, the impact of two clues is equal to the product of the impact of each clue. Conditional independence simplifies the number of estimates needed for measuring the joint impact of several pieces of information. Without this assumption, evaluating the joint impact of two pieces of information requires more than two estimates. With it, the likelihood ratio of each clue will suffice.

With the assumption of conditional independence, the Bayes' formula can be written as:

$$\text{Posterior odds} = (\text{Likelihood ratio for } C_1) \times \ldots \times (\text{Likelihood ratio for } C_n) \times (\text{Prior odds})$$

This formula differs from the earlier Bayes' odds form in the sense that the joint impact of characteristics C_1 through C_2 is written as the product of the likelihood ratios associated with each characteristic.

The chances for conditional dependence increase along with the number of clues, so clues are likely to be conditionally dependent if the model contains more than six or seven clues. When clues are conditionally dependent, either one clue must be dropped from the analysis or the dependent clues must be combined into a new cluster of clues. If age and computer literacy were conditionally dependent, then either could be dropped from the analysis. As an alternative, we could define a new cluster with these levels:

- Below 30 and programs a computer
- Below 30 and frequently uses a computer
- Below 30 and uses computer output
- Below 30 and does not use a computer
- Above 30 and programs a computer
- Above 30 and frequently uses a computer
- Above 30 and uses computer output
- Above 30 and does not use a computer

The new clue is constructed by combining the levels of age and computer literacy. There are statistical procedures for estimating conditional dependence; however, the following process works quite well (Gustafson et al. 1973a):

- Write each clue on a 3 × 5 card.
- Ask each expert to assume a specific population (in our example, the population of people who join the HMO). Ask each expert to describe the population so that we can verify that he or she is thinking about the right one.
- Ask the expert to pair the cards according to this rule: If knowing the value of one card will make it considerably easier to estimate the value of the other, place the two cards together.
- Repeat these steps for other populations (in our example, the employees who do not join).

- If several experts are involved, ask them to present their clustering of cards to each other.
- Have experts discuss any areas of disagreement, and remind them that only major dependencies should be clustered.
- Ask panelists to modify their clusters as they wish.
- Use majority rule to choose the final clusters. (To be accepted, a cluster must be approved by the majority of experts.)

Experts will have in mind different, sometimes wrong, notions of dependence, so the words *conditional dependence* should be avoided. Instead, we focus on whether, in a specific population, one clue tells us a lot about another. We find this form of questioning easier to understand.

Step 6. Estimate the impact of clues

In previous steps, we defined the forecast event and organized a set of clues that could be used in the forecast. Since we intend to use the Bayes' formula to aggregate the effects of various clues, the impact of each clue should be measured as a likelihood ratio. This section explains how to estimate likelihood ratios, but other approaches are possible (Huber 1974).

To estimate likelihood ratios, experts should think of the prevalence of the clue in a specific population. The importance of this point is not always appreciated. A likelihood estimate is conditioned on the forecast event, not vice versa. Thus, the impact of being young (age less than 30) on the probability of joining the HMO is determined by finding the number of young employees among joiners. There is a crucial distinction between this probability and the probability of joining if one is young. The first statement is conditioned on joining the HMO, the second on being young. The definition of likelihood must be kept in mind—it is conditioned on the forecast event, not the presence of the clue. The likelihood of joining if younger than 30 is p(younger than 30|join), while probability of joining the HMO for a person younger than 30 is p(join|younger than 30).

A likelihood is estimated by asking questions like:

Q: Of 100 people who do join, how many are younger than 30?

The question is repeated but now conditioned on not joining:

> Q: Of 100 people who do not join the HMO, how many are younger than 30?

The ratio of the answers to these two questions determines the likelihood ratio associated with being younger than 30. This ratio could be estimated directly with questions like:

> Q: Imagine two employees, one who will join the HMO and one who will not. Who is more likely to be younger than 30? How many times more likely?

We estimate the likelihood ratios by relying on experts' opinions, but the question naturally arises about whether experts can accurately estimate probabilities. Before answering we need to emphasize that accurate probability estimation does not mean being correct in every forecast. For example, if we forecast that an employee has a 60 percent chance of joining the proposed HMO but the employee does not join, was the forecast inaccurate? Not necessarily. The accuracy of probability forecasts cannot be assessed by the occurrence of a single event. A better way to check the accuracy of a probability is to check it against observed frequency counts. A 60 percent chance of joining is accurate if 60 of 100 employees join the proposed HMO. A single case reveals nothing about the accuracy of probability estimates.

Systematic bias may exist in subjective estimates of probabilities (Lichtenstein and Phillips 1977; Slovic, Fischhoff, and Lichtenstein 1977). Research shows that subjective probabilities for rare events are inordinately low, while they are inordinately high for common events. These results have led some psychologists to conclude that cognitive limitations of the assessor inevitably flaws subjective probability estimates. For example, Hogarth (1975) concludes: "Man is a selective, sequential information processing system with limited capacity, he is ill-suited for assessing probability distributions."

Alemi, Gustafson, and Johnson (1986) argue that accuracy of subjective estimates can be increased through three steps. First, experts should be allowed to use familiar terminology and decision aids. Distortion of probability estimates can be seen in a diverse group of experimental subjects, but not among all real experts. For example, meteorologists seem to be fine probability estimators (Winkler and Murphy 1973). Weather forecasters are special because they assess familiar phenomena and have access to a host of relevant and overlapping objective information and judgment aids (such as computers and satellite photos). The point is that experts can reliably estimate likelihood ratios if they are dealing with a familiar concept and have access to their usual tools. In this regard, Edwards writes:

> If substantive experts are indeed allowed the time and the necessary tools (e.g., paper and pencil), they can accurately assess probabilities. Granted that assessed probability is not precise to the third digit, it nevertheless is a systematic and coherent assessment of the individual's belief. (See Edward's comments following Hogarth's 1975 article.)

A second way of improving experts' estimates is to train them in selected probability concepts (Lichtenstein and Fischhoff 1978). In particular, experts should learn the meaning of a likelihood ratio. Ratios larger than 1 support the occurrence of the forecast event; ratios less than 1 oppose the probability of the forecast event. A ratio of 1-to-2 reduces the odds of the forecast by half; a ratio of 2 doubles the odds.

The experts also should be taught the relationship between odds and probability. Odds of 2-to-1 mean a probability of 0.67; odds of 5-to-1 mean a probability of 0.83; odds of 10-to-1 mean a probability of an almost certain event. The forecaster should walk the expert through and discuss in depth the likelihood ratio for the first clue before proceeding. We have noticed that the first few estimates of probability can take four or five minutes each, as many things are discussed and modified. Later estimates often take less than a minute.

A third step for improving experts' estimates of probabilities is to rely on more than one expert and on a process of estimation, discussion, and reestimation. This method can reduce inaccuracies by as much as 33 percent compared to individual estimates (Gustafson et al. 1973b). Relying on a group of experts increases the chance of identifying major errors. In addition, the process of individual estimation, group discussion, and individual reestimation reduces pressures for artificial consensus while promoting information exchange among the experts.

Step 7. Estimate prior odds

According to Bayes' formula, forecasts require two types of estimates: likelihood ratios associated with specific clues, and prior odds associated with the target event. Prior odds can be assessed by finding the prevalence of the event. In a situation without a precedent, prior odds can be estimated by asking experts to imagine the future prevalence of the event. Thus, the odds for joining may be assessed by asking:

Q: Out of 100 employees, how many will join?

The response to this question provides the probability of joining, p(J), and this probability can be used to calculate the odds for joining:

$$\text{Odds for joining} = \frac{p(J)}{[1 + p(J)]}$$

When no reasonable prior estimate is available, we prefer instead to assume arbitrarily that the prior odds for joining are 1-to-1, and allow clues on the nature of the employee population to alter these odds as we proceed.

Step 8. Develop scenario forecasts

Decision makers use scenarios to think about alternative futures. The purpose of forecasting with scenarios is to make several numerical predictions, not one. Some future events result from self-fulfilling prophecies—a predicted event happens because we take steps to increase the chance it will happen. In this circumstance, predictions are less important than choosing the ideal future and working to make it come about. Scenarios help the decision maker choose a future and make it occur.

Scenarios are written as coherent and internally consistent narrative scripts. The more believable they are, the better. Scenarios are constructed by selecting various combinations of clue levels, writing a script, and adding details to make the group of clues more credible. An optimistic scenario may be constructed by choosing only clue levels that support the occurrence of the forecast event; a pessimistic scenario combines clues that oppose the event's occurrence. Realistic scenarios, on the other hand, are constructed from a mix of clue levels. In the HMO example, scenarios could describe hypothetical employees who would join the organization. A customer most likely to join is constructed by assembling all characteristics that support joining:

> A 29-year-old male employee earns more than $60,000. He is busy and values his time; he is familiar with computers, using them both at work and at home. He is currently an HMO member, though not completely satisfied with it.

A pessimistic scenario describes the employees least likely to join:

> A 55-year-old female employee earning less than $35,000. She has never used computers and has refused to join the firm's existing HMO.

More realistic scenarios combine other clue levels:

> A 55-year-old female employee earning more than $60,000 has used computers but did not join the firm's existing HMO.

A large set of scenarios can be made by randomly choosing clue levels and then asking experts to throw out impossible combinations. To do this, first write each clue level on a card and make one pile for each clue. Each pile will contain all the levels of one clue. Randomly select a level from each pile, write it on a piece of paper, and return the card to the pile. Once all clues are represented on the piece of paper, have an expert check the scenario, and discard scenarios that are wildly improbable (for example, a 20-year-old earning more than $80,000).

If experts are evaluating many scenarios (perhaps 100 or more), arrange the scenario text so they can understand them easily and omit frivolous detail. If experts are reviewing a few scenarios (perhaps 20 or so), add detail and write narratives to enhance the scenarios' credibility.

Because scenarios examine multiple futures, they introduce an element of uncertainty and prepare decision makers for surprises. In the example, the examination of scenarios of possible customers helped the decision makers understand that large segments of the population may not consider the HMO desirable. This led to two changes. First, a committee was assigned to make the proposal more attractive to segments not currently attracted to it. This group went back to the drawing board to examine the unmet needs of people unlikely to join. Second, another committee examined how the proposed HMO could serve a minority of a firm's employees and still succeed.

Sometimes forecasting is complete after we have examined the scenarios, but if the decision makers want a numerical forecast, we must take two more steps.

Step 9. Validate the model

Any subjective probability model is in the final analysis just a set of opinions, and as such should not be trusted until it passes vigorous evaluation. The evaluation of a subjective model requires answers to two related questions:

- Does the model reflect the experts' views?
- Are the experts' views accurate?

To answer the first question, design about 100 scenarios, ask the expert to rate each, and compare these ratings to model predictions. If the two match closely, then the model simulates the expert's judgments. For example, we can generate 100 hypothetical employees and ask the expert to rate the probability that each will join the proposed HMO. We would also ask the expert to arrange the cases from more to less likely, to review pairs of adjacent employees to see if the rank order is reasonable, and to change the rank orders of the employees if needed. We would then use the Bayes' model to forecast whether each hypothetical employee will join and compare this prediction to the expert's ranking. If the rank order correlation is higher than 0.70, we would conclude that the model simulates many aspects of the expert's intuitions (see Table 8-1).

Figure 8-1 shows the relationship between predictions and expert's ratings. The straight line shows the expected relationship. Correlation shows the degree of agreement between expert and model. (Correlation = 0.45.)

Some differences between the model and the expert should be expected, as the expert will show many idiosyncrasies and inconsistencies not found in the model. But the model's predictions and the expert's intuitions should not sharply diverge. If the correlation is lower than 0.5, then perhaps the expert's intuitions have not been effectively modeled, in which case the model must be modified. The likelihood ratios might be too high, or some important clues might have been omitted.

The above procedure leaves unanswered the larger and per-

Table 8-1 Agreement between Expert and Model

Case	*Model Predictions*	*Expert's Intuition*
1	80	40
2	60	20
3	80	40
⋮	⋮	⋮
98	40	30
99	90	80
100	10	20

Figure 8-1 Agreement Between Expert Judgment and Model Prediction

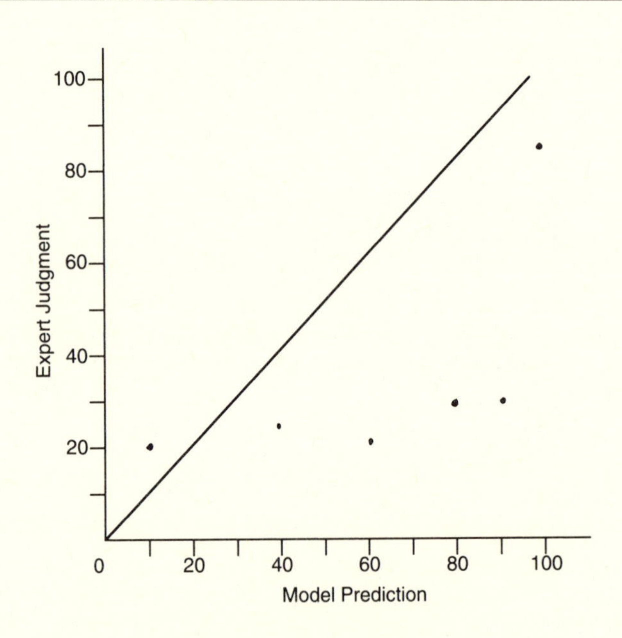

haps more difficult question of the accuracy of the expert's intuitions. Intuitions can be validated if they can be compared to observed frequencies, but this is seldom possible (Howard 1980). In fact, if we had access to observed frequencies, we would probably not bother consulting experts to create subjective probability models.

One way to increase our confidence in expert opinions is to use several experts. If experts reach a consensus, then we feel comfortable with a model that predicts that consensus. Consensus means that experts, after discussing the problem, independently rate the hypothetical scenarios close to one another. One way of checking the degree of agreement among experts' ratings of the scenarios is to correlate the ratings of each pair of experts. Correlation values above 0.75 suggest excellent agreement; values between 0.50 and 0.75 suggest more moderate agreement. If the correlations are below 0.50, then experts differed, and it is best to examine their differences and redefine the forecast. See Table 8-2.

Figure 8-2 shows the relationship between ratings of experts A and B. A straight line shows the expected relationship between the two experts. (Correlation = 0.67.)

Some investigators feel that a model, even if it predicts the consensus of the best experts, is still not valid because they think that only objective data can really validate a model. According to this rationale, a model provides no reason to act unless it is backed by objective data. While we agree that no model can be fully validated until its results can be compared to real data, we nevertheless feel that in many circumstances expert opinion is sufficient grounds for action. In some circumstances (surgery is an example), we must take action on the basis of experts' opinions. These same critics of subjective models would doubtless trust two surgeons who suggested they risk their lives on the operating table. Along these same lines, we should be willing to accept expert opinion as a basis for business and policy action.

Table 8-2 Two Experts' Ratings of 100 Cases

Case	*Expert A's Rating*	*Expert B's Rating*
1	50	60
2	40	30
3	20	10
⋮	⋮	⋮
98	90	80
99	40	20
100	80	30

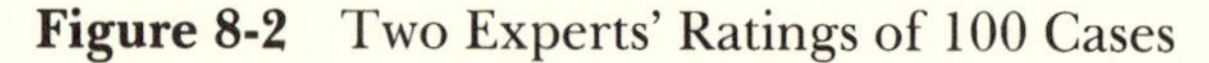

Figure 8-2 Two Experts' Ratings of 100 Cases

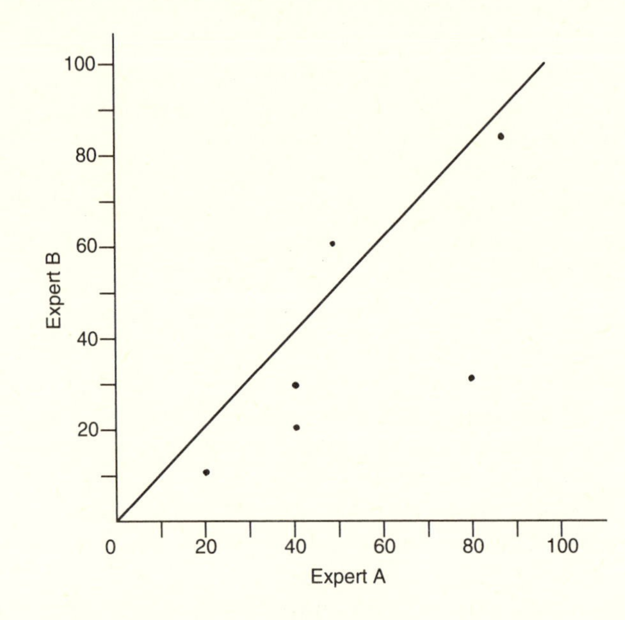

Step 10. Use the model to forecast

To make a forecast, we should begin by describing the characteristics of the employees. Suppose an expert believes, or we have observed, that the employees at the firm that may offer the proposed HMO have the characteristics described in Table 8-3. We use the Bayes' formula and the likelihood ratios associated with each characteristic to calculate the probability of the employees' joining the HMO.

In our example, suppose we evaluate a 29-year-old man earning $60,000 who is computer literate but not an HMO member. Suppose the likelihood ratios associated with these characteristics are 1.2 for being young, 1.1 for being male, 1.2 for having a high hourly rate, 3.0 for being computer literate, and 0.5 for not being a member of an HMO. Likelihood ratios greater than 1.0 increase odds of joining, while ratios less than 1.0 reduce the odds. Assuming an equal prior, this employee's odds of joining are:

$$\text{Odds of joining} = 1.1 \times 1.2 \times 3 \times 0.5 \times 1 = 1.98$$

The probability of joining can be calculated with this formula:

Table 8-3 Employee Characteristics

	Small Minority	*Typical A*	*Typical B*
Age	29 or younger	30 or older	30 or older
Income	$60,000+	$30–60,000	$60,000+
Gender	Male	Male	Female
Computer literacy	Yes	Yes	Yes
Member of an HMO	No	Yes	No
Observed count	50	700	200

$$\text{Probability} = \text{Posterior odds}/(1 + \text{Posterior odds})$$

In the above example, the probability of joining is:

$$\text{Probability of joining} = 1.98/(1 + 1.98) = 0.66$$

The probability of joining can be used to estimate the number of employees likely to join the new HMO (in other words, demand for the proposed product). If we expect to have 50 of the above type of employee, we can expect 33 = (50 × 0.66) to join. If we do similar calculations for other types of employees, we can calculate the total demand for the proposed HMO.

Readers might like to know that, in fact, analysis of demand for the proposed HMO showed that most employees would not join but that 12 percent of the employed population might join. Careful analysis allowed the planners to identify a small group of employees who could be expected to support the proposed HMO, showing that a niche was available for the innovative plan.

Summary

Forecasts of unique events are useful, but they are difficult because of the lack of data on which to calculate probabilities. Even when events are not unique, frequency counts are often unavailable, given time and budget constraints. However, the judgments of people with substantial expertise can serve as the basis of forecasts.

In tasks where many clues are needed for forecasting, experts may not function at their best, and as the number of clues increases, the task of forecasting becomes increasingly arduous. Bayes' theorem is a mathematical formula that can be used to aggregate the impact of various clues. This approach combines the strength of human expertise (estimating the relationship between the clue and the forecast) with the consistency of a statistical model. Validating these models poses a thorny problem because no objective standards are available. But once the model has passed scrutiny from several experts from different backgrounds, we feel sufficiently confident about the model to recommend action based on its forecasts.

References

Alemi, F., D. H. Gustafson, and M. Johnson. 1986. "How to Construct a Subjective Index." *Evaluation and the Health Professions* 9 (1): 45–52.

Bayes, T. 1963. "Essays toward Solving a Problem in the Doctrine of Changes." *Philosophical Translation of Royal Society* 53: 370–418.

Beach, H. B. 1975. "Expert Judgment about Uncertainty: Bayesian Decision Making in Realistic Settings." *Organizational Behavior and Human Performance* 14: 10–59.

DeFinetti, B. 1964. "Foresight Its Logical Laws, Its Subjective Sources." Translated and reprinted in *Studies in Subjective Probabilities,* edited by R. Kyburg and G. Smokler, pp. 93–158.

Gustafson, D. H., R. Ludke, J. J. Kestly, and F. Larson. 1973a. "Probabilistic Information Processes: Implications and Evaluation of Semi-PIP Diagnostic System." *Computers and Biomedical Research* 6.

Gustafson, D. H., R. K. Shukla, A. Delbecq, and G. W. Walster. 1973b. "A Comparative Study of Differences in Subjective Likelihood Estimates Made by the Individual, Interacting Groups, and Nominal Groups." *Organizational Behavior and Human Performance* 9: 280–91.

Hogarth, R. M. 1975. "Cognitive Processes and the Assessment of Subjective Probabilities." *Journal of American Statistical Association* 350.

Howard, R. A. 1980. "An Assessment of Decision Analysis." *Operations Research* 28 (1): 4–27.

Huber, G. 1974. "Methods for Quantifying Subjective Probabilities and Multi-Attribute Utilities." *Decision Science* 5: 430.

Lichtenstein, S., and B. Fischhoff. 1978. "Training for Calibration." Army Research Institute Technical Report TR-78-A32.

Lichtenstein, S., and L. D. Phillips. 1977. "Calibration of Probabilities: The State of the Art." In *Decision Making and Change in Human Affairs,* proceedings of the 5th Research Conference on Subjective Probability, Utility, and Decision Making, Darmstadt, edited by H. Jungermann and G. de Zeeuw. Dordrecht: Reidel.

Ramsey, P. P. 1950. "Truth and Probability." In *The Foundation of Mathematics and Other Logical Essays,* edited by Braithwaite. New York: Humanities Press. Also reprinted in *Studies in Subjective Probability.* New York: John Wiley Press, 1976.

Savage, L. 1954. *The Foundation of Statistics.* New York: John Wiley.

Schum, D. A. 1965. "Inferences on the Basis of Conditionally Non-independent Data." Army Research Laboratory Technical Report No. 65–161.

Slovic, P., B. Fischhoff, and S. Lichtenstein. 1977. "Knowing with Certainty the Appropriateness of Extreme Confidence." *Journal of Experimental Psychology: Human Performance and Perception* 3 (4): 552–64.

Snyder, M., and W. B. Swann. 1978. "Behavioral Confirmation in Social Interaction: From Social Perception to Social Reality." *Journal of Experimental Social Psychology* 14: 148–62.

Winkler, R. L., and A. H. Murphy. 1973. "Experiments in the Laboratory and the Real World." *Organizational Behavior and Human Performance* 10: 225–70.

9

Option Generation

The health care financing dilemma is fundamentally a search for options. We have searched for years to find a mechanism that joins incentives to improve quality of care with incentives to control costs. HMOs, preferred providers, second opinions, and incentives to employees to seek low-cost care have all been tried with virtually no observable impact on spiraling health care costs. More recently, two promising options have reentered the debate: variants of national health insurance and Japanese management principles, that is, continuous quality improvement (CQI) and, ultimately, total quality management (TQM). While we believe TQM and CQI hold promise for controlling costs and improving quality, their implementation is very difficult.

We know that to successfully transform an organization to TQM or CQI, we must (as stated in Chapter 2) create a tension for change, show that TQM is a superior alternative, provide social support for those attempting to implement it, build skills, develop a plan, and create a feedback system. The issue is not what but how to do each of these elements of change. The health field has so little experience in this transformation that there is no good set of options on how to create these elements of change. The options must be generated.

Developing a comprehensive set of options is a fundamental difficulty in formulating policy. Many policymakers can sympathize with the expression "between a rock and a hard place" because they know all too well the feeling of having to choose one of two alternatives when neither is appealing. In reality, other options could often be far more rewarding, but they are not even considered because the process of option generation is atrophied or absent.

Often the available options seem diametrically opposed. Legislators who must vote on bills that would require family planning agencies to inform parents that their teenagers have requested birth control services have basically two options: yes and no. Moreover, options are often undimensional in nature. For example, we might negotiate parental notification and then negotiate sex education requirements in schools. This process would be suitable if the two topics were totally unrelated, but they, and many other issues, are actually part of a multidimensional problem. Furthermore, options that leave little room for compromise set the stage for confrontation which one side will win and the other will lose. Far better to think of something that will allow both sides to leave the table smiling—or at least not frowning.

We do not deny that defining issues in this manner may be good political strategy, but such confrontation can lead to disaster as a problem lurches from an excess on one side to an excess on the other.

If options are important, they are not always so easy to think up, even if the constituencies are receptive to new options. Research has identified some roadblocks to creative option generation. Dunker (1945) and Abelson (1977) found that some decision makers become fixated on certain features of a problem and ignore others. People learn a set of rules, and those rules generate options, even though rules can result in unbridgeable gulfs between the parties. For example, these rules are irreconcilable when it comes to abortion: (1) nothing we do shall interfere with a woman's right to control her body, and (2) nothing we do shall interfere with an unborn child's right to live. The challenge is to get decision makers to deviate from those rules for a while and take the time to focus on generating options that will give each side what it really needs.

This chapter reviews research on how people create options, presents an option generation process that benefits from the central findings of this research, and shows a variation allowing that process to be done quickly. Finally, it discusses certain constraints on option generation.

A number of studies have examined how people could generate better options. Many of these studies sought to help people break free from well-learned, restrictive rules on the way they perceive problems. Demonstrating alternative rules or problem descriptions can help decision makers break out of these ruts. Hayes, Watermon, and Robinson (1977) found that decision makers solve a problem more rapidly if they can fit it into a familiar category. Unfortunately, such familiarity can reduce the number of options developed. It would only be a slight exaggeration to argue that the task of policy analysts is finding other ways of seeing problems.

For instance, if we believe nursing home quality assurance should be classified as a "regulatory problem," we naturally assume that the solution is to write regulations. One nursing home regulatory system actually applied 1,547 regulations to all homes (Gustafson et al. 1980). If we classify the problem as a "statistical quality control problem" (Deming 1988), we naturally believe we can solve it by rapidly screening a random sample of nursing home residents. If too many residents fail the screening, all residents would be examined.

Another way to create options is to examine objectives (Humphreys 1983). In the example, one set of objectives might be to increase a nursing home's commitment and ability to continuously improve quality of care without increasing costs. Pitz, Sach, and Heerboth (1980) found a substantial increase in option quantity when subjects were asked to generate several for each objective. (It can be assumed that in most policy contexts many objectives must be served at once, such as quality improvement along with cost containment.) Interestingly, subjects created more options by working on objectives singly rather than as a group. The number of options was also reduced if subjects were given examples of options. We conclude that defining objectives and using them as prompts can enhance option generation if the objectives are addressed singly. On the other hand, using examples tends to inhibit creativity. (Suggesting examples is not the same as suggesting new ways of seeing a problem. Examples are options. New ways of seeing a problem are new models, each of which may contain or suggest several options.)

Unfortunately, options generated by addressing single objectives may serve only that objective, so we still need a set of composite options that can serve multiple objectives. The process of creating a composite option seems to require:

1. A clear understanding of objectives and their relative importance
2. A list of options to meet each objective
3. An assessment of how each option affects each objective

4. An identification of how options can be modified or combined to create composites that satisfy at least the crucial objectives

Instead of focusing on objectives, we can examine purpose—in other words, examine why the system under study exists. Often focusing on the reason for existence yields a substantially different set of options (Nadler 1981). For example, suppose the problem is late payment of Medicaid claims. If a claims processor that takes 150 days to pay a Medicaid claim can pay Medicare claims in just 32 days, something is wrong with the Medicaid system. If we took an objectives approach, we might decide our objective is to reduce time until payment. If we take a reason-for-existence approach, we might seek to eliminate the system altogether. In the claims payment example, we might examine purpose and search for redundant systems by this routine:

Problem	Delay in claims payment
Reason for existence of claims payment system	To reimburse providers for services rendered
How is payment made?	A claim is submitted and paid each time service is provided
Are there other ways of reimbursing?	(a) A fixed, prospective payment (b) Hiring providers directly on salary

The point is that we can generate new ideas by understanding the present system's purpose and by questioning the need for it. Several processes address option generation from this perspective (Volkema 1983; Nadler 1981).

Option generation research has also focused on system requirements, also called essential elements of the solution. Utterbach (1974) argues that instead of generating options immediately after we gain our understanding of the problem, we should discern the essential elements of any solution. Once these elements (which we call essential elements) are understood, they will help free the mind to find more effective solutions.

In the Medicaid payment example, typical essential elements might include:

- An ability to reimburse providers no later than one month after date of service
- Incentives to deliver only needed services

- Incentives to monitor services provided and ascertain the differences in service patterns across different physicians treating the same problem

How can we identify these essential elements? Often, key elements are best identified by seeking help from theoretical and practical experts outside the organization or even the field in which the problem exists. A theoretical expert is a person who has studied identical or similar problems in other settings. In the Medicaid example, people who have studied auto insurance or security systems might be considered theoretical experts who could offer innovative perspectives on health insurance issues. A practical expert is someone who has designed or operated a system with a similar purpose.

It is important not to ask detailed questions about the other system's operation because we are now concentrating on the necessary elements. The reason for focusing on the essential elements is that they are probably the same regardless of the setting in which the solution will be implemented. We want to avoid imitating systems that operate in different environments because solutions that are familiar to our experts may have aspects that are inappropriate to our setting. So Utterbach would argue that we ask the experts to focus on characteristics of the solution that are essential to success. These elements, in turn, give us clues that have broader application.

Option Generation Strategy

The literature provides important clues to a superior process for option generation.

- Specify objectives and ask people to generate options for each objective in sequence.
- Specify why the system we seek to improve exists and ask if there is a way to satisfy this reason without having the system at all.
- Ask the theoretical and practical experts to specify the essential elements of successful solutions they have seen to related problems.
- Clearly understand the process or means by which the existing system functions. In doing so, we can identify sources of waste, rework, duplication, and needless complexity (Deming 1988) that should be addressed.

In the following sections, we will build upon these findings to create a strategy to generate options. Although we present this strategy as a sequence of steps, in reality most users will cycle in and out of these steps in various orders.

Systematic and perceptual analyses

Often, the first step is to understand the problem from both systematic and perceptual viewpoints. A systematic viewpoint allows us to compare the actual performance of the system to our expectations. The components would include documenting and assessing the system's purpose, its clientele, its specific products, the procedures used to produce those products, the information used and collected in that process, the personnel and equipment employed, and the unique characteristics of the environment in which the system functions. In each case, we should collect facts, not perceptions: What kind of clients are served? What kind and number of products are delivered? In the process, we identify shortcomings in the system's performance (sources of waste, rework, duplication and needless complexity).

While a systematic analysis gathers facts, perceptions may be the real problem. This is because even if a system performs as intended, a problem exists if the public thinks its performance is inadequate. In a perceptual analysis, opinion leaders must be identified and their perceptions of the system explored. The outline above can be used as long as perceptions replace facts as the focus. In perceptual analysis, we ask: What does each opinion leader believe the purpose of the system should be? What kinds of clients should be served? After eliciting each expectation, we would gain the expert's opinion about system performance with questions like: "How well do you think this purpose is being met?" After systematic and perceptual analyses are completed, the PIMS will understand the issue quite well.

Creating preliminary options

The next phase is to independently use systematic and perceptual analyses to create preliminary options. First we should attempt to eliminate the need for the system. If after these attempts it is clear that the system is essential, then we should (1) have outside practical and theoretical experts identify and prioritize necessary elements of any workable solution, and (2) incorporate those essential elements into several options (Figure 9-1).

The PIMS can prioritize the problems and identify solutions to key problems (Figure 9-2). Using both the purposes and problem descriptions in the list above, the PIMS can search for different ways of searching for solutions. For regulatory problems, we might examine accepted models of regulatory systems. For statistical quality control problems, we might look at accepted models of similar systems. These models will lead us to identify a new set of options (Figure 9-3).

Next, we identify and prioritize the objectives of influential constituencies. What does each one want to accomplish by solving the problem? The most important objectives are used independently (following Pitz's advice) to identify options (Figure 9-4).

For instance, in the Medicaid payment example, the constituencies might be the physicians who want to be paid and the regulators who want to (1) pay a fair price (2) for appropriate care (3) that was actually delivered. Pitz would argue that the physicians' objective should be used to generate one set of options and each regulators' objective would similarly generate a separate set of options.

Finally, we create composite solutions and compare the problems, purposes, objectives, models, essential elements, and solutions. A group of creative people try to invent a few options that:

1. Satisfy the major purposes
2. Eliminate the most important problems
3. Contain the essential elements
4. Reach the crucial goals of the key constituencies

Figures 9-1 through 9-4 show that in this hypothetical situation 23 independent options were generated and synthesized into a smaller set of composite options. As this synthesis begins, two things are likely to become apparent:

1. The 23 options contain three to five fundamental concepts that should serve as the basis of the composite options.
2. The options are simply components of some composite options.

Suppose one option that satisfies a physician's objective is a voucher system (similar to food stamps) that patients could use to

Figure 9-1 Options Generated from Purposes

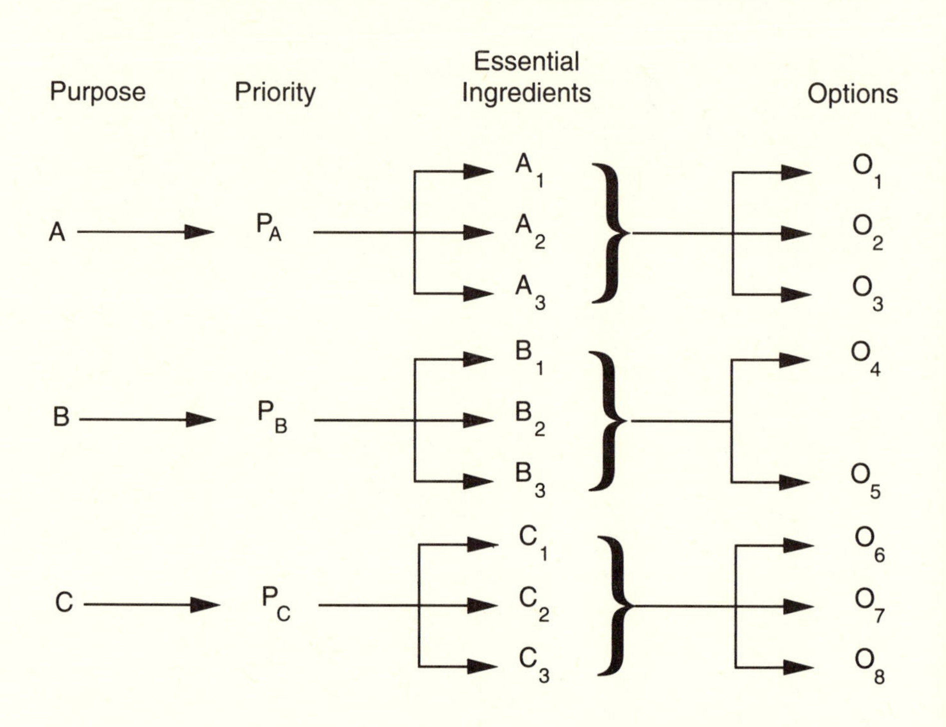

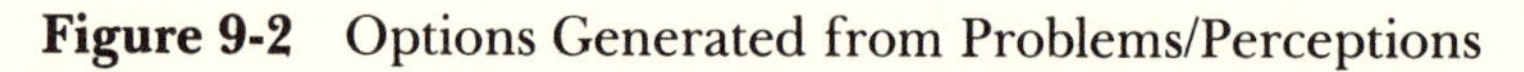

Figure 9-2 Options Generated from Problems/Perceptions

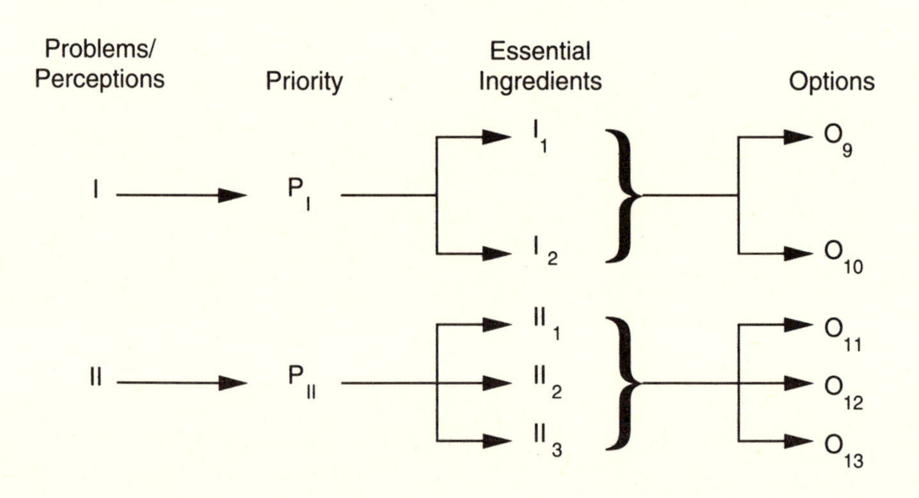

Figure 9-3 Options Generated from Problems and Purposes

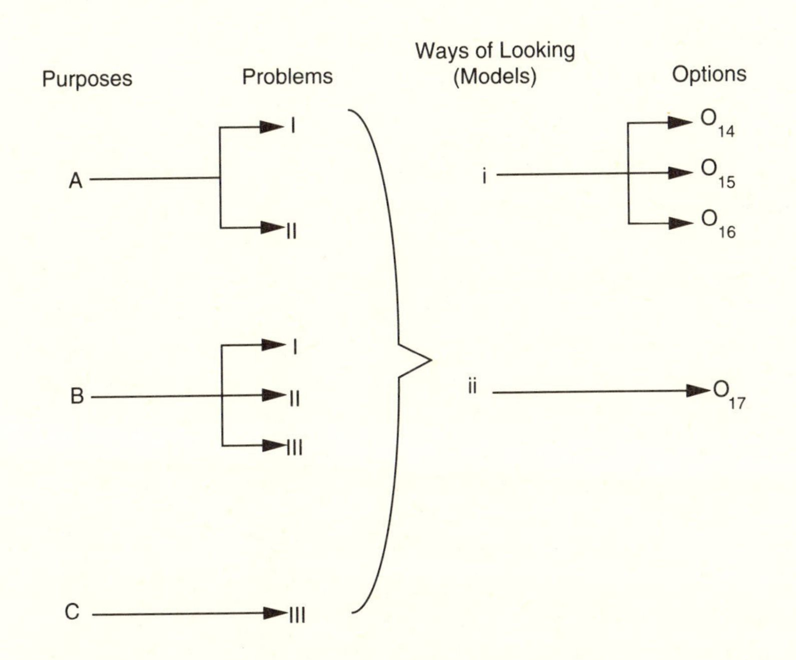

Figure 9-4 Options Generated from Constituency Objectives

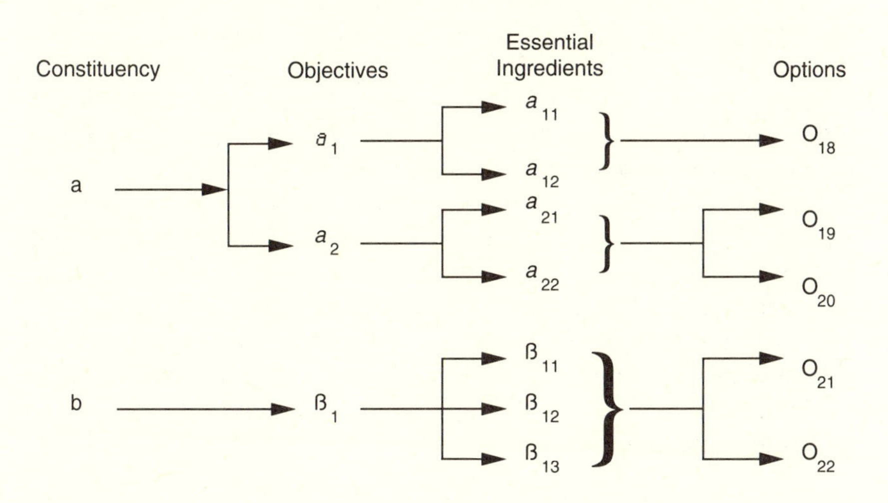

pay for care. Suppose a regulator option to ensure appropriateness is that any physician found performing inappropriate surgery can face the full legal penalties for "assault with a deadly weapon." A composite option is voucher payments coupled with inappropriate care constraints. While neither component of the option is acceptable by itself, together they might be the start of an effective option.

It is important to remember that the original set of options need not be included in the composites. Often, the original set sparks new and better ideas that are more appropriate to the situation.

One important conclusion of research conducted by Utterbach (1975), Bryson and Delbecq (1979), and others is that different people should be key actors during the various steps of the process. A systems engineer might be best at examining the operation of the existing system. People directly affected by the system (using our example, patients and providers) may be the best source of perceptions about the system. Outside theoretical and practical experts are best for identifying the essential elements. A group of creative people who understand the environment of the new system might be suited to generating the options. A group of practical people can refine those options to make them more realistic. In any event, a PIMS needs the flexibility to call on different people at various stages of the process.

Ideally, a PIMS would have several months and thousands of dollars to address issues like these, but often time constraints prevent us from involving the proper people or exploring options from a sufficient number of perspectives. Even though this is not enough time to convene groups of experts, it may be long enough to briefly involve crucial experts and go rapidly through the critical steps. When time is pressing, telephone interviews with a few creative and trusted people may be our primary method of collecting data. When resources and time are limited, it is important for the PIMS to use both very carefully and adhere strictly to the budget.

We should recognize that some problems are important enough to oblige us to devote two or three days of work to them. If three working days were available for option generation, the following schedule could be used:

Stage I. Acquiring expert advice (2 hours)

A. Telephone interviews with three trusted colleagues who know the problem. Goal: to identify a list of actors, opinion leaders, and theoretical and practical experts.
B. Classify players as follows:
   1. Actors: people with extensive knowledge of the problem and system
   2. Opinion leaders: people whose views matter to those who will eventually decide
   3. Experts: People representing different approaches
C. Ask an assistant to find out if and when each is available for phone interviews.
D. Based on quality and accessibility, contact at most three actors, three opinion leaders, and three experts.

Stage II. Identifying the purpose and the problems (9 hours)

A. Identify three principal documents on the system's purpose, operation, and problems (1 hour).
B. Have a staffer review and brief you alone on the contents (3 hours).
C. Interview at least two actors and two opinion leaders by telephone to learn reasons for the system's existence and its problems—waste, rework, duplication, and needless complexity (2 hours).
D. Using interviews and written materials, write lists of purposes and perceived problems (30 minutes).
E. By telephone, ask at least two actors and two opinion leaders to rate the importance of items on each list. A 0-7 Likert scale can be used. Then average the scores and refine them to reflect your impressions (2 hours).
F. Select no more than three purposes and four problems for further development (30 minutes).

Stage III. Identify essential elements (8 hours)

A. Contact at least two practical and two theoretical experts. Describe the system, its purposes, and its problems. Ask them to:
   1. List elements they consider essential to any successful solution.

2. Suggest written materials that might be useful for understanding those elements and their importance (3 hours).

B. Have a staff person read that material, highlight key points, prepare a written summary, and brief you on these matters (3 hours).

C. Based on the interviews and briefing, select at most five essential elements (1 hour).

D. On large flip charts, prepare Figures 9-1 through 9-4, but stop before reaching the options (30 minutes).

E. Dictate your rationale for the choices (30 minutes).

Stage IV. Option generation (3.5 hours)

A. Convene a working group of no more than three trusted and creative colleagues.

B. Display Figures 9-1 through 9-4, and talk about how they were developed and what they mean (30 minutes).

C. Ask each person independently to develop (without discussion) at least one option for each figure (15 minutes).

D. Ask each to present his or her options. Allow others to ask questions to clarify (20 minutes).

E. Ask each person to independently list the principal advantages, disadvantages, and modifications he or she would make to each option and present those ideas without discussion (45 minutes).

F. Adjourn the meeting. Use your notes and their lists to create the list of options. Next to each option, list what you believe are its principal advantages, disadvantages, and necessary modifications (1 hour).

Stage V. Developing composite options (4 hours)

A. Ask no more than two creative, trusted colleagues to join you for a two-hour meeting. At least one person should not have been at the meeting in step III.

B. Introduce the system and give background on the task at hand, then present and discuss Figures 9-1 through 9-4. Allow the group to ask clarifying questions (30 minutes).

C. Ask colleagues to create (without discussion) at least two options that accomplish the purposes, eliminate the problems, include the essential elements, and satisfy as many constituent objectives as possible (15 minutes).

D. Ask each person (including yourself) to present options, allowing time for questions. Then ask each person to silently generate a list of advantages, disadvantages, and modifications for each option (45 minutes).
E. Have each person present his or her list, while the others remain silent (30 minutes).
F. Using notes and the lists, you generate a final set of no more than five options. Each option should be described in terms of the following aspects (you can estimate) (2 hours):
   - Purpose
   - Clients served and not served
   - Specific products of the option
   - Method of operation
   - Personnel required
   - Equipment and other resources needed
   - Rough estimate of relative cost of the options (absolute cost is not needed)
   - Unique environmental characteristics addressed by the option
   - Information used and to be gathered from system operation

Discussion

The essence of this discussion is that two important initiatives can improve option generation. The first is to take the time to independently generate options from several perspectives. These options result from questions like: What options will satisfy the purposes of the system? What will correct problems with the system? What options do different models of the situation suggest? What options include the essential elements listed by outside experts?

The second theme is to involve a broad spectrum of experts in finding creative and acceptable options. For example, users of a system can almost always identify its problems. External experts can identify essential elements of options. Creative insiders can generate the actual options. Opinion leaders can determine whether the solution is politically acceptable.

We have stressed that outside practical and theoretical experts should be contacted for their advice on the essential elements of a solution. But what can be done if such experts are not available?

Again, we plead that you look deeper. Even if you cannot get someone who has designed an outstanding nursing home regulatory system, you may find someone who has carefully considered the regulation of public utilities. Thus, the manager of a local electric utility might have excellent ideas about the strengths, weaknesses, and possible modifications of the utility regulatory system. Alternatively, perhaps an excellent high school principal's knowledge of discipline can be considered expertise on regulation. If so, can you transfer some of the principal's ideas to nursing home regulation?

Do not assume that expertise or creativity is unavailable. If we understand what is meant by creativity and expertise, and if we expand our characterization of the problem, the assistance we seek may become not only apparent but available.

After Option Generation

Once the set of options has been generated, we must begin selecting among them. Obviously this is an important task. Start by identifying the selection criteria. In critical situations, you might assign weights to the criteria and rate each option against each criterion. The MAV model (Chapter 7) could help in this task. Some options that sound exciting will appear impractical in the cold light of reality. While some unrealistic options can be revised, others must be discarded. By "realistic option," we mean one that satisfies technical, political, and practical constraints.

Technical constraints exist if tools do not yet exist to implement the option. Suppose an option requires a computer chip with yet-unattainable storage capacity. Although this technical constraint may disappear in the future, it renders the option unrealistic for the present. Political constraints exist if powerful people are not expected to approve the option in the current climate. Practical constraints exist if resources are inadequate for implementation.

While these constraints cannot be ignored, the real question is when to address them. Some argue that they should be imposed from the start of the option generation process. Others argue that the constraints should be used to prune a list of freely generated options. Little empirical evidence seems to favor one approach at the expense of the other. The important point is to be sure the constraints are real before letting them influence option generation or refinement. Far too often, constraints that are flexible are accepted as fixed, leading to an inferior set of options.

References

Abelson, R. P. 1977. "Script Processing in a Future Formative and Decision Making." In *Cognition and Social Behavior,* edited by J. Carroll and J. Payne. Hillsdale, NJ: Earlbaum.

Bryson, J. M., and A. L. Delbecq. 1979. "A Contingent Approach to Strategy and Tactics in Project Planning." *APA Journal* (April): 167–79.

Deming, W. E. 1988. *Out of the Crisis.* Boston: MIT Press.

Dunker, K. 1945. "On Problem Solving." Translated by L. S. Lees from the 1935 original *Psych Monograph* 58 (270).

Gustafson, D., C. Fiss, J. Fryback, P. Smelzer, and M. Hiles. 1980. "The Wisconsin System: Quality of Care in Nursing Homes." *American Health Care Association Journal* 6, no. 5 (September).

Hayes, J. R., E. Waterman, and C. Robinson. 1977. "Identifying Relevant Aspects of a Problem Text." *Cognitive Science* 9: 297–313.

Humphreys, P. 1983. "Use of a Problem Structuring Technique for Option Generation: A Computer Choice Case Study." In *Analyzing and Aiding Decision Processes,* edited by P. C. Humphreys, O. Svenson, and A. Vari. Amsterdam: North Holland Press.

Nadler, G. 1981. *The Planning and Design Approach.* New York: John Wiley.

Pitz, G. F., N. J. Sach, and M. R. Heerboth. 1980. "Eliciting a Formal Problem Structure for Individual Decision Analysis." *Organizational Behavior and Human Performance* 26: 396–408.

Utterbach, J. 1974. "Innovation in Industry and Diffusion of Technology." *Science* 183 (15 February): 620–27.

Volkema, R. 1983. "Problem Formulation in Planning and Design." *Management Science* 29 (6): 639–52.

10

Expected Utility and Decision Trees

This chapter introduces decision trees, a tool for choosing between alternatives that will lead to uncertain outcomes. We have already introduced tools for measuring a decision maker's utility and uncertainty. These tools are useful for many problems, but their usefulness is limited when a series of intervening events is likely. When a sequence of events must be analyzed, decision trees provide a means to consider both utility and uncertainty.

The first part of this chapter defines decision trees, shows how they are constructed, and describes how they can be analyzed using mathematical expectations. The second part introduces additional tools for analysis of decision trees.

The first step in looking at decision trees may be to review the limitations of the MAV and Bayesian models. As we have shown, uncertainty can be considered in the multiattribute value model. For example, in deciding about contraceptive methods, uncertainty about the effectiveness of each option can be modeled as one dimension in a multidimensional decision that also includes ease of use, need to interrupt sexual activity, and cost. In this scheme, utility scores would be assigned to different levels of each dimension, and the importance of various dimensions would be assessed. The overall utility of an option would be the sum of the utility of its individual attributes weighted by their relative importance. Note that an inherently uncertain phenomenon is being modeled without resort to the concept and calculus of probability. Such simplifications reduce the need to introduce unfamiliar terminology to decision makers and decrease the likelihood that the analysis will be distracting. (Parsimonious use of exotic terminology helps decision makers focus on the problem, not the jargon.)

But there are occasions when such simplifications do not make sense, especially if several events will intervene before the outcome of interest occurs. For example, a benefits manager interested in how the choice of physicians will affect the cost of care must analyze a sequence of events. A change in physician leads to a change in frequency of visits and referrals, which alters the frequency and place of hospitalization, which modifies the cost of hospitalization. If the analysis measures uncertainty about cost of care while ignoring these intervening events, then the sequence of the events and certain important relationships are lost. Losing the sequence is like reading the beginning and ending of a novel: it may be effective at getting the message across but not at communicating the story. And even if the sequence of events is irrelevant to the analysis, it may be important to communicating the results.

If an MAV model were used to analyze the above decision by a benefit manager, we would probably be concerned with four attributes: cost of hospitalization, cost of clinic visits, and two uncertainties associated with each cost dimension. But such a picture obscures the web of relationships among deductible levels, copayments, rates of hospitalization, rates of clinic visits, costs, and discounts. For this reason, we feel the MAV approach results in an ineffective analysis. One of us feels decision trees are useful only for portraying an analysis, while the other two authors feel that trees and the theory behind them are also useful as methods of analysis. Although we do not fully agree, we will present both sides so you can decide for yourself.

Before describing trees in detail, it will be helpful to outline the example to be used in this chapter.

The Benefit Manager's Dilemma

Throughout the remainder of the chapter we will return to the dilemma faced by the benefits manager of a corporation with 992 employees, all of them covered by an indemnity health insurance program. Employees can seek care from any physician and, after satisfying an annual deductible, must pay only a copayment, with the employer paying the remainder.

A preferred provider organization (PPO) has approached the benefits manager and offered to discount services to employees who use its clinic and hospital. As an inducement, the PPO wants the company to increase the deductible and/or copayment required of employees who use other providers. Employees would still be free to seek care from any provider, but it would cost them more.

The logic of the arrangement is simple—the PPO can offer a discount because it expects a high volume of sales. Nevertheless, the benefits manager wonders what would happen if employees start using the preferred provider. In particular, an increase in the rate of referrals and clinic visits could easily eat away the savings on the price per visit. Change of physicians could also alter the employees' place and rate of hospitalization, which would likewise threaten the potential savings.

We should clarify some terminology before proceedings. Discount refers to proposed charges at the PPO compared to what the employer would pay under its existing arrangement with the current provider. Deductible is a minimum sum that must be exceeded before the health plan begins to pick up the bill. Copayment is the portion of the bill the employee must pay after the deductible is exceeded.

Describing the Problem as a Tree

Imagine a tree with a root, a trunk, and many branches. Lay it on its side, and you have an image of a decision tree. The word *tree* has a special meaning in graph theory. No branch of the tree is ever connected to the root, trunk, or branch leading to it. Thus, a tree is not circular; you cannot begin at one place, travel along the tree, and return to the same place. Because a decision tree shows the temporal sequence—events to the left happen before events to the right—we begin describing a decision tree with its trunk.

The origin of a decision tree is a decision, shown as a small square node with two or more lines emanating from it. Each line corresponds to one option. In our example, two lines represent the options of signing a contract with the preferred provider or continuing with the status quo.

The second component of a decision tree is a chance node. This node shows the occurrence of events over which the decision maker has no direct control. From a chance node several lines are drawn, each showing a different possible event. Suppose that joining a PPO will change the utilization of hospital and outpatient care. Figure 10-1 shows a portrayal of these events.

Figure 10-1 The Beginning of a Decision Tree is a Decision Node

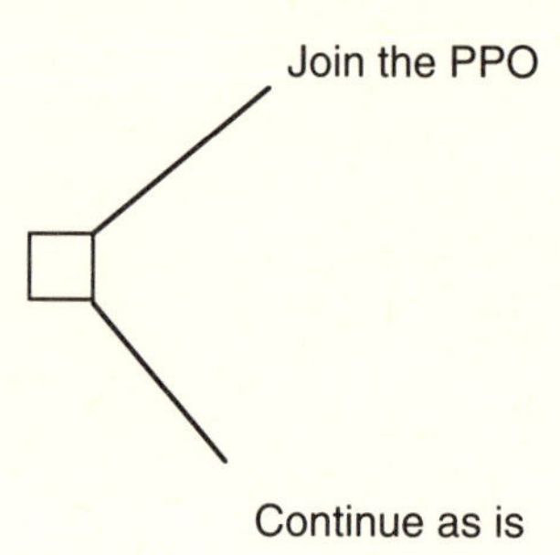

Figure 10-2 Events are Placed to the Right of a Decision Node

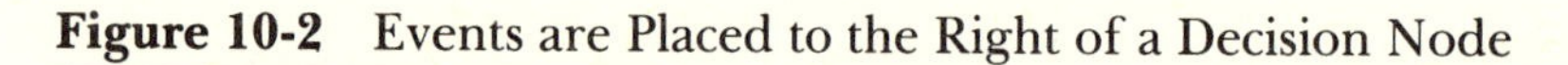

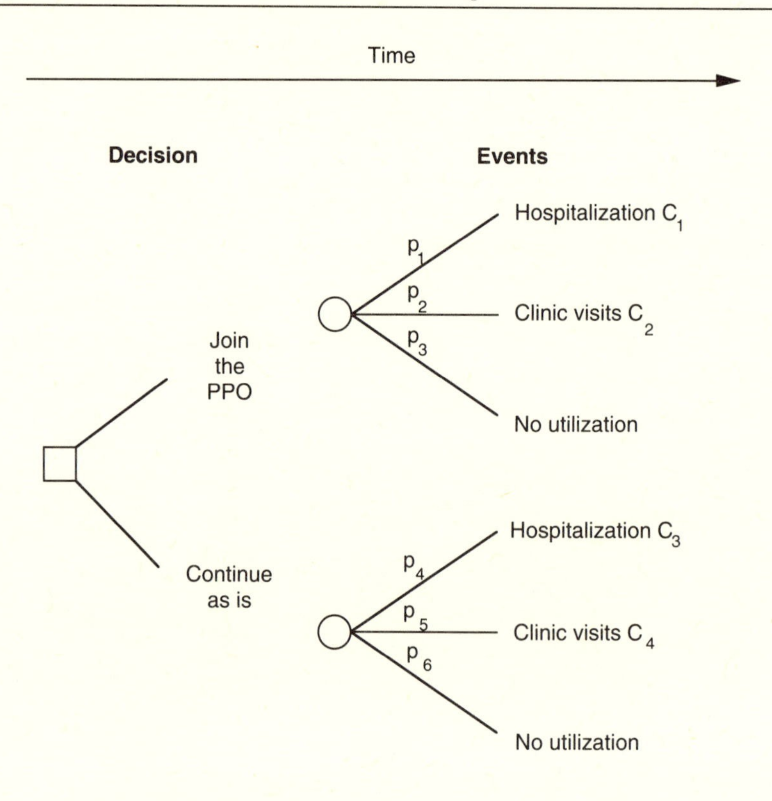

Note that the chance node is identified by a circle. The distinction between circles and boxes indicates whether the decision maker has control over the events that follow a node. Figure 10-2 suggests that, for people who join the preferred provider, there is an unspecified probability of hospitalization, outpatient care, or no utilization. We mark these probabilities p_1, p_2, . . . , p_6. (It is the practice to place probabilities above the lines leading to the events they are concerned with.)

The third element in a decision tree is the consequences. While the middle of the tree shows events following the decision, the right side, at the end of the branches, shows the consequences of these events. Suppose, for the sake of simplicity, that the benefits manager is only interested in costs, and not just costs to anybody but costs to the employer, which exclude copayments and deductibles paid by the employee. Since we do not yet know these costs, we label hospital and clinic charges C_1, C_2, . . . , C_4 and show them at the right.

Figure 10-3 represents the major elements of a decision tree: decisions, chance events, and consequences (in this case, costs). In addition to these elements, a tree contains a temporal sequence—events at the left precede events on the right. One use of the tree would be to elucidate our ideas about whether the employer should encourage employees to use the preferred provider. Then we approached the client to review our ideas in their present, simplistic state.

Use the Process and the Product

A decision tree, once analyzed and reported, indicates a preferred option and the rationale for choosing it. Such a report communicates the nature of the decision to other members of the organization. The tree and the final report on the preferred option are important organizational documents that can influence people, for better or worse, long after the original decision makers have left.

While the analysis and the final report are important by themselves, the process of gathering data and modifying the tree are equally important—perhaps more so. The process helps in several ways:

- It informs all parties that a decision is looming and they must articulate their concerns before it is completed.

Figure 10-3 Consequences Are Placed to the Right of a Decision Tree

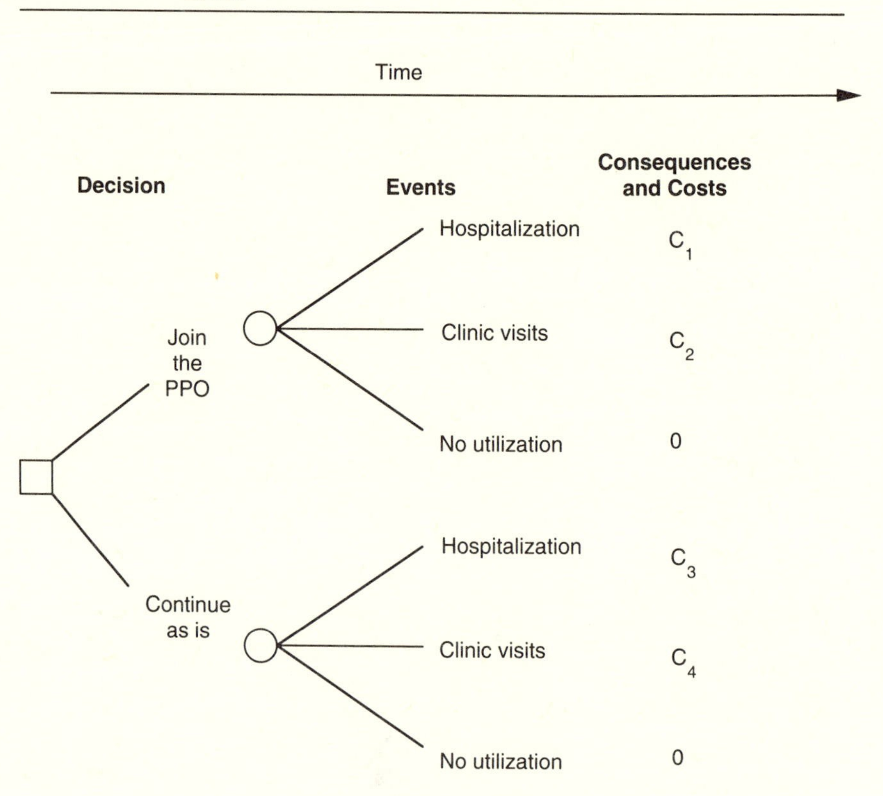

- As these concerns are added to the analysis, the process reassures clients that the analysis is fair and open.
- It provides new insights while facilitating discussion of the decision.
- It removes decision makers at various levels from day-to-day concerns, allowing them to ponder the impending changes. As decision makers put more thought into the decision, they develop more insight into their own beliefs. And the discussions among decision makers will develop further information and insights. If the analysis was done without their involvement, the positive atmosphere of collaboration would be lost.

- It signals the moment to close ranks behind the action after the analysis and decision are made. No business can afford continual self-study and doubt, and the analysis process helps distinguish the time for questioning from the time for action.

When describing a decision tree, the goal is to grow progressively more realistic as the work proceeds. The first attempt is a simple description of the problem, a sort of straw man to be worked on and improved. We will use this approach in helping the benefits manager decide about the PPO.

The first step would be to ask the client to review a very preliminary version of the decision tree (Figure 10-3). At this point, the client is likely to see more questions than answers. Not only is the concept of decision analysis unclear, but the client also has many concerns not reflected in our analysis. An important element in this meeting is to convince the client that the analysis is not intended as an independent review of the problem but rather to reflect the client's concerns. We explain that some decision makers commission studies and do not want to hear about them until the final reports are finished, but then we deem this procedure unacceptable and insist on client involvement throughout the process.

We also stress that the analysis is only a vehicle to the more complicated process of negotiating, locating alternatives, and deciding to act. We show how analysis can simulate the monetary consequences of various contracts to help the vice president for operations understand the cost of the proposed contracts. We acknowledge that many ongoing activities may affect our analysis, which we see as a small but important component of these negotiations.

Despite these reassurances, it's not easy to get a busy benefits manager to examine a foreign concept like a decision tree. We convince the manager of the usefulness of our approach by rearranging the tree and showing how it can be altered to reflect the manager's concerns. When he or she says that joining the PPO might increase hospitalization rates, we show that this means he or she expects the probability of hospitalization to increase after signing the contract. Continuing in this fashion, we demonstrate how other concerns are, or can be, reflected in the analysis. If we find a concern that is not already incorporated in it, we agree to refine the model to reflect it.

Among the issues that arise from this meeting are:

- The analysis should separate mental health and general outpatient care, as payments for mental health care could not surpass a fixed amount.
- The analysis should separate general outpatient care from mental health care because payments for the latter are capped while payments for the former are not.
- The analysis should concentrate on employees who file claims, because only they incur costs to the employer.

We revise the model to reflect these issues, and in subsequent meetings the client adds still more details, particularly about the relationship among the copayment, discount, and deductible. This is important because the order in which these terms are incorporated changes the value of the different options. Negotiations between the employer and the preferred provider have suggested that the discount is on the first dollar, before the employee pays the deductible or copayment. Hospitalization has no copayment and a $200 deductible for individuals. Outpatient care has a $200 individual and a $500 family deductible. Both the mental health and general clinic visits count toward outpatient deductible and copayment. The insurance plan required meeting the deductible before the copayment. Once these considerations were incorporated, we reached the revised model presented in Figure 10-4.

Now the client receives a summary of the comments from meetings and an updated version of the tree. We explain to the client that some of the above issues are too complicated to be shown in detail on the tree but that they will be incorporated in our computer model. We do not want to trap the decision makers in the details of how these changes enter the analysis, but we do assure the client that the analysis reflects them.

In summary, development of the decision tree proceeds toward increased specification and complexity. The earliest model is simple, later models are more sophisticated, and the final one may be too complicated to show all elements graphically, and is used primarily for analytical purposes. Each step toward increasing specification involves interaction with the client—an essential element to a successful analysis.

Figure 10-4 Revised Model for Consequences of Joining the PPO for an Average Employee Family Unit

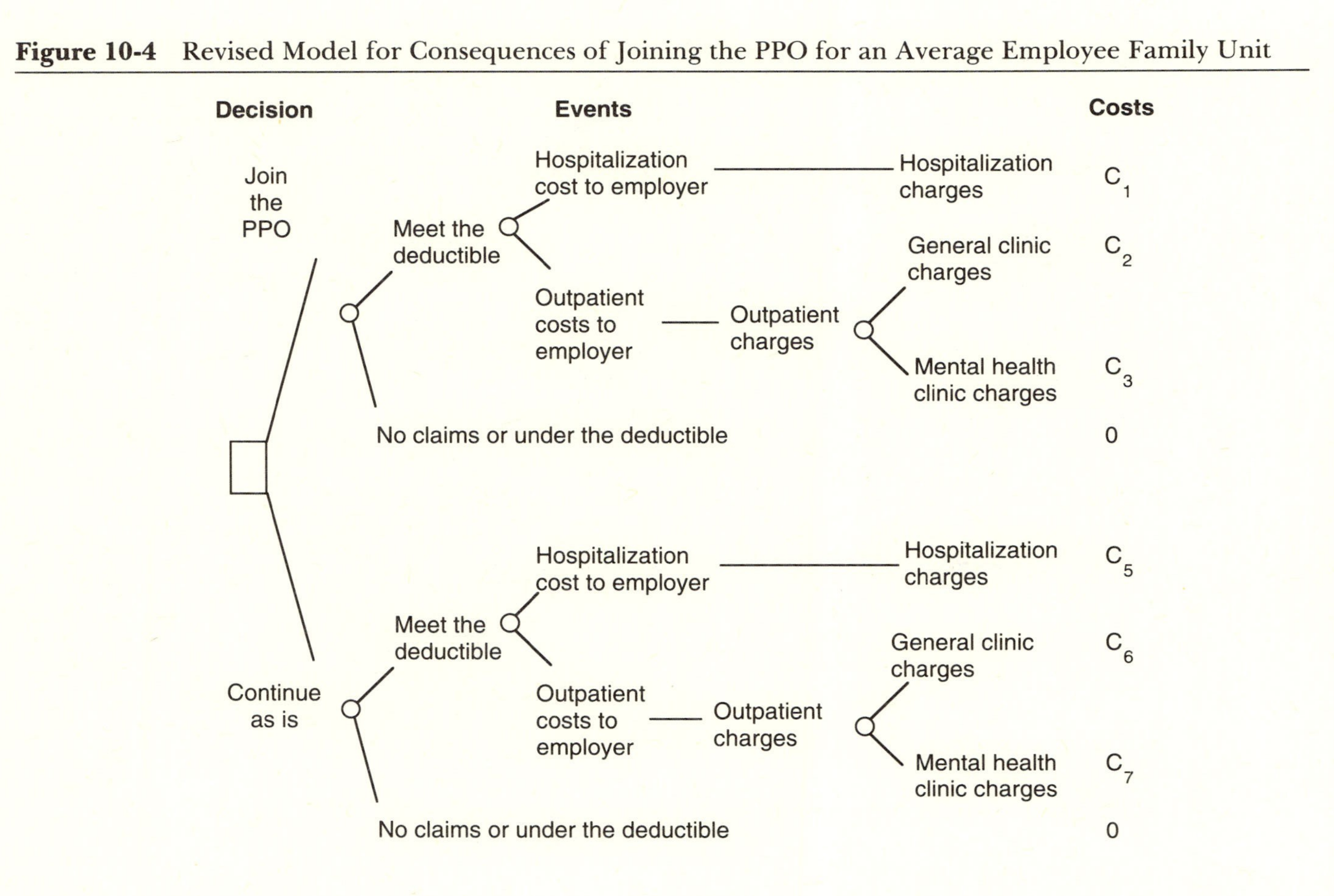

Estimating Probabilities and Costs

The numerical designation of values and uncertainties in decision trees permits quantitative evaluation of various courses of action. The measurement of preferences was discussed in Chapter 7; measurement of uncertainty was discussed in Chapter 8. If you are not familiar with these chapters, it may help to review them now. Here, we will focus on the use of these estimates in a tree.

In the example we are discussing, it is relatively simple to obtain the costs and probabilities for the current situation, shown on the lower part of the decision tree. After reviewing a 10 percent sample of employee records, we calculate the average costs and probabilities for hospitalization and clinic visits (see Figure 10-5). But what basis and method can be used to predict changes in costs and probabilities after offering the preferred provider?

As we have throughout this book, for this example we advocate a mix of objective and subjective estimates. To save money, data are collected on variables only if additional precision would make a difference to our conclusions. To highlight how objective and subjective information can be combined, we show below how we estimated "probability of hospitalization" and "cost per hospitalization."

We project probability of hospitalization by interviewing an expert. According to the research literature on health services, the utilization of hospitals is affected by characteristics of both the provider and the insurance plan. Our expert suggested that four characteristics could alter utilization patterns in the setting under investigation (see Table 10-1):

- The limited availability of beds could reduce the frequency of hospitalization (the PPO hospital has a 5 percent higher occupancy rate than average for the community).
- Physician-generated demand could lead to increased clinic visits and increased hospitalization because physicians in the preferred clinic see 5 percent fewer patients than doctors at nearby clinics.
- A move to the preferred practice—a group practice—could increase the frequency of referrals for the 20 percent of employees who now see solo physicians.

Figure 10-5 The Decision to Offer a PPO and its Consequences on the Average Employee Family Unit

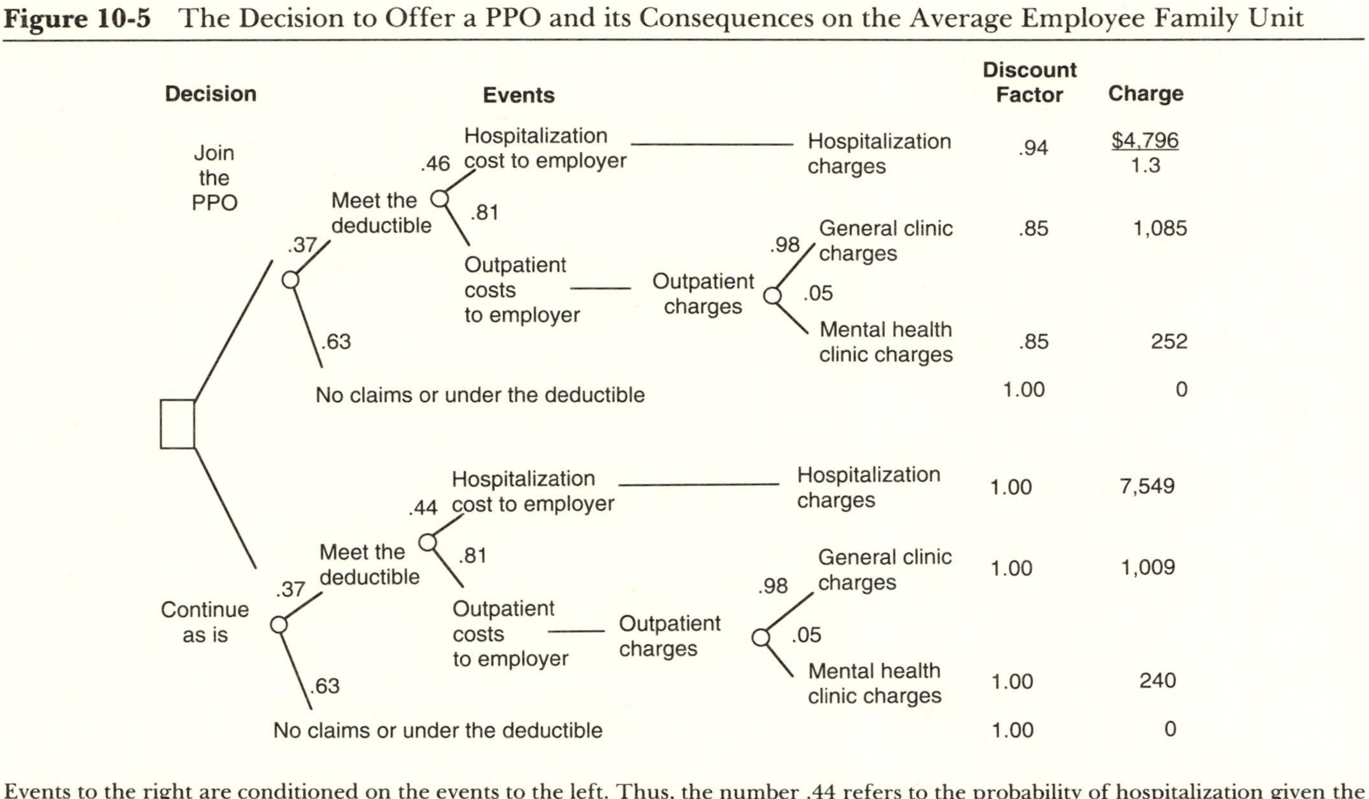

Events to the right are conditioned on the events to the left. Thus, the number .44 refers to the probability of hospitalization given the employee family unit has met the deductible. (Note that probabilities around each node do not add up to one when the events around the node are not mutually exclusive.)

Table 10-1 Increase in Hospitalization Rates Projected at the Preferred Clinic

	Differences between PPO and Other Clinics	*Impact of 1% Increase*	*Net Effect*
Occupancy rate of primary hospital	+5%[a]	−0.43[b]	−2.15
Number of patients seen per day	−5%[a]	−0.65[b]	+3.15
Group vs. solo practice	+20%[a]	+0.007[b]	−0.14
Effect of copayment reduction	−10%[c]	+0.1[d]	−1.00
Net projected change in hospitalization rate			−2.14

[a]From an expert's opinions.
[b]From Rosenblatt and Moscovice (1984).
[c]From proposed PPO contract.
[d]From Newhouse et al. (1981).

- The frequency of clinic use and eventually the frequency of hospitalization could both increase because of reductions in copayments.

This estimate shows how we gauge the impact of joining the preferred provider by combining the expert's concerns with the research literature. Although these estimates are at best guesses, they are sufficient unless the conclusion of the analysis is affected by their precision. This assumption can be tested with a process known as sensitivity analysis. In this process, described more fully later, the expert's estimates are changed to see whether small variations will radically alter our conclusions. In this case, sensitivity analysis identifies a few variables (especially cost per hospitalization) for which precise measurement is important. Therefore, additional data are collected for these variables, and the tree is reanalyzed in light of them.

In estimating cost per hospitalization, we assume that the employees will incur the same charges as current patients at the preferred hospital. However, this is misleading because the provider, as a large referral center, treats patients who are extremely ill. Company employees are unlikely to be as sick, and thus will not incur equally high charges, so we adjust charges to reflect this difference.

We make this adjustment using a system developed by Medicare to measure differences in mix of illness. In this system, each group of diseases is assigned a cost relative to the average case. Patients with diseases requiring more resources have higher costs and are assigned values greater than one. Similarly, patients with relatively inexpensive diseases receive a value less than one. The case mix for an institution is the sum of these costs weighted by their frequency of occurrence at the institution. If C_i indicates the relative cost for disease i, and $p_{i,j}$ measures the frequency of occurrence of disease i at hospital j, then:

$$\text{Case mix for institution } j = \Sigma_i C_i p_{i,j}$$

Data on case mix came from a review of employee records and patient records at the preferred hospital. In 1982, employees had a case mix of 0.90; the case-mix index at the preferred hospital that year was 1.17. This suggests that the diseases treated at the preferred hospital are about 30 percent more costly than those typically faced by employees, so we proportionally adjust the average hospitalization charges at the PPO.

We present this method of estimating probability of hospitalization and cost per hospitalization to highlight the ease and the importance of using a mix of experts' opinions and objective data. To save money and time, analysts should rely on expert opinion, unless additional precision is needed to ensure an accurate conclusion.

The Great Expectation and Analysis of Trees

The analysis of decision trees is based on the concept of expectation. In everyday use, the word suggests some sort of anticipation about the future rather than an exact formula. In this everyday sense, one may ask, "What do you expect a preferred provider to cost our company?" and you are free to answer as your assumptions and intuitions suggest. In mathematics, expectation is precisely defined. If you believe that costs $c_1, c_2, \ldots, c_n$ may happen with probabilities $p_1, p_2, \ldots, p_n$, then the mathematical expectation is

$$\text{Expected cost} = \Sigma_i p_i c_i$$

The expected cost is the sum of costs weighted by the probability of their occurrence. If 40 percent of employees will incur only outpatient charges of $1,009, 10 percent will incur hospital and outpatient charges totalling $8,549, and the remaining 50 percent of employees will incur no charges, the expected cost of care is:

$$0.40 \times \$1{,}009 + 0.10 \times \$8{,}549 + 0.50 \times \$0 = \$1{,}258.50$$

The theory of expectation is widely used in physical sciences to describe uncertain events; wide use, of course, does not imply accuracy, just popularity. In fact, when you talk about future costs you may not be thinking about the mathematical notion of expected cost. But for now we assume that you accept mathematical expectation as a good way of thinking about future events. Later in this chapter we will reevaluate this assumption and present alternative definitions of expectation.

Using Mathematical Expectation

Employees would have two options in the example under discussion: to join the preferred provider and take advantage of its discounts, or to continue with the current system. If they decide to join, the change of provider will alter both utilization rates and costs to the employer. To use mathematical expectation to project the cost of each option, we start from the extreme right-hand nodes and replace uncertain costs by expected costs; costs at the node are multiplied by the probability of their occurrence and totalled. This total then replaces the entire node.

For example, under the current plan (lower part of Figure 10-5), the expected cost to the bank per family unit is calculated as:

1. The *expected clinic charge* is the sum of mental health charges and general clinic charges (each times their probability of occurrence):

$$\$240 \times 0.05 + \$1{,}009 \times 0.98 = \$1{,}012$$

2. The *expected deductible* paid by employees was calculated similarly: cost times expected rate of usage. Eight percent of employees claimed family deductibles of $500, and the rest claimed individual deductibles of $200. (This information was not shown on the tree but included in the analysis.)

$$\$500 \times 0.8 + \$200 \times 0.92 = \$244$$

3. The *copayment* was calculated using usage data. Employees paid for 20 percent of charges after the deductible was paid.

Figure 10-6 The Mental Health and Clinic Charges Are Replaced With Their Expected Value

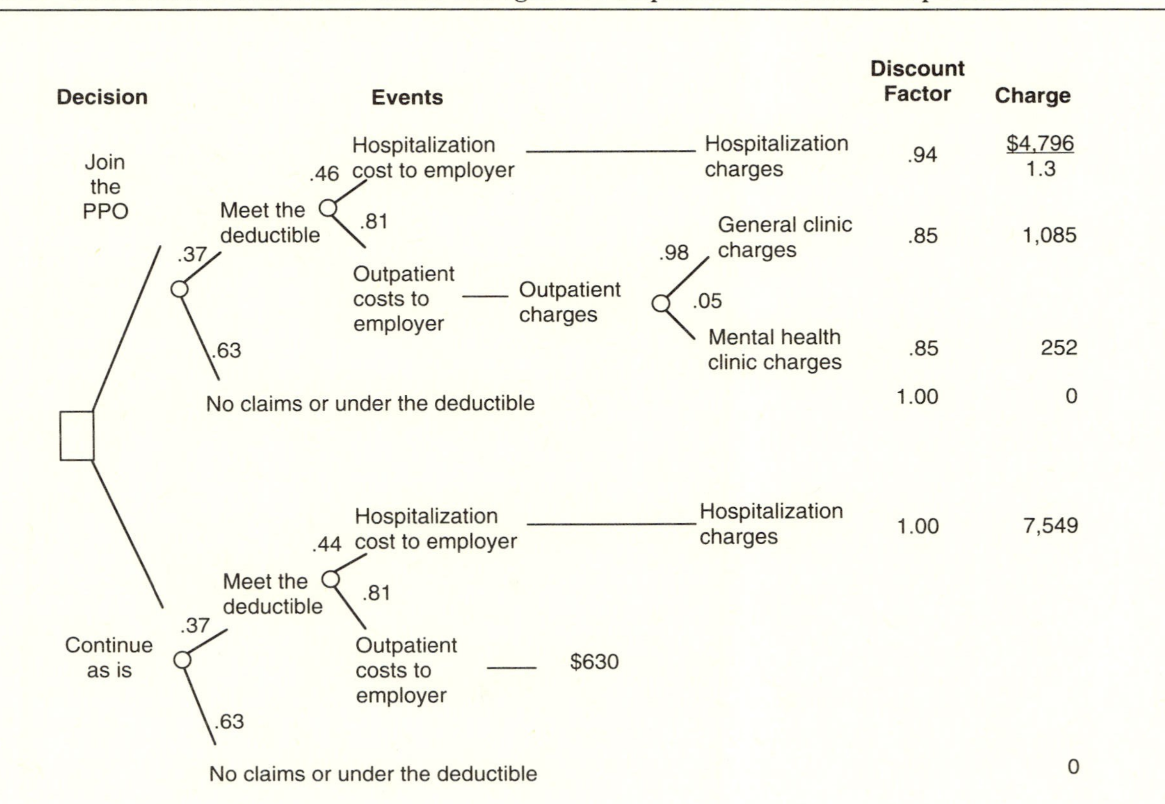

Figure 10-7 Hospitalization and Clinic Costs Replaced With Their Expected Value

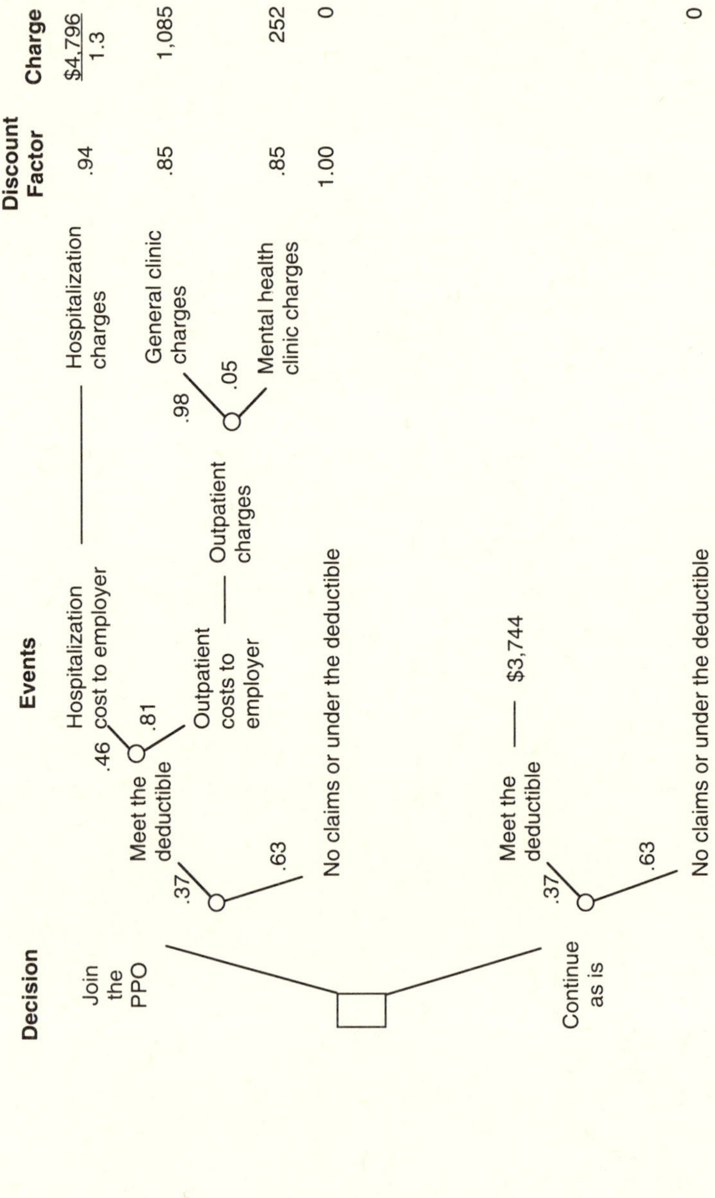

(This information was not shown on the tree, but was made part of the analysis.)

$$(\$1{,}012 - \$244) \times 0.20 = \$158$$

4. The expected deductible and copayment were subtracted from clinic charges to find the *expected clinic costs to the employer:*

$$\$1{,}012 - \$244 - \$158 = \$630$$

   This amount replaced the node for mental health and general clinic charges in the modified tree shown in Figure 10-6.

5. No copayment was required on the hospitalization portion, but there was a $200 deductible. The *net hospitalization cost* to the employer for each employee was:

$$\$7{,}549 - \$200 = \$7{,}349$$

6. The *expected total costs for hospitalization and clinic visits for employees who meet the deductible* was calculated by multiplying each cost at the node by its probability of occurrence:

$$0.44 \times \$7{,}349 + 0.81 \times \$630 = \$3{,}744$$

   This amount replaces the node for hospitalization and clinic visits in the modified tree shown in Figure 10-7.

7. The *expected cost for all employees* (regardless of whether they met the deductible) was calculated by multiplying the above cost by the probability of meeting the deductible.

$$0.37 \times \$3{,}744 + 0.63 \times \$0 = \$1{,}385$$

   This figure replaces the final chance node in the tree and suggests that the expected cost of the current situation if $1,385 per family unit. Joining the preferred provider would be more cost-effective if its expected cost is below this. Using mathematical expectation, and analyzing the top part of the tree, we calculate expected costs after joining the preferred provider as $743, for an expected saving of $642 per family unit.

Figure 10-8 The Savings Persisted Despite an Additional 5 Percent Change in the Hospitalization Rate

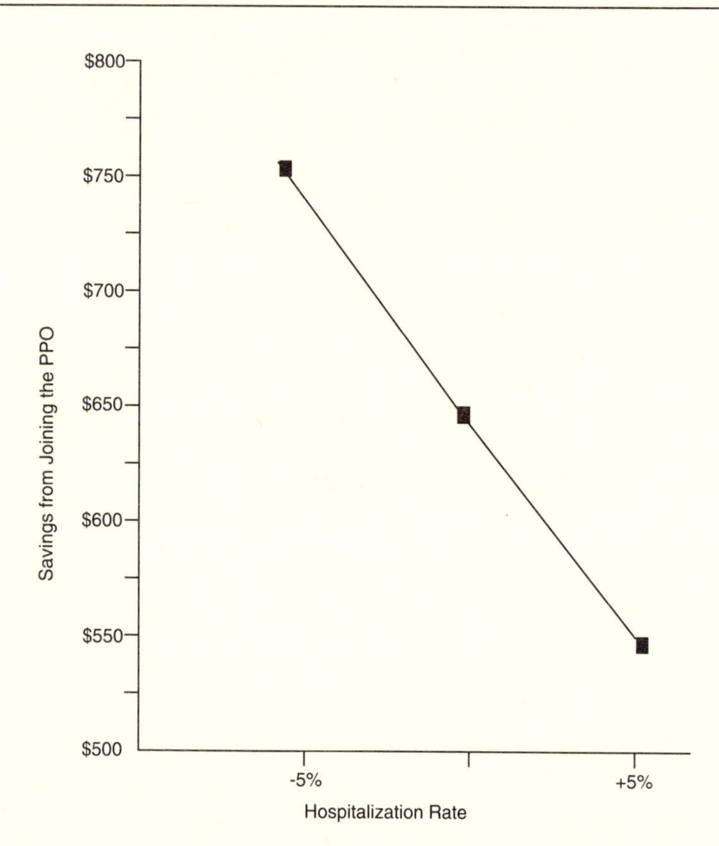

Sensitivity to Conclusions and Assumptions

The point of constructing a decision tree is not only to recommend or reject an option but also to show which factors tip the scale toward or against it. Sensitivity analysis is the repeated examination of two options while varying some cost or probability to examine how conclusions depend on assumptions. In this case we can examine several scenarios using different assumptions about costs and utilization to find out which factors are most important and hence must be estimated most accurately.

The first step in a sensitivity analysis is determining a reasonable range for the variable of interest. Next, use the decision tree to

Figure 10-9 The Savings Persisted Despite a 35 Percent Increase in Clinic Costs

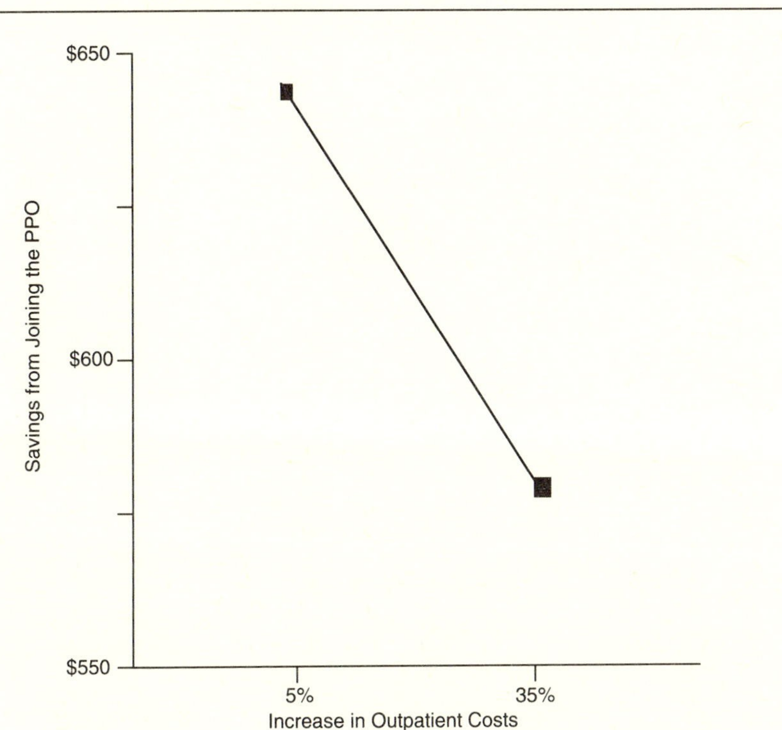

calculate the expected cost of the proposed PPO at the low and high ends of the range. Graph these values as in Figure 10-8: the variable of interest is on the *x* axis and the expected savings of the PPO is on the *y* axis. Assuming only one variable has been changed, draw a straight line between these two points. If the expected cost savings becomes negative along the line (meaning PPO costs are higher than current costs), the conclusion is sensitive to the variable. If, over the specified range, changes in the variable do not alter the conclusions, the conclusions are not sensitive to changes in this variable.

Figure 10-8 shows that the conclusions are not sensitive to a 5 percent increase in hospitalization, because joining the preferred provider still would save the company money.

In calculating the cost for clinic visits per person, we assumed that the cost of ancillary services at the preferred clinic would be similar to present costs. Figure 10-9 shows the sensitivity of this con-

Figure 10-10 The Savings Disappear When in Calculation of Average Hospitalization Charges the Two Most Expensive Cases Were Ignored

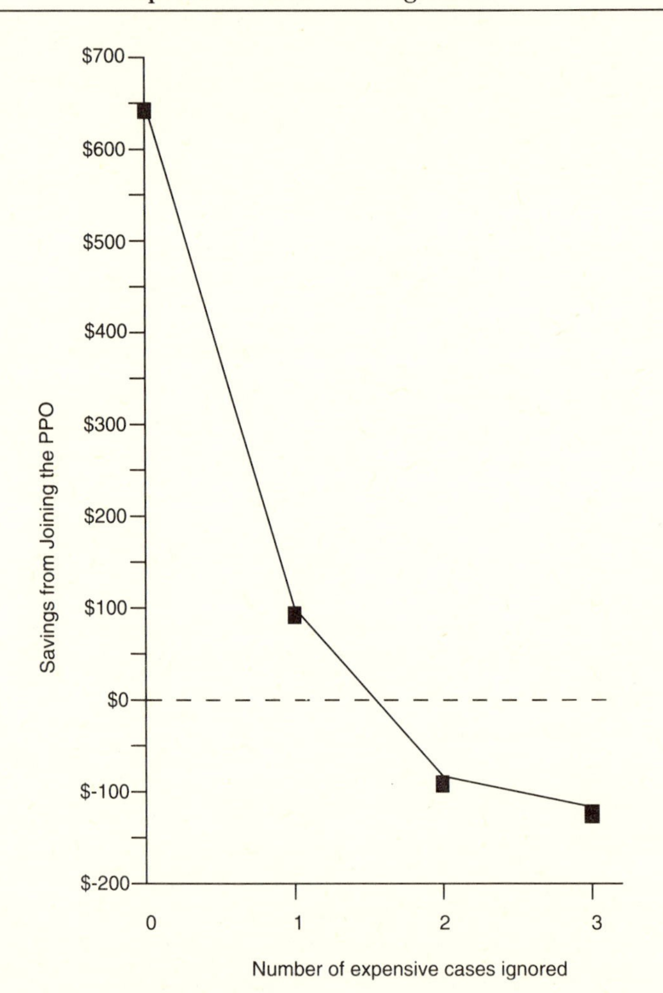

clusion to a 35 percent increase in average outpatient charges. The preferred provider remains cost-effective even under these substantially higher charges.

Among all the assumptions we tested, the conclusions were only sensitive to estimates of case-mix differences. Since the employee population was small (992 employees and their dependents), their past consumption of resources may not indicate future charges very well. In particular, a few cases extremely expensive during the year studied may be unlikely to recur each year. To evaluate the impact of the costly cases, we excluded them one at a time and reevaluated the estimates in the tree. This process changed not only the current average hospitalization charges but also the estimate of case-mix differences (see Figure 10-10). While at the beginning the current plan seemed more expensive, as more cases were excluded it became less expensive.

Because the analysis is sensitive to both cost per hospitalization and case mix, we must estimate both variables more accurately. From an additional year of data, we collect information on the health care costs of 10 percent of employees and calculate the average hospitalization charge and the case-mix difference. Apparently, employees used more expensive services in this year, and the new data suggest a higher average charge and a slightly larger case-mix difference. This indicates that our original estimates of charges under the current plan may have been conservative. When these data were presented to the client, he decided that the current year's experience was not materially different from what he expected in the future, and despite the sensitivity of the conclusions to case-mix calculations, there was no need for additional data from a third year.

The sensitivity analysis helped this decision maker see which factors would affect the cost of the proposed changes in the health plan. It helped him separate the wheat from the chaff and look at the important items. Understanding that the cost per hospitalization would affect overall cost helped the client negotiate for hospitalization discounts in addition to the clinic discounts existing in the proposed contract.

Presentation and Option Generation

One of the most productive uses of the decision tree is in generating new ways of looking at a problem. We saw evidence of this in the comments we received upon finishing this analysis. First, the clients told us, "We told you so." Since their original opinions contrasted with our conclusions, this comment suggested that careful thought and communication during the analysis had brought them to a conclusion that seemed correct to them, and that our presentation of the results reinforced what they had learned during the process.

Second, we also received criticism that indicated that we had overlooked certain perspectives. One decision maker believed that the potential savings were insufficient to counterbalance the political and economic costs of instituting the proposed change. It turned out that the present health care providers were customers of the client, and that signing a contract with the PPO might alienate them and induce them to take their business elsewhere. Incorporating the risk of losing customers would improve our calculation and help the client decide whether the savings would counterbalance the political costs.

Finally, a third critical perspective emerged: Would it be better to await a better offer from a different provider? We could have reflected the consequences of waiting by placing an additional branch from the decision node, and clearly this would have provided a better basis for the decision.

New avenues often open up when an analysis is completed. It is important to remember that one purpose of this activity is to help decision makers understand the components of their problem and to devise increasingly imaginative solutions to it. Therefore, there is no reason to act defensively if a client begins articulating new options and considerations while you present your findings. Instead, encourage the client to discuss the option, and consider modifying the analysis to include it.

A serious shortcoming with decision trees is that many clients believe they show every possible option. Actually, there is considerable danger in assuming that the problem is as simple as a tree makes it seem. In our example, many other options may exist for reducing health care costs aside from joining the preferred provider, but perhaps because they were not included in the analysis, they can be ignored by the decision maker, who, like the rest of us, is victim to the "out of sight, out of mind" fallacy (also see Fischhoff et al. 1978).

The "myth of analysis" can explain why things not seen are not considered. This is the belief that analysis is impartial and rests on proper assumptions and that a small change will not affect the outcome. Perpetuating this myth prevents further inquiry and imaginative solutions to problems. Decision trees could easily fall into this trap because they appear so comprehensive and logical that decision makers fail to imagine any course of action not explicitly included in them.

We can satisfy an important function of analysis—helping generate ideas by increasing our insight about the problem—by breaking the final report presentation into two segments. First we summarize the results of the tree and the sensitivity analysis, then we ask the clients to share their ideas about options not modeled in the analysis. If we do not explicitly search for new alternatives, the decision tree might do more harm than good. Instead of fostering creativity, it can allow the analyst and decision maker to hide behind a cloak of missed options and poorly comprehended mathematics.

This completes our presentation of constructing, analyzing, and reporting a decision tree. Before we leave this chapter, however, we will introduce several other tools that can improve analysis of trees by changing the way expectations are calculated (also see Schoemaker 1982).

Expected Utility Theory

Expected utility is an alternative method of evaluating an outcome that attempts to take the decision maker's point of view and has the advantage of more accurately reflecting a decision maker's preference.

Recall that we asked you to suppose that mathematical expectation is a good way of predicting future events, and that the benefits manager is concerned only about expected cost. We now show how the analysis might be improved by relaxing these two assumptions.

When in 1738 Bernoulli was experimenting with the notion of expectation, he noticed that people did not prefer the alternative with highest expected monetary value, that people are not willing to pay a large amount of money for a gamble with infinite expected return. In explanation, Bernoulli suggested that people maximize utility rather than monetary value, and costs should be transformed to utilities before expectations are taken. He named this model expected utility.

According to expected utility, if an alternative has n outcomes with costs $c_1, \ldots, c_n$, associated probabilities of $p_1, \ldots, p_n$, and each cost has a particular utility to the decision maker, say $u_1, \ldots, u_n$, then:

$$\text{Expected utility} = \Sigma_i \, p_i \, u_i$$

Bernoulli resolved the paradox of why people would not participate in a gamble with infinite return by arguing that the first dollar gained has a greater utility than the millionth dollar. A beauty of a utility model is that it allows the marginal value of gains and losses to decrease with their magnitude. In contrast, mathematical expectation assigns every dollar the same value. When the costs of outcomes differ considerably, say, when one outcome costs $1,000,000 and another $1,000, we can prevent small gains from being overvalued by using utilities instead of costs.

Utilities are also better than costs in testing whether benefits meet the client's goals. Using costs in the preferred provider analysis, we found that joining the PPO would lead to expected savings of about $70,000. Yet, when the client had not acted six months after completing the analysis, it became clear that this saving was not sufficient to cause him to act because his target expectations were much higher. Thus, he chose to await a better opportunity. We could have uncovered this problem if, instead of monetary returns, we had used utility of gains relative to the client's target expectation. Compared to the client's target return, $70,000 had little utility.

The expected utility model is also better than mathematical expectation at incorporating nonmonetary considerations. In Chapter 7, we described how to measure utility over many dimensions, monetary and nonmonetary. When money is not a crucial factor in the decision, the decision tree outcomes should be measured in terms of utilities rather than costs. In this chapter's example, cost was not the sole concern—the firm had many objectives for changing its health care plan. If it wanted only to lower costs, it could have ceased providing health care coverage entirely, or it could have increased the copayment. The firm was concerned about employees' reactions, which it anticipated would be based on concerns for quality, accessibility, and, to a lesser extent, cost to employees.

The utility model was also useful for another client concern—that some employees would object to changing physicians. We surveyed employees to assess their utility for switching physicians, then used utility models to predict how many survey participants would join the preferred provider. This study portrays a situation in which nonmonetary factors are more important than cost, and therefore, the analysis is best done with utility.

Utility is also preferable for clients who must consider attitudes toward risk. This is because expected utility, in contrast to expected cost, reflects attitudes toward risk. A risk-neutral individual bets the expected monetary value of a gamble. A risk taker bets more on the same gamble because he or she associates more utility to the high returns. A risk adverse individual cares less for the high returns and bets less. Research shows that most individuals are risk seeking when they can choose between a small gain and a gamble for a large gain and are risk adverse when they must choose between a small loss and a gamble for a large loss (Kahneman and Tversky 1979). The mathematical model of expectation does not reflect these types of preferences. A utility model does.

A client, especially when trying to decide for an organization, may exclude personal attitudes about risk and request that the analysis of the decision tree be based on expected cost and not expected utility. Thus, the client may prefer to assume a risk-neutral person and behave as if every dollar of gain or loss were equivalent. The advantage of making the risk attitudes explicit is that it leads to insights about one's own policies; the disadvantage is that such policies may not be relevant to other decision makers.

Transformation of costs to utilities is important in most situations. But when the analysis is not done for a specific decision maker, monetary values are paramount, the marginal value of a dollar seems constant across the range of consequences, and attitudes toward risk seem irrelevant, then it may be reasonable to explicitly measure the cost and implicitly consider the nonmonetary issues.

Prospect Theory

The notion that monetary values should be transformed before taking expectations has led to speculation that probabilities should be transformed as well. Experiments by Kahneman and Tversky (1979) showed that low probabilities are perceived lower than their objective value and high probabilities are treated as certainties. These and other experimental results suggest that decision makers distort probabilities when evaluating uncertain alternatives. Kahneman and Tversky suggested a method of analysis called prospect theory, in which probabilities are transformed to reflect the decision maker's understanding of them. These transformed probabilities are then used to weight the utility of a decision's consequences.

Analysts generally have different uses in mind when they model a decision. Sometimes, as when an expert's judgment is modeled, the analyst tries to influence the users of the decision tree. Such a model predicts an expert's judgment and can serve as a "prescriptive" norm for a novice. Other times the analyst is trying to understand how a decision is made; hence the approach is descriptive. When an analyst constructs a model to predict employee choices, the purpose is to describe behavior, not to prescribe it. But when an analyst constructs a model about what will happen if an employee joins the PPO, the purpose is to prescribe. This type of model addresses whether the employer should offer a PPO. When the analysis is prescriptive, there is no need for a transformation of probabilities.

References

Bernoulli, D. 1738. "Specimen Theoriae Novae de Mensura Sortis." *Commetarri Academiae Scientaiarum Imperialis Petropolitanae* V: 175–92. Also see "Exposition of a New Theory on Measurement of Risk," Louise Sommer (trans.), *Econometrica* 22 (1954): 23–26.

Fischhoff, B., P. Slovic, and S. Lichtenstein. 1978. "Fault Trees: Sensitivity of Estimated Failure Probabilities to Problem Presentation." *Journal of Experimental Psychology: Human Perception and Performance* 4 (2): 330–34.

Kahneman, D., and A. Tversky. 1979. "Prospect Theory: An Analysis of Decision Under Risk." *Econometrica* 47 (2): 263–91.

Rosenblatt, R. A., and I. S. Moscovice. 1984. "The Physician as Gatekeeper: Determinants of Physicians' Hospitalization Rate." *Medical Care* 22 (2): 150–59.

Schoemaker, P. J. H. 1982."The Expected Utility Model: Its Variants, Purposes, Evidence and Limitations." Center for Decision Research, Graduate School of Business, University of Chicago. Also see *Journal of Economic Literature,* 1983.

11

Decision-Oriented Program Evaluation

Large-scale evaluations of health and social service programs are commonly initiated to help policymakers make decisions on topics such as:

- How to improve productivity in hospitals, particularly regarding materials management, nurse staffing, and general management methods
- How to structure the process of forming health policy
- Whether to adopt an innovative preventive program of community care (for the elderly, for example)
- Whether to adopt specialized programs to fight a disease such as tuberculosis
- How to manage long-range planning of resource allocation

Not surprisingly, with this kind of interest, program evaluation has become a big business and an important field of study (Wilderman 1979; Polivka and Steg 1978). Program evaluations are requested and funded by virtually every department of health and social services, not to mention many legislatures, governors, and city administrations.

The basic concept of evaluating social and health care programs, as shown in Figure 11-1, is straightforward. A program is expected to meet certain performance standards (a). The program actually performs at a level (b) that may equal or exceed the standards or fall short because of flaws or unexpected environmental influences. Actual performance is compared to expected perfor-

Figure 11-1 Schematic Representation of Program Evaluation

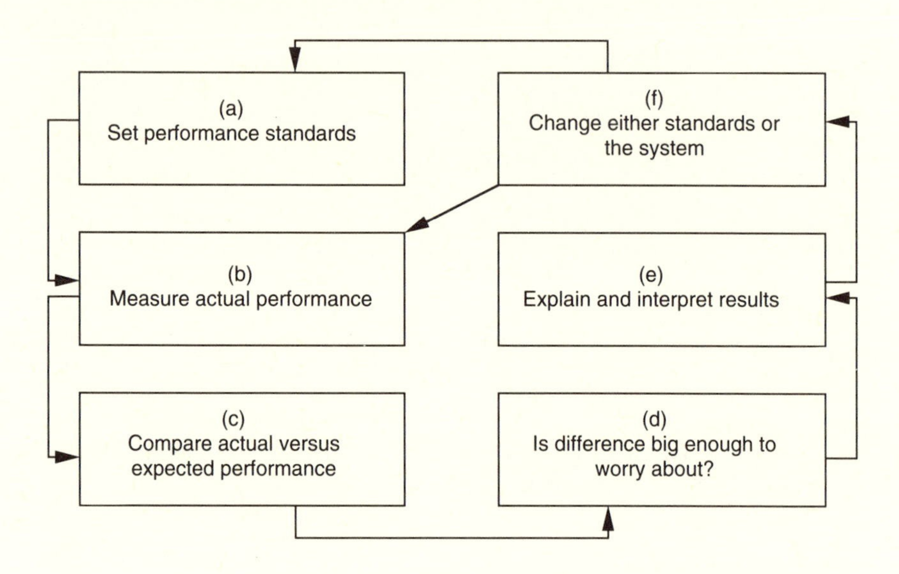

mance (c), and decisions are made about which, if any, of the discrepancies are worrisome (d). The findings are explained and interpreted to decision makers (e), and changes are introduced in either system performance or the expectations of it (f).

While the basic concepts of evaluation are simple, actual implementation can be quite complex, and numerous evaluation techniques and philosophies have been introduced over the years (Alemi 1988). The major approaches are categorized as experimental, case study and cost-benefit analysis.

Some researchers have advocated an *experimental* approach, with carefully designed studies using experimental and control sites, random assignment of subjects, and pre-and posttests. These people prefer clearly defined research hypotheses that examine objective measures of performance (Scheirer 1978; Campbell and Stanley 1966; Boruch 1975). Variations on this experimental theme often remove the random assignment criterion. These "quasi-experimental" designs interject alternative explanations for findings. Over the years we have used these designs and have almost always regretted it. We now believe it is possible to assign randomly in almost any evaluation. It just requires a lot of careful, creative thinking. We don't believe an experimental evaluation is always necessary, but when it is, random assignment must be an essential element of it.

Another school advocates examining *case studies* (Weiss and Rein 1970; Heilman 1980), arguing that case studies are superior to experiments because of the difficulty of identifying criteria for experimental evaluation. Further, they say that experiments require random subject assignment and pre- and posttests, both of which are impractical because they interfere with program operation. This school prefers using unobtrusive methods to examine a broad range of objectives and procedures and which are sensitive to unintentional side effects. This approach, they contend, helps administrators improve programs instead of just judging them. Some case studies report on services offered and characteristics of their use, while others are less concerned with the physical world and emphasize the values and goals of the actors. Their reports tend to contain holistic impressions that convey the mood of the program. We believe that any evaluation must contain case studies to help people really understand and act on the conclusions.

Cost-benefit analysis evaluates programs (Thompson 1980) by measuring costs and benefits in monetary terms and calculating the ratio of costs to benefits. This ratio is an efficiency statistic showing what is gained for various expenditures. There are many types of benefit analysis. Some analyses assume that the market price of services fairly and accurately measures program benefits (Watson 1979); others measure benefits on the basis of opinion surveys. Variations on the cost-benefit theme involve comparisons that don't translate everything into a dollar equivalent. The critical characteristic of those studies is an ability to compare what you get against what it costs to get it.

Many Evaluations Are Ignored

Although program evaluations take a good deal of time and money, their results are often ignored. Even if interesting data are collected and analyzed, evaluations have no impact if their results are not directly relevant and timely to a decision (Agarwala 1977; Cox 1977; Cook and Gruder 1978). Often, evaluation reports present a variety of unrelated findings and thus confuse rather than clarify the decision maker's choices (Wright 1974; Dickson, Senn, and Chervany 1977).

It is our opinion that evaluation studies with little impact generally began with a poor design. We believe an evaluation can gain impact if the evaluator understands and focuses on the options, values, and uncertainties of the decision makers. To provide the kind of evaluation that supports policy formation, relevance to the decision must be designed in at the start, not tacked on at the end. Edwards, Gutentag, and Snapper (1975) wrote: "Evaluations, we believe, exist (or perhaps only should exist) to facilitate intelligent decision making ... an evaluation research program will often satisfy curiosity. But if it does no more, if it does not improve the basis for decisions about the program and its competitors, then it loses its distinctive character as evaluation research and becomes simply research" (p. 140).

Decision-Oriented Evaluation Design

Certain design factors can increase the relevance of an evaluation to the actual decision making. The evaluators should:

- Identify the primary users of the evaluation. Often, evaluators say their findings are intended for "policymakers," not an individual. The evaluators do not contact an individual policymaker because either they consider such person's views irrelevant or they perceive that such individuals hold their positions temporarily while the policy issue and the evaluation task remain more or less permanently. We prefer naming the decision makers before designing evaluation studies to meet their needs, because, when asked, decision makers usually express needs dictated by their positions, not their idiosyncrasies. Even though decision makers change, their needs remain stable, and identifying the needs of a single decision maker is not a waste of effort.
- Identify the decision-making needs of decision makers, and provide information to meet their needs. Gustafson and Thesen (1981) found that once decision maker information needs were prioritized in order of importance, the top 65 percent of the priorities had nothing to do with the type of questions addressed by typical evaluation data. Decision makers tend to seek help on issues that depend more on values and expectations than on statistics. Although program statistics cannot be ignored, policymakers should provide information that will actually influence decisions.

- As part of the evaluation, suggest options that creatively address the issues. Too often evaluations produce evidence about the strengths and weaknesses of the existing system, touching only briefly on what improvements could be made. An effective evaluation must devote considerable effort to identifying improvements that will remove the weaknesses.
- Identify and attempt to reduce the most important uncertainties involved in the system being evaluated. Frequently a system performs well under some conditions and less well under others. Attempts to improve system performance are limited by our ability to predict when true differentiating conditions will arise. Evaluators should identify the differentiating conditions, develop means to predict when they will arise, or propose system improvements that are less susceptible to variations in those conditions.
- Explain how they reached their results. Statistical analysis can satisfy detached, rational decision makers, but many policymakers find that examples underscore the justification for conclusions in a more persuasive manner than rational and statistical arguments alone. Such examples can be constructed by careful evaluation of the program. To explain the evaluation results, we must design a protocol that permits us to observe, experience, and describe how the system actually operates. Succinct examples can make an evaluation report "come alive" and help decision makers feel and understand at several levels the consequences of their decisions. The use of examples has three benefits: (1) it permits us to describe how the system actually functions; (2) it permits us to compare actual operation to intended operation; and (3) if done by qualified personnel, it omits the need to assess the adequacy of the system's operation.
- Examine the sensitivity of the evaluation findings to assess their practical significance. With a sufficiently large data base, almost any difference can be statistically significant. What we want at this point is to determine how erroneous our assumptions can be before they would cause the decision makers to act mistakenly. Evaluation studies can conduct sensitivity analyses but rarely do.
- Present results more quickly and when they can be most useful. Too often, decisions must be made before the evaluation

is complete (see Chapter 12). Some people argue that it is misleading to present preliminary results, but we contend that evaluators, while they are designing the study, should be sensitive to timing, particularly in terms of knowing the critical moments in the policy process. Reliable information that could influence policy almost always surfaces during an evaluation, and if evaluators know the timing, they can give input when it is most useful. The decisions will be made anyway, and policymakers will act with whatever information they have.

In summary, we suggest that program evaluation should be tied to the decision-making process. The remainder of this chapter presents a nine-step strategy for such a decision-oriented evaluation design. We will clarify our presentation by referring to our evaluation of the nursing home quality assurance process (Gustafson et al. 1990).

Continued rises in nursing home costs in the United States have stimulated increasing debate about how regulation can improve the industry. Some critics find the government ineffective at evaluating nursing homes (Moreland Act Commission 1976). These critics argue that current surveys of nursing home quality are too frequent, are too intensive, and have little relation to the health status and functional ability of nursing home residents. Gustafson et al. (1981) were asked to evaluate the process of surveying nursing home quality, and we use their experience to illustrate how a decision-oriented evaluation is done.

Step 1. Identify the decision makers

The first step in planning an evaluation is to examine the potential users and invite them to help devise the plan of action. In the example, three groups might be expected to use the evaluation results: (1) the state government, to decide what program to implement; (2) the federal government, to decide whether to support the state's decision and whether to transfer aspects of the project to other states; and (3) several lobbying groups (nursing home associations), to choose their positions on the topic. We identified individuals from each group, asked them to collaborate with our evaluation team, and kept them informed of progress and preliminary conclusions throughout the study.

Step 2. Examine the decision makers' concerns and assumptions

Next we talked to the chosen decision makers (the program administrator and experts on the program, for example) to determine their concerns and assumptions, and to identify the potential strengths, weakness, and intended operations of the program.

In the example, a decision maker who was concerned about the paperwork burden for quality assurance deemed the effort excessive and wasteful. A second decision maker was concerned with the cost of the quality assurance process and worried that it would divert money from resident care. A third was more concerned that quality assurance funds be distributed in a way that helped not only to identify problems but also to facilitate solutions. This person preferred to find solutions that would bring nursing homes into compliance with regulations. A fourth decision maker felt that the quality assurance process should attend not just to clients' medical needs but also to their psychological and social ones. All these divergent objectives were important because they suggested where to look while designing a quality assurance process to address the decision makers' real needs.

We also helped identify and clarify each decision maker's assumptions. These assumptions are important because, regardless of accuracy, they can influence decisions if not challenged. One decision maker believed the state must play a policing role in quality assurance by identifying problems and penalizing offending homes. Another person believed the state should adopt the role of change agent and take any necessary steps to raise the quality of care, even if it had to pay a home to solve its problems. We examined arguments for and against these philosophies about the role of government, and while we did not collect new data on these issues, our final report reviewed others' research on the matter.

Step 3. Add your observations

Another important method of examining a program is to use one's own observations. The perceptions of decision makers, while very useful for examining problems in detail, do not prove that problems exist, only that they are perceived to exist. Thus, it is important to examine reports of problems to see that they are, indeed, real problems. We suggest that members of the evaluation team watch the system from beginning to end to create a picture of its functioning.

A system analyst should literally follow the quality assurance team through a nursing home and draw a flowchart of the process.

While observational studies are not statistically valid, they can add substantial explanatory power and credibility to an evaluation and allow us to explain failure and suggest improvements. A valuable side effect of such observations is to gather stories describing specific successes and failures. These stories have powerful explanatory value, often more than the statistical conclusions of the evaluation. The observations not only suggest how and where to modify the program, they also indicate areas that should be targeted for empirical data collection.

Step 4. Conduct a mock evaluation

The next step is performing a mock evaluation, which is a field test to refine the evaluation protocol and increase efficiency. The mock evaluation keeps the decision maker informed and involved. Too often, decision makers first see the results of the evaluation when reading the final report. While this sequence probably allows enough time to produce a fine product, time alone guarantees neither quality nor relevance. We prefer informing the decision maker about our findings as the project proceeds, because, by definition, we are gathering information that could influence a decision. Decision makers will want access to this information. The mock evaluation lets the decision maker tell us which areas require more emphasis, allowing us to alter our approach while we have time.

A mock evaluation is similar to a real one except that experts' opinions replace much of the data. This "make-believe" evaluation helps estimate how much money and time are needed to complete the evaluation. It also changes the data collection procedures, sample size requirements (because we gain a more realistic estimate of variance in the data), and analysis procedures. Finally, the mock evaluation gives a preview of likely conclusions, which allows decision makers to tell whether the projected report will address the vital issues, as well as identify weaknesses in the methodology that still can be corrected.

Critics of such previews wonder about the ethics of presenting findings that may be disproven by careful subsequent observation. But supporters counter by questioning the ethics of withholding information that could inform policy. These questions represent two extreme positions on a difficult issue. It is true that preliminary results may receive more credibility than they deserve. Moreover, decision makers may press to alter the evaluation design to prevent reaching embarrassing conclusions. We feel those dangers may be outweighed by the alternatives of producing irrelevant data, missing critical questions, or failing to contribute valuable information when it can help the policy debate.

After the decision makers have read the mock report, we ask them to speculate about how its findings might affect their actions. As our evaluation team describes its preliminary findings, the decision makers will explain their possible courses of action and list other information that would increase the evaluation's utility.

It is important to make sure that evaluation findings lead to action. Decision makers can react in many ways to various findings. Some consider negative findings sufficient basis for changing their opinions and modifying the system, while others continue adhering to existing opinions. If our findings are unable to motivate the decision makers to change the system, this is a signal that we could be collecting the wrong data. At this point, we can decide to collect different data or analyze it more appropriately. The goal remains to provide information that really influences the decision maker's choices.

In the nursing home study, we observed several nursing home surveys, talked with interested parties, developed a flowchart of the process, and then asked the group to consider what they would do differently if the evaluation suggested that current efforts were indeed effective. We then repeated the question for negative findings on various dimensions. The discussion revealed that our experts, like others in the field, believed that existing quality assurance efforts were inefficient and ineffective, and these people expected the evaluation to confirm their intuition. But they felt evaluation findings would make a difference in the course of action they would follow. In other words, they were certain about the effectiveness of the current system but uncertain how to improve it. This is an important distinction, because an evaluation study that only gauged the effectiveness of the current system would confirm their suspicions but not help them act. What they needed was a study to pave the way for change, not just to criticize a system that was clearly failing.

Our evaluation team and its advisory group at this point developed an alternative method of nursing home quality assurance that helped to reallocate resources by focusing on influencing the few problematic homes, not on the majority of adequate ones. We designed a brief nursing home survey to identify a problem home and target it for more intensive examination. Then we designed an evaluation to contrast this alternative approach to the existing method of evaluation. Thus, the mock evaluation led us to create an alternative system for improving nursing home quality, and instead of just evaluating quality of the current system, we compared and contrasted two evaluation systems.

The mock evaluation is a preview that helps the decision makers see what information the evaluation will provide and suggested improvements that could be made in the design. And "showing off" the evaluation makes the decision makers more likely to delay their decision making until the final report is complete.

Step 5. Pick a focus

Focus is vital. In the planning stage, our discussions with decision makers usually expand the scope of the upcoming evaluation, but fixed resources force us to choose which decision makers' uncertainties to address, and how to do so. For example, further examination of the potential impact of the evaluation of nursing home quality assurance revealed a sequence of decisions that affected whether evaluation findings would lead to action. The state policymakers were responsible for deciding whether to adopt the proposed changes in quality assurance. This decision needed the approval of federal decision makers, who relied on the opinions of several experts as well as our evaluation. Both state and federal decisions to modify the quality assurance method depended on a number of factors, including public pressure to balance the budget, demand for more nursing home services, the mood of Congress toward deregulation, and the positions of the nursing home industry and various interest groups. Each of these factors could have been included in our effort, but we didn't have the money to include all and we needed to select a few.

Some factors in the decision-making process may be beyond the expertise of the evaluation team. For example, the evaluators might not be qualified to assess the mood of Congress. Although the evaluation need not provide data on all important aspects of the decision process, the choice not to provide data must be made consciously. Thus, we must identify early in the process which components to include and which to exclude as a conscious and informed part of evaluation planning.

To those who think we are advocating sloppy analyses, we answer that evaluations often operate on limited budgets and thus must allocate resources to produce the best product without "busting the budget." This means that specificity in some areas must be sacrificed to gain greater detail elsewhere.

Step 6. Identify criteria for an evaluation

Now the evaluation team and decision makers set the evaluation criteria, based on program objectives and proposed strengths and weakness of the program. (See Chapter 7 for a discussion of how to identify evaluation criteria, or "attributes"; see Chapter 5 to see how the analysis can be done using a group of decision makers.)

The nursing home evaluation focused on a number of questions, one of which was the difference between the existing method of quality assurance and our alternative method. We used these criteria to evaluate this issue:

- Relation to regulatory action: The quality assurance effort should lead to consistent regulatory actions.
- Ease of use: Administering quality assurance should interfere with delivering nursing home care no more than necessary.
- Reliability: The quality assurance effort should produce consistent findings, no matter who does the reviews.
- Validity: The findings should correlate with adverse outcomes of nursing home care (such as an increasing rate of deterioration in residents' ability to attend to their daily activities).
- Impact: Quality assurance should change the way long-term care is delivered (see Chapter 15).
- Cost: The cost of conducting quality assurance must be measured to allow us to select the most cost-effective method.

We created an evaluation design that divided the state into three regions. In the lower half of the state (and the most populous), nursing homes were randomly assigned to control and experimental conditions, after ensuring an equal number of proprietary nursing homes and nonprofit nursing homes, of similar sizes, treating similar patients, would be placed in each group. The northern half of the state was divided into two regions, one receiving the new regulatory method and one not. This was done to observe how the management of the regulatory process would rate. Such random assignment greatly increased the credibility of the evaluation.

A second aspect of design was the measures used. Previously a nursing home's quality was judged on the basis of the number of conditions, standards, and elements found out of compliance. However, it was apparent that radical differences in severity of violations could take place within a level (e.g., element). It was decided to convene a panel of experts to rate numerically the severity of different violations. Tests of reliability between experts and over time demonstrated that the measures were good.

Step 7. Set expectations

Once the evaluation design is completed, we ask decision makers to predict the evaluation findings. This request accomplishes two things. First, it identifies the decision makers' biases so we can design an evaluation that responds to them. Second, it gives a basis for comparing evaluation findings to the decision maker's expectations, without attributing them to specific people. The impact of evaluation results is often diluted by hindsight. Reviewers might respond to the results by saying, "That is what I would have expected, but.... " Documenting expectations in advance prevents such dilution.

In the nursing home example, decision makers expected that the alternative method would be slightly better than the current method, but they were surprised at how much better it performed. There were substantial cost savings as well as improvements in effectiveness. Because we had documented their expectations, their reaction was more akin to "Aha!" than to "That's what we expected."

Step 8. Compare actual and expected performance

In this phase we collect data to compare actual and expected performance. For more information on data collection and statistical comparison, consult the many books on evaluation that cover these topics in detail. There are a variety of ways this can be done. Chapter 15 reports on our use of change theory as the basis for our measure of impact for the evaluation of nursing home regulatory systems. In this case, "actual" and "expected" performance comparisons are replaced by comparisons of the control (or currently operating) and experimental (new) methods of nursing home quality assessment. The evaluation reported in Chapter 15 was one of several comparisons made. Another, not reported there, compared the number and severity of violations detected. The control method detected many more violations, but most were violations in paperwork reporting requirements. The experimental methods detected more severe violations. Moreover, the new method proved to be more effective in promoting actual improvements in quality.

Step 9. Examine sensitivity of actions to findings

Sensitivity analysis allows us to examine the practical impact of our findings. In this step, we asked decision makers to describe the various course of actions they would have taken if they had received specific evaluation findings. Then we asked them to consider what they would have done upon receiving different findings. Once we knew the threshold above which the actions would change, we calculated the probability that our findings could contain errors large enough to cause a mistake. Using the nursing home example, the evaluation might have revealed that one method of quality assurance was 5 percent more expensive than another. Decision makers might tell us that savings of 20 percent would induce them to change their decision. In this case, the decision makers would be asked identify a threshold, say 15 percent, above which they would change their decision. The evaluation team would then calculate the probability that the findings contained an error large enough to exceed the threshold. In other words, the team would state the chance that a reported 5 percent difference in cost is indeed a 15 percent difference in cost.

Sensitivity analysis allows decision makers to modify their confidence in the evaluation findings. If the findings suggest that the reported practical differences are real and not the result of chance, then confidence increases. Otherwise, it decreases.

References

Agarwala, R. R. 1977. "Why is Evaluation Research Not Utilized?" *Evaluation Studies Review Journal* 2: 27.

Alemi, F. 1988. "Subjective and Objective Methods of Program Evaluation." *Evaluation Review* 11 (6): 765–74.

Boruch, R. F. 1975. "On Common Contentions about Randomized Field Experiments." In *Experimental Testing of Public Policy; The Proceedings of the 1974 Social Science Research Conference of Social Experiments* by R. F. Boruch and H. W. Riecken. Boulder, CO: Westview Press.

Campbell, D. T., and J. C. Stanley. 1966. *Experimental and Quasi Experimental Design for Research.* Chicago: Rand McNally.

Cook, T., and C. Gruder. 1978. "Meta-evaluation Research." *Evaluation Quarterly* 2 (1): 5.

Cox, G. B. 1977. "Managerial Style: Implications for the Utilization of Program Evaluation Information." *Evaluation Quarterly* 1 (3).

Dickson, G. W., J. A. Senn, and N. L. Chervany. 1977. "Research in Management Information Systems: The Minnesota Experiments." *Management Science* 23 (9): 913–23.

Edwards, W., M. Gutentag, and K. Snapper. 1975. "A Decision-Theoretic Approach to Evaluation Research." In *Handbook of Evaluation Research,* edited by E. L. Streuning and W. Gutentag. London: Sage Publications.

Gustafson, D. H., and A. Thesen. 1981. "Are Traditional Information Systems Adequate for Policy Makers?" *HCM Review* (Winter).

Gustafson, D. H., C. J. Fiss, Jr., and J. C. Fryback. 1981. "Quality of Care in Nursing Homes: New Wisconsin Evaluation System." *Journal of Long Term Care Administration* 9 (2).

Gustafson, D. H., F. C. Sainfort, R. V. Konigsveld, and D. R. Zimmerman. "The Quality Assessment Index (QAI) for Measuring Nursing Home Quality." *Health Services Research* 25 (1): 97–128.

Heilman, J. G. 1980. "Paradigmatic Choices in Evaluation Methodology." *Evaluation Review* 4 (5): 693–712.

Moreland Act Commission. 1976. *Assessment and Placement: Anything Goes.* Prepared by the New York State Moreland Act Commission on Nursing Homes and Residential Facilities.

Polivka, L., and E. Steg. 1978. "Program Evaluation and Policy Developments: Bridging the Gap." *Evaluation Quarterly* 2 (4): 696.

Scheirer, M. A. 1978. "Program Participants, Positive Perception: Psychological Conflict of Interest in Social Program Evaluation." *Evaluation Quarterly* 2 (1).

Thompson, M. S. 1980. *Benefit-Analysis for Program Evaluation.* Beverly Hills, CA: Sage Publications. Watson, S. R. 1979. "Decision Analysis as a Replacement for Cost/Benefit Analysis." University of Cambridge, Department of Engineering.

Weiss, R., and M. Rein. 1970. "The Evaluation of Broad Aim Programs: Experimental Design, Its Difficulties and An Alternative." *Administrative Science Quarterly* 15 (1).

Wilderman, R. 1979. "Evaluation Research and the Socio-Political Structure: A Review." *American Journal of Community Psychology* 7 (1).

Wright, P. 1974. "The Harassed Decision-Maker: Time Pressures, Distractions, and the Use of Evidence." *Journal of Applied Psychology* 59 (5): 555–61.

Section III

Policy Support Tools

12

Determining Information Requirements

Policy information management systems (PIMS) for policymakers, executives, and planners must rest on a clear understanding of the decisions these individuals will make. Although most PIMSs are designed by studying the past and present decisions and functioning of the organization requesting the system, identifying future information needs by such extrapolation does not guarantee adequate decision support . Indeed, the cost of collecting, analyzing, and maintaining data on all potential issues is generally prohibitive, especially because some issues will probably not recur. But if history is not a good guide, what can inform us in choosing which information to provide—and when—to the policymaker? The answer to this question forms the bulk of this chapter. Briefly, the information should reflect the probability of needing a specific item of information; the cost of collecting, maintaining, and analyzing data; and the likelihood of having time to collect and analyze data before a decision must be made.

A related issue is whether information should be collected and analyzed before or after it is needed. In addressing the question of timing, the following trade-offs should be considered:

- *Relevance versus timely availability*. Data collected and analyzed after the need arises are more likely to be relevant to the decision maker's task, but this timing increases the chance that the data will not be available when needed.
- *Periodic or continuous data gathering*. Collecting data as the need arises allows analysts and decision makers to define precisely

what must be collected. Regular data collection often leads to data categories that are too narrow or too broad. Periodic collection—collecting data only when the need arises—allows us to assemble the most appropriate data but may not give us a basis for comparing trends. In general, the frequency of collection should increase along with the turbulence of the environment.

By its very nature, building a PIMS requires us to predict which issues a decision maker will face and to identify which data are essential to analyzing and resolving those issues. The temptation to collect data on every conceivable contingency must be tempered by a recognition that collection is expensive. In practically every case, the number of data items must be minimized. One method of doing so is described below.

Why Use a Methodology to Determine Information Needs?

The process of finding out exactly what information is needed is called the information requirements determination, or IRD. It's not an easy challenge to design a PIMS to meet the information needs and expectations of policymakers. Various obstacles prevent simple determination of information requirements. For example, users often will not differentiate true information needs from wishes (Taggart and Tharp 1977). Further, memory limitations affect the perception of need. Managers' information needs are usually determined by interviewing them, but if they cannot remember instances in which information was lacking for a decision, interviews will not be very useful. If the manager has a good memory for these instances, the analyst will have a good basis for specifying the information requirements. Otherwise, the list of information needs will be incomplete.

Various severe and systematic cognitive biases (IBM Corporation 1981) also impair the ability to define information needs. Users' lack of expertise is a common obstacle to formulating and stating true needs (Ellis 1982; Hogarth 1981). This is not to suggest that analysts can identify users' needs better than users themselves. Although analysts have tools to identify needs, they may be ignorant of the subject matter, so users and analysts must collaborate to determine information needs.

Despite its importance to developing a PIMS (Rockart 1979),

IRD is fraught with difficulties (Gustafson and Thesen 1981; Munro and Davis 1977). Two key steps in developing a list of information needs are identifying who knows what information will be needed to solve foreseeable problems (Alemi et al. 1984) and assuring that information requirements are relevant and valid (Mason and Mitroff 1973; Wack 1985). In addition, if the PIMS serves multiple users, all must agree on the information requirements (Dubois et al. 1982).

Dubois et al. (1982) have suggested three categories of difficulties in the IRD phase:

- A well-defined set of requirements does not exist or is unstable.
- The user is unable to specify requirements.
- The analysts are unable to elicit those requirements and/or evaluate them for correctness and completeness.

Many IRD approaches have been suggested to improve our ability to understand information needs, as in Taggart and Tharp (1977) and Yadav (1983). It seems that a task analysis followed by a detailed data flow analysis (Hira and Mori 1982; Mintzberg 1975) was most common in the past. Other approaches have included: asking the decision maker about needs (Huysmans 1970; Ross and Schoman 1977); deriving requirements from the existing information system (Valusek 1985); looking at strategic goals and concerns (Checkland 1981); and doing input-process-output analysis (Lundeberg 1979).

All of these approaches have appealing features, but each suffers from one or more of these weaknesses:

- A focus on past and present issues and information needs. This may be acceptable in a stable environment, but not in a dynamic one.
- A focus on a single issue, task, or decision. Again, the resulting information system could be useful in a highly repetitive operational setting but would be too narrow to deal with the diverse set of issues faced by most policymakers and planners.
- A focus on the information currently processed instead of on the issues. This technique may not significantly improve the quality of decision support.
- A focus on the decision makers' personal goals, which may be inconsistent with the organizational goals.

- A focus on observing users' behavior without seeking their insight or helping them examine their experience to assess needs more creatively.

Until now, IRD techniques have emphasized structuring need statements into formats compatible with computers, not on finding mechanisms to *formulate needs* from the user's perspective (Gustafson et al. 1982). While a few user-oriented IRD tools are described in the literature (King 1978; Mitroff 1974), the overwhelming majority of IRD tools developed in the past decade concentrate on structuring and representing the needs expressed by users rather than on helping users formulate their needs more effectively.

We believe the most critical problems with present PIMS construction are

- The black-or-white style of including or excluding information items, which omits intermediate options
- The separation of data collection from analysis—as if the value of data is independent of their analysis
- The assumption that only empirically observable events are data, and that experts' opinions are irrelevant to policy decisions.

Methodology

This chapter presents an alternative method of IRD which we developed to overcome the above weaknesses. The methodology starts by having a panel of experts generate a list of important upcoming issues and create a generic taxonomy of information items that are needed to deal with those issues. Then a panel of users selects a relevant subset of the issues and prioritizes the information items so they will best address the chosen issues. Finally, a resource allocation is created to govern the collection and analysis of information on the basis of the new priorities.

By developing a comprehensive list of upcoming issues, the methodology minimizes the dangers of determining future information needs by extrapolating from current needs. This type of "issue-driven" methodology also minimizes the potential for miscommunication between analysts and users. By using panels of experts and users rather than an individual from each group, the methodology minimizes the cognitive and behavioral limitations of having a single person define information needs. By using a consistent scale for prioritizing the information needs for all users, the methodology minimizes conflict over setting priorities. (See Table 12-1.)

The ultimate purpose of our methodology is to identify a core of generic information needs that will address the widest possible spectrum of issues. The product of our efforts is (1) a prioritized list of information items that are relevant to a large set of issues, and (2) a resource allocation plan listing which items of information will be collected at what time. The methodology comprises five steps and requires access to a panel of experts on issues for which the PIMS will be used.

Table 12-2 summarizes the methodology. Each step is discussed below and illustrated by a case study that shows how information needs would be set for a commission charged with setting mental health manpower policy.

Step 1. Identify policy issues

The first step is to identify and prioritize issues likely to confront the policymakers served by the information system. A number of mechanisms can be used for anticipating policy issues. Techniques like the Delphi, Nominal Group, or Integrative Group Process have been used to capture the opinions of groups of experts on the likelihood that issues will occur. "Experts" include users of the information (the policymakers) as well as outsiders with extensive experience in predicting the emergence of relevant issues. (See Chapter 11 and Chapter 3.) Systems analyses (to identify weaknesses in system operation), literature reviews, and observing the environment can also be helpful. In any case, the identification process should result in a list of policy issues along with priorities that reflect both the intrinsic importance (i.e., magnitude of the consequences) of the issue and the likelihood it will become salient and thus require a response (IBM Corporation 1981).

In the case study reported here, 48 experts in mental health policy from several states participated in a three-month-long Delphi study to identify and set priorities on key issues likely to face the mental health field in the next five years. The experts rated the issues in terms of probability of occurrence and intrinsic importance. The importance and probability estimates were multiplied to yield an expected value score for each issue. Later, the 37 members of a commission responsible for setting mental health manpower policy

Table 12-1 Features and Benefits of the IRD Methodology

Feature	*Benefit*
Comprehensive list of upcoming issues	Reduces danger of extrapolating from current needs; improves communication
Panels of experts and users	Minimizes cognitive and behavioral limitations of a single person
Consistent scale for prioritizing needs	Minimizes conflicts in setting priorities

Table 12-2 Summary of the Methodology

Step	*Objective*	*Performed by*
1	Identify issues	Panel of experts
2	Develop a generic taxonomy of information needs	Panel of experts and systems analysis team
3	Contextualize taxonomy	Systems analysis team
4	Prioritize information items	Panel of users
5	Develop resource allocation plan for data collection and analysis strategies	Systems analysis team

in one state (the users of the system) selected from that list six issues best reflecting the goals of their commission. For illustrative purposes, three of the six issues selected are shown in Table 12-3.

Step 2. Develop a taxonomy of information requirements

The panel of 48 experts was asked, in a second three-month-long Delphi study, to identify information items that would substantially reduce their uncertainty about how to deal with the issues identified in step 1. They were also asked to name the three most important policy issues they had recently addressed and to identify which information items made, or could have made, a difference in analyzing and resolving each issue.

Table 12-3 Three Upcoming Issues in Mental Health Manpower

1. How to reimburse the growing number of paraprofessionals who treat mental health clients?
2. What educational programs or policies should the state adopt to upgrade knowledge of mental illness among primary care physicians, who deliver many services to rural, mentally ill clients?
3. Many chronically mentally ill patients live in nursing homes that cannot provide adequate care. What policies should the state adopt to improve care for these patients?

After these questions were answered, the analysis team eliminated redundancies and combined the responses into a single list. The team also culled issue-specific components of each question and incorporated suggested clarifications. The remaining questions were grouped into categories on the basis of topic similarity. For example, all questions dealing with aspects of the clients were grouped in the "client population" category. Finally, the panel of experts reviewed and modified the full list of questions, which were now phrased generically, and relabeled the categories as necessary. The analysis team again consolidated the responses into a list of 69 general items arranged in ten categories. We believe the set of 69 information needs shown in Appendix 12-A are applicable to a wide range of issues and thus represent a valuable contribution to a master plan for developing many types of information systems. However, while the set of information items is already relatively small, it can be further focused by prioritizing its items.

Step 3. Make the information items specific to the topic

The third step is to modify the items so they are clearly relevant to the specific issue. In the case study, the intended users—the mental health manpower commissioners—focused on the chronically mentally ill, so the analysis team adapted the generic list to that population. One change was to substitute "chronically mentally ill client" for "client" in the descriptions. Another change was to substitute specific designations ("social worker" or "nurse practitioner") for the generic "personnel."

Step 4. Rate the importance of information requirements

No information system can collect and analyze data to address every possible information requirement. Even if one could guarantee that the format in which the data were collected was appropriate to a particular policy analysis, the cost would be prohibitive. It is imperative to establish priorities on what information is to be collected and analyzed.

In step 4 we ask the policymakers to rate the importance of each information item for specific policy issues. When finished, the information items rated highest are considered most responsive to the policymaker's needs. In this study, each commissioner was asked to choose the three issues on which they were most informed from the six previously selected (see Table 12-3 for a partial listing). (The analysis team ensured that each of the six issues was addressed by at least three commissioners.) Using the instrument shown in Table 12-4, each commissioner rated the relevance of all 69 information items for each issue they selected. The scores for each information item are shown in Appendix 12-A. The right side of the appendix lists the priority scores of each item. This priority score was calculated by averaging the scores given for each item by all analysts on all six issues. For example, item 1 in the appendix (how clients differ from each other) has an average score of 1.8 (out of a possible 3) across all six issues for all raters.

Step 5. Write a resource allocation plan for data collection and analysis

By plotting the average ratings (an item's importance across all issues) against the range of those ratings (i.e., how consistently important it was across issues), we determine what direction our resource allocation should take. The following four sets of information show how we can establish categories of information based on data we have collected so far.

Set	*Name*	*Characteristics*	*Important to*
1	Essential information	High average importance and low range	All issues

Table 12-4 Instrument to Collect Priority Ratings on All Information Items

Issue 1: The growing role of paraprofessionals in both treatment and maintenance of mental health clients has necessitated considerations about how the professionals should be reimbursed.	Assign a score between 1 and 3 to each information item: 3—Absolutely essential to address the issue. 2—Nice to have but not essential to address the issue. 1—Not needed to address the issue.

Information Item 46:

What do we know about the *different* funding mechanisms for allocating financial resources to paraprofessionals?

Rating: ______________________________

Comment: ______________________________

____

2	Rapid collection and analysis	High average importance and high range	Most issues
3	Periodic information	Moderate importance and high range	A few issues
4	Low priority	Low importance and low to moderate range	No issues

The essential information set. Items in this set have high average importance and a small range (denoting uniform importance across issues). These items should be collected and analyzed regularly so a policymaker can access them immediately when needed. These information items are found in the lower right of Figure 12-1. In this case study, the essential information set includes 14 items consisting of a description and analysis of:

Item	*Data item*
5	How expensive is it to provide services that fully meet clients' needs?

Figure 12-1 Average and Range of Perceived Information Importance Strategies for Data Collection and Analysis

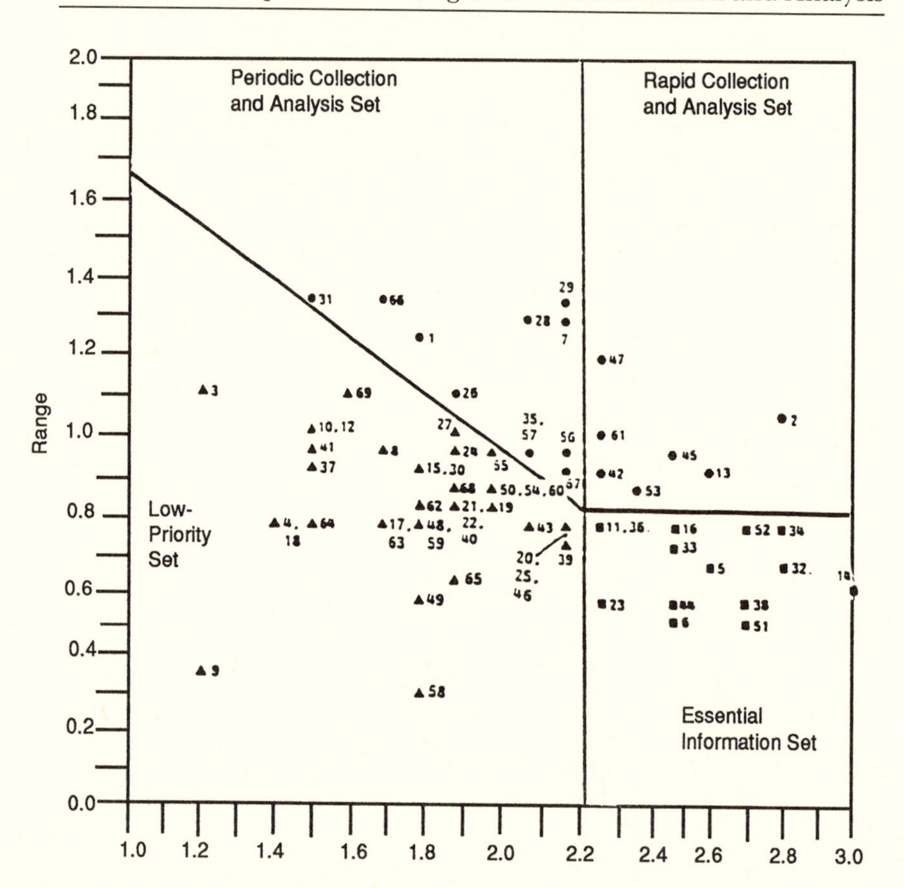

6	What kind of clients need what kind of services?
11	What services are actually used by each type of client, and how satisfied is the client with them?
14	How are services coordinated?
16	Are there gaps in the system's ability to meet client needs and expectations?
32	What resources are required to meet clients' service needs?

33	What are the total funds available for serving the client?
34	How are funds allocated?
36	What are the constraints on funding?
38	Are additional sources of funding available?
44	What is the existing system (e.g., client flow, personnel, facilities)?
51	What is the cost of caring for different clients within the current system?
52	What is the cost of providing resources to meet all the needs of the clients?

The practice of routinely providing these descriptions and analyses has interesting resource and information system implications. For example, items 6, 14, 16, 32, and 44 certainly belong within the broad framework of policy analysis, yet they require systems and decision analyses not typically found in computerized information systems. Items 5, 33, and 51 are cost analyses, at least two of which require modeling and simulation, not just simply reporting the data. Hence, the character of the policy information management system needed here is likely to be quite different from the typical information system that supports policy analysis.

The rapid collection and analysis set. Items in this set are, on the average, only slightly less important than those in the essential information set, but they have a larger range, indicating that their relevance across issues varies substantially. These information items are likely to be needed on short notice for some but not most policy issues. Rather than collecting (and analyzing) data that may never be used, the best use of resources here may be to construct a data collection, modeling, and analysis system that can be quickly activated. The system should identify who will collect the data; where, from whom, and within what period it will be collected; and what instruments will be used to do the collecting. The models and files for analyzing the data should be specified, and personnel qualified to implement the analysis should be identified. In the case study, the PIMS selected this rapid collection data set:

Item	*Data item*
2	The size and character of the client population
13	The number, type, and distribution of services provided
42	The organization of the existing system
45	The monitoring and evaluation methods used with the system
47	Options to the existing system
53	The costs of implementing alternatives to the existing system
61	Legislative and regulatory changes needed to alter the system

This set includes two types of information. Items 2, 13, 42, and 45 require more elaborate description of the existing system compared to item 44 in the essential information set. Items 47, 53, and 61 deal with alternatives to the existing delivery system.

The periodic information set. These information items are unimportant for most policy issues but vital to a few. The items cannot justify regular data collection or analysis, and might not even justify detailed planning. The best strategy may be to allocate resources for special studies (perhaps conducted by vendors). In the case study, the periodic information set consists of 11 items:

Item	*Data item*
1	How clients differ from one another
7	Who gets what priority for different types of care
26	How different providers serve the underserved areas
28	Constraints specifically limiting substitutability of providers
29	The existing provider certification and education systems
31	Training that might be needed by families and the community
35	Possible funding sources and methods of obtaining funds

56	Legislative/regulatory constraints on the existing/ optional systems
57	How legislation defines responsibility for the client population
66	How and whether to mobilize power groups
67	How the community is affected by the care delivery system

Many of these (28, 31, 35, 56, 66, and perhaps 7 and 29) seem relevant primarily if a change is planned in the system, so the nature of the information needed would depend on the options under consideration. The information required is not likely to be present in a typical information system. In addressing this set, the PIMS should find resources to conduct quick response studies that draw on the judgments of experts rather than empirical data.

The low-priority set. Information items in this set have low average importance and moderate to low range and would only rarely be used in most policy analyses. Even when they are used, they would not typically be crucial to the analyses. We suggest that no planning or preparation be directed toward these items. The following items are examples of the low-priority set:

Item	*Data item*
4	Episodes of illness: frequency and responsiveness to treatment
9	The role of the family in caring for client
46	What knowledge is needed to correct the problem
58	How existing laws and regulations duplicate each other
64	The position, attitude, and agenda of affected interest groups

Items 4, 9, 46, and 58 require a level of detail that may not be needed in the selection and design of policy planning, although they may be important during implementation. It surprised us that item 64 received such a uniformly low score because interest groups typically have a great effect on policy formation. Possibly policymakers felt they already knew the positions of these groups and therefore needed no further information.

The boundaries between sets should be drawn according to the availability of resources to collect and analyze the data, and by comparing the cost of collecting data in advance with the cost of collecting it as needed. In the case study, resource limitations had not been determined, so the boundaries shown in Figure 12–1 were set by finding "natural" clusters of information items (numbers in Figure 12-1 refer to the items in Appendix 12-A).

Discussion

The methodology we have discussed provides a means with which to identify a core of information items that will support the analysis of a wide range of issues. It also provides a means of determining which items to collect and analyze and what priority to assign them. While some data should be routinely collected and analyzed, in other cases resources should be invested not in *doing* but in *planning* the collection and analysis. Finally, this analysis suggests that much of the information needed by policymakers can come from the minds of experts rather than from observing and documenting events or facts.

The methodology requires that the analysis team identify and gain access to a group of experts who will generate and update a list of future issues or potential problems in the field. The selection of experts is crucial since their collective vision will drive the design of the PIMS. It is also important that the panel be available on a regular basis (the frequency depends on the degree of environmental turbulence) to determine which issues are emerging and whether they require additional items of information. It is our experience, however, that the generic taxonomy tends to be highly robust and that, while issues come and go, the taxonomy seldom changes.

The average importance and range of items, on the other hand, may change, so they should be reassessed periodically. For example, in this era of tight budgets, information items that deal with costs, resources, and funding are usually important. And because the boundaries of the four information sets depend on resources, changes in resources for collecting and maintaining the information systems will alter the boundaries. A loss of resources indicates that we should adopt a stricter definition of importance, so the low-priority set would grow at the expense of the other three.

Certain refinements could improve the accuracy of the ratings. For instance, it might make sense to use an importance rating scale with more benchmarks (e.g., 1 to 7 instead of 1 to 3) and to use the variance rather than the range to classify the items. Also, a different subset of issues and a different set of commissioners could be used to test the stability of the ratings.

In some cases, expert judgments will suffice as the source of information. In other cases, empirical data would be useful if available, but it is unrealistic to collect a great deal of empirical data on short notice. Fortunately, you can draw upon the judgments of experts as surrogates for this data (see Chapter 16 for an example of this process).

The principal idea behind this methodology is to create in advance a flexible information system that includes a core of critical data, an ability to rapidly access sources of other data as new issues arise, and the resources to "hire out" special studies in rare cases.

Appendix 12-A. Generic Taxonomy of Information Needs for Setting Mental Health Policy and Their Average Importance and Range

Category 1. Client population

Number	*Information item*	*Rating*
1.	How clients differ from each other?	1.80
2.	The size of each of the different client populations.	2.80
3.	The problems caused by the client's illness and how serious these problems will be.	1.20
4.	Do the frequency of episodes change under different treatments?	1.40
5.	The disparity between where the clients come from and where they go to receive services.	2.60
6.	What kinds of clients need what kind of services?	2.50
7.	Who receives priority for care?	2.20
8.	The characteristics of a client who is treatable and those of one who needs maintenance assistance.	1.70
9.	The family role in caring for the clients.	1.20
10.	The clients who need services but who do not utilize them.	1.50

Category 2. Utilization of services

11.	The services actually used by each type of client and the clients' satisfaction with them.	2.30
12.	The services needed by the family of the client.	1.50
13.	The services and how they are/should be distributed.	2.60
14.	The coordination of services.	3.00
15.	The eligibility criteria for receiving different services.	1.80

16.	The gaps in the system's ability to meet client needs and expectations	2.50
17.	The utilization patterns of different clients and services	1.70
18.	Why some services are utilized differently than others?	1.40
19.	The potential demand for different services.	2.00
20.	The degree to which different services are interchangeable.	2.20
21.	The services available through the client's family.	1.90
22.	The extent to which client needs are being met.	1.90

Category 3. Skills/training

23.	The kinds of skills needed to manage client's behavior.	2.30
24.	The kinds of skills available and their distribution.	1.90
25.	The number, location and training of professionals delivering direct patient care.	2.20
26.	How different personnel are to serve the underserved areas.	1.90
27.	The sources or skills and their flexibility.	1.90
28.	The constraints that limit the use of substitutes to compensate for the shortage of skills.	2.10
29.	The current educational and certification processes.	2.20
30.	The internal training programs that should be advocated.	1.80
31.	The kinds of training the family and the community need.	1.50

Category 4. Resources

32.	The resource requirements for meeting the client needs.	2.80
33.	The total funds available for serving the client?	2.50
34.	How funds are allocated?	
35.	What are the funding sources and percents of each source?	2.10
36.	The limits or constraints on funding.	2.30
37.	The likelihood of continued financial support.	1.50
38.	Additional sources of funds.	2.70

Category 5. Organization and operation

39.	The system and what is outside the system. (What parts of the system treating the clients do we have control over?)	2.20
40.	What services are provided outside governmental control?	1.90
41.	The existing system readiness for change.	1.50
42.	The existing organization of the system.	2.30
43.	The existing management of the system.	2.10
44.	The existing flow within the system.	2.50
45.	The monitoring and evaluation methods of the system.	2.50

Category 6. Options

46.	What is the technology/knowledge needed to correct problem?	2.20
47.	What options exist that would reduce the problems.	2.30
48.	The advantages and disadvantages of each option.	1.80

49.	How each option would function (provide scenarios).	1.80
50.	The effectiveness of each option in other settings.	2.00

Category 7. Cost of operation

51.	The cost of caring for the client in the present situation.	2.70
52.	The cost to provide services that actually meet all the needs of the client population.	2.70
53.	The cost to implement/operate each alternative.	2.40
54.	Who bears the cost of providing services?	2.00
55.	Who would bear the cost under each alternative system?	2.00

Category 8. Legislation

56.	How existing legislation, administrative rules and governmental policies constrain the ability to provide services in other ways.	2.20
57.	How existing legislation, etc., defines/assigns legal responsibility for the client population.	2.10
58.	How existing laws duplicate, interact or conflict.	1.80
59.	The experience and implications of litigation that has challenged existing legislation.	1.80
60.	The potential costs and benefits of deregulation.	2.00
61.	Legislative policy or other regulatory changes needed to implement a proposed solution.	2.30
62.	How legislation currently being drafted/ considered is likely to affect the problem.	1.80

Category 9. Power groups

63.	Who is being hurt, or helped by the proposed option.	1.70
64.	The position, attitudes and the agenda of interest groups.	1.50
65.	Coalitions that should/could be built among interest groups.	1.90
66.	How can power groups be mobilized? How have they been used elsewhere in similar situations?	1.70

Category 10. Social costs

67.	How the community is affected by the current system.	2.20
68.	The attitude and the tolerance level of the community regarding the client.	1.90
69.	How the community will be affected by the alternative systems for dealing with the problem	1.60

References

Alemi, F., W. L. Cats-Baril, and D. H. Gustafson. 1984. "Integrative Group Process." Center for Health Systems Research and Analysis Working Paper, University of Wisconsin-Madison.

Checkland, P. 1981. *Systems Thinking, Systems Practice.* New York: John Wiley.

Dubois, E., J. P. Finance, and A. VanLamsweerde. 1982. "Towards a Deductive Approach to Information System Specification and Design." In *Requirements Engineering Environments,* edited by Y. Ohno. Tokyo: Ohmsha.

Ellis, H. C. 1982. "A Refined Model for Definition of System Requirements." *Database Journal* 12 (3): 2–9.

Gustafson, D. H., and A. Thesen. 1981. "Are Traditional Information Systems Adequate for Policymakers?" *Health Care Management Review* (Winter): 51–63.

Gustafson, D. H., W. Cats-Baril, and F. Alemi. 1982. "Predicting The Emergence of Policy Issues: Bayesian and MAU Models." Center for Health Systems Research and Analysis Working Paper, University of Wisconsin-Madison.

Hira, H., and K. Mori. 1982. "Customer-Needs Analysis Procedures: C-NAP." In *Requirements Engineering Environments,* edited by Y. Ohno. Tokyo: Ohmsha.

Hogarth, R. 1981. *Judgment and Choice.* New York: John Wiley.

Huysmans, J. H. B. M. 1970. "The Effectiveness of the Cognitive Constraint in Implementing Operations Research Proposals." *Management Science* 17 (1): 92–104.

IBM Corporation. 1981. *Business Systems Planning—Information Systems Planning Guide,* Application Manual, GE20-0527-3.

King, W. R. 1978. "Strategic Planning for Management Information Systems." *MIS Quarterly* 2 (1): 27–37.

Lundeberg, M. 1979. "An Approach for Involving Users in the Specifications of Information Systems." In *Formal Models and Practical Tools for Information Systems Design,* edited by H. J. Schneider. Amsterdam: North-Holland.

Mason, R. O., and I. I. Mitroff. 1973. "A Program for Research on Management Information Systems." *Management Science* 19 (5): 475–87.

Mintzberg, H. 1975. "The Manager's Folklore and Fact." *Harvard Business Review* 53 (4): 49–61.

Mitroff, I., and T. Featheringham. 1974. "On Systematic Problem Solving and Error of the Third Kind." *Behavioral Science* 19 (3): 383–93.

Munro, M. C., and G. B. Davis. 1977. "Determining Management Information Needs—A Comparison of Methods." *MIS Quarterly* 1 (2): 55–67.

Rockart, J. F. 1979. "Critical Success Factors." *Harvard Business Review* 57 (2): 81–91.

Ross, D. T., and K. E. Schoman. 1977. "Structured Analysis for Requirements Definition." *IEEE Transactions on Software Engineering* SE-3 (1): 6–15.

Taggart, W. M., and M. O. Tharp. 1977. "A Survey of Information Requirements Analysis Techniques." *Computing Surveys* 9 (4): 273–90.

Valusek, J. R. 1985. "Information Requirements Determination: An Empirical Investigation of Obstacles Within an Individual." Unpublished doctoral dissertation, University of Wisconsin-Madison.

Wack, P. 1985. "Scenarios: Uncharted Waters Ahead." *Harvard Business Review* 63 (5): 72–89.

Yadav, S. B. 1983. "Determining an Organization's Information Requirements: A State-of-the-Art Survey." *Data Base* 14 (3): 3–20.

13

Setting Priorities for Policy Analysis

Previous chapters have discussed the importance to organizations of monitoring and adapting to environmental change. Organizations always have difficulty adapting to change, but the inevitable problems are multiplied if a surprise forces them to adapt on short notice. Time pressures almost inevitably increase the likelihood of technical or political errors.

Millions of dollars are spent attempting to predict changes, in the state of the economy, the attitudes of voters, or other vital matters. To policymakers, one of the vital environmental influences is a change in the attitude of a key constituency. For instance, a family planning service will likely be quite concerned if a parents association suddenly decides to oppose its policy of providing birth control devices to adolescents without notifying parents. If an organization knows such change is impending, it can initiate timely efforts to adjust.

But predicting the emergence of issues is only half the battle. Because of resource and time limitations, the policymaker and the PIMS can analyze only a handful of issues in any detail. Thus, we advocate a two-step process to forecast environmental changes: first, identify the set of potential issues; and second, prioritize issues within that set for more extensive examination. A summary of issue identification follows; then the remainder of this chapter will address issue prioritization.

Several well-developed issue identification processes (such as Delphi, nominal group techniques, and the integrative group process) have already been discussed (see Chapter 5). The critical point in using these processes is to seek the people most qualified to identify issues, such as clients, administrators, researchers, journalists, and pollsters. Using formal group processes should not replace other means of identifying issues, such as reading newspapers and journals or listening to legislators and their aides. We do suggest that decision makers periodically use a group process to create an exhaustive list of issues, complete with a brief analysis of each one. The list of issues can also serve as the starting point for a later issue identification session, in which the decision makers may just update the existing list or create an entirely new one. We now turn to using decision-analytic techniques to prioritize issues.

Issue Prioritization

Many group processes can be used to prioritize issues, generally by asking judges to select a small number of issues (often seven) which they, as individuals, consider important. The judges then rank order these issues by assigning the value 7 to the most important issue, 6 to the next most important, and so on. By simply adding the votes, we obtain an overall ranking of issue priority. This voting method allows us to add votes directly.

Significant variation among judges in assigned priorities can be the result of three causes: legitimate disagreement about priorities, differing interpretations, and rating priority on differing dimensions.

If the experts legitimately disagree on the priorities, you have a conflict that may be addressed by techniques discussed in Chapter 14.

If the experts interpret the issues differently, several procedures can be used to increase the judges' understanding and allow priorities to be set more accurately. Probably the most common approach is the talk-prioritize (TP) procedure, wherein experts discuss the alternatives before voting. Because discussion before voting encourages the dominating person to force consensus, additional group processes may be needed to compensate.

Facilitators have at least two other options for dealing with differing interpretations. One is a process used periodically in the Delphi technique, called prioritize-feedback-prioritize (PFP). In PFP, judges establish a preliminary set of priorities and share them anonymously with fellow judges. Sometimes each judge is asked to justify on paper the priority assignments. After reviewing the priorities of all judges (and possibly the reasons for them), each judge is asked to reprioritize. These values are added to yield the final score.

A similar option is prioritize-talk-prioritize (PTP), in which the time for face-to-face discussion is limited to a few minutes. This reduces the pressure to change opinions but retains the pressure to understand each other. After the discussion, each judge silently and secretly prioritizes again. As with PFP, priorities are added to yield the final score.

Experiments to evaluate these processes have found that PTP performs significantly better than PFP, TP, or prioritizing without feedback. PTP's superiority may stem from the compulsion for each judge to take a position on the issue before justifying that stand. (PTP differs from PFP, which lacks a face-to-face discussion of priorities, and in which feedback usually consists of showing how, and briefly why, the other judges score the various items.) A judge with an established position in PTP may be curious about why others disagreed and may tend to reconsider his or her position before voting again. Because the final vote is anonymous, judges need not change their votes in public. One concern with PTP is that judges may become defensive and entrenched in poorly considered priorities. The group facilitator can guard against this by encouraging panelists to explain why they voted as they did and limit (but not prohibit) debate on the merits of that explanation.

A third cause of differences in priorities lies in differences in importance judges attach to the dimensions determining priorities. Our experience suggests that two dimensions commonly determine which issues will be addressed: political ripeness and intrinsic importance.

Political Ripeness

An issue is politically ripe if it has developed to the stage where it attracts attention in the form of public interest, media exposure, and/or pressure from interest groups. Any of these factors can create a demand for action on the issue. Pertinent catastrophic events can cause issues to ripen in a hurry, forcing the policymaker and policy analyst into a crisis management situation. However, issues usually ripen gradually. Sensitizing a policy analyst to indicators that an issue is ripening will enable the analyst to begin addressing the issue early on in its life cycle, and thus lessen the time constraints associated with public demand for an issue's resolution. (See Chapter 3.)

Our model of issue ripeness was developed by convening a panel of respected legislators, journalists, and administrators from the health and social service fields. We used the integrative group process described in Chapter 5 to:

- Identify which factors should be measured to gauge issue ripeness
- Identify possible levels for each factor
- Assign values to each level
- Assign weights to each factor

A MAV model format was used to aggregate these data into an indicator of ripeness. Table 13-1 lists all the factors generated by the panel. A full description of each factor is found in Appendix 13-A.

Table 13-2 presents the actual model. The first column presents the factor weight; only 12 factors were weighted by the panel. The second column lists the factor number from Table 13-1. The third column lists the value on each factor level. The final column lists the levels on each factor. Note that each factor has one level assigned 0 and another 100. These scores were assigned to the lowest and highest levels associated with each factor. The actual issue of interest may not fit any of these descriptions perfectly and will need to be placed somewhere in between. For instance, on factor 5 (priority assigned by bureau-level policymakers) an issue may be of moderate priority. It doesn't fit as low (value 50) or high (value 100). A judge will need to decide what value to assign (e.g., 75).

According to Table 13-2, the panel felt the most important factor (1) was "legislative activity" about an issue. The panel seemed to feel that the introduction of a bill with reasonable chance of passage would strongly substantiate the argument that the issue is ripe. A legislative call for study of the issue (factor 3) indicates possible ripening. On the other hand, a decisive defeat of legislation indicates the issue is not ripe. Our expert panel believed that because elected representatives are particularly sensitive to the timing of issues, passage by a legislature indicates their consensus that the issue is timely or, in our terms, ripe.

The active involvement of "special interest groups" also seems an important indicator of ripening. Interest groups, by their nature, have a way of calling public attention to, and mobilizing support for,

Table 13-1 Factors That Predict When a Potential Issue Will Ripen

1. Legislation has been introduced.
2. A legislator is interested in the potential issue.
3. Legislature has requested a study.
4. Potential issue might surface as campaign issue.
5. Priority given to potential issue by bureau-level policymakers.
6. Priority given to potential issue by county-level policymakers.
7. Priority given to potential issue by bureau-level program people.
8. Priority given to potential issue by county-level program people.
9. Priority given to potential issue by public and private providers.
10. An alternative, implemented in other states, is being advocated in this state.
11. Criticism about the potential issue is coming from professionals outside the system that is supposed to deal with the issue.
12. Other systems are dealing with issues that are linked to the potential issue.
13. A technological development has impact on the potential issue.
14. Related issues in the same political arena are ripening.
15. Potential issue is repeatedly identified through existing, formal grievance mechanisms.
16. Potential issue has been addressed at the federal level.
17. Potential issue has been addressed by the governor or secretary responsible for the system in which the potential issue is emerging.
18. Potential issue is more likely to occur because of a change in funding.
19. Court cases may affect the potential issue.
20. Special groups have shown concern about the potential issue.
21. The public has been exposed to the potential issue by mass media.
22. First-line, grass-roots field workers identify the potential issue as important.
23. Potential issue is generating informal discussions among lay persons or policymakers.

an issue. The more resources they commit to an issue and the more influential the groups, the more likely the issue is to ripen.

The third factor of key importance is activity in the courts. If court cases have already fundamentally changed the issue, we can consider this an indication of ripeness, even if the court was in an-

Table 13-2 Index for Rating Ripeness

Factor Weight	*Factor Number from Table 13-1*	*Value*	*Factor Level Scale*	*Factor Level*
.15	1 (Legislation)	0	a.	Has been introduced, decisively defeated and dropped
		33	b.	Has not been introduced
		66	c.	Has been introduced, was given serious consideration, but eventually defeated
		100	d.	Has been introduced with reasonable chance of passage
.14	20 (Interest groups)	0	a.	Have been lobbying strongly against the issue
		25	b.	Have not been involved in the issue
		50	c.	Have shown substantial concern for the issue
		100	d.	Have been lobbying strongly for the issue
.13	19 (Litigation)	0	a.	Cases that dealt with the issue have been thrown out
		100	b.	Cases have been ruled in favor of the issue
.10	17 (Status)	0	a.	Has not raised the issue
		100	b.	Has raised the issue
.08	5 (Bureau priority)	0	a.	It is not considered an issue at all
		50	b.	The issue has low priority
		100	c.	The issue has high priority
.08	6 (County priority)	0	a.	Does not consider the issue at all
		50	b.	The issue has low priority
		100	c.	The issue has high priority
.08	21 (Media attention)	0	a.	Has not made the issue visible
		100	b.	Has made the issue visible
.06	4 (Campaign issue)	0	a.	Is not a campaign issue
		100	b.	Is a campaign issue

Continued

Table 13-2 Continued

Factor Weight	*Factor Number from Table 13-1*	*Value*	*Factor Level Scale*	*Factor Level*
.05	10 (Promptly in other states)	0	a.	Do not consider it an issue at all
		50	b.	Have assigned a low priority to the issue
		100	c.	Have assigned a high priority to the issue
.05	3 (Legislative study)	0	a.	A study requested by the legislative branch has been dropped
		50	b.	A study has not been requested by the legislative branch
		100	c.	A study requested by the legislative branch has been initiated
.04	11 (Professional attention)	0	a.	There has not been significant professional criticism of the existing system as it relates to the issue
		100	b.	There has been significant professional criticism of the existing system as it relates to the issue
.04	14 (Related issues)	0	a.	The issue is not closely related to other high-priority issues being considered now
		100	b.	The issue is closely related to other issues currently being addressed with high priority

other state. However, if a court clearly signals that it is not interested in the issue, the issue moves away from ripeness.

Key leaders in the executive branch can also indicate ripeness, especially if the interest comes from the chief administrative officer (such as the head of a state department of health and social services). When such an executive publicly raises the issue, it indicates the timing is right to attack the issue. However, the executive-branch official directly responsible for the system and city and county officials also should be watched for indications of ripeness.

Media interest is an important indicator of ripeness, especially if a feature story or series is published on the issue. The panel felt the media, like special interest groups, can hasten the ripening process by calling public attention to the issue.

Issues that are addressed directly in a political campaign can be ripening. However, because of the transient nature of campaigns, the panel did not give this factor as much weight as might be expected. Issues raised before an election can be dropped after the contest is settled.

According to our panel, two groups that seem to have little influence or association with ripeness are the professionals the issue affects and the policy analysts involved with the issue. Vested interest may cause professionals to overstate the importance of "their" issues. But the complexion changes if a group of professionals has an effective interest group touching on the issue. The lack of association between policy analysts and ripeness may result from the tendency of analysts to focus on a few issues, and possibly to do so too late in the game. Policy analysts may be better as indicators of current ripeness than as predictors of future ripeness.

The model generated by the expert panel for determination of issue ripeness could have several uses for readers involved with policy analysis in general and social services policy in particular. You may want simply to review the factors, factor levels, and factor weights to see how experts in this field feel about these predictors. You may wish to judge interest among legislators or to identify the positions and activities of interest groups. Regardless of the extent to which this model is taken, an understanding of the inherent thought process will help you gain proficiency in examining issue ripeness.

This ripeness model has been used to score the relative ripeness of issues in several experiments. In one experiment, the judgment of a panel of mental health policymakers was compared to the model's evaluation of the ripeness of 21 issues. The average correlation between the panel and model ratings was .77, suggesting that the model has enough validity to warrant its discussion just described.

The issue ripeness model is a multiattribute value (MAV) similar to those discussed in Chapter 7. Each issue is assigned a value on each factor. The factor level score is weighted by the factor importance, and those weighted scores are added. An example of the use of the ripeness model follows. A hypothetical issue is described in terms of the levels shown in Table 13-3. Values were taken from Table 13-2.

Table 13-3 Example Ripeness Model

Weight	*Factor Level*	*Value*
.15	Has not been introduced as legislated	33
.14	But interest groups have shown concern	100
.13	And no court cases have addressed the issue	15
.10	Secretary, governor, city/county have not raised the issue	0
.08	But the mass media has run a series	100
.06	It is not a campaign issue	0
.05	And a legislative study has not been requested	0
.04	Yet professionals consider the issues central	100
.05	Policy analysts assign it a low priority	50
.04	It stands alone	0

The score for this example issue would be 35.40, calculated as follows:

$$.15\,(33) + .14\,(100) + .13\,(15) + .10\,(0) + .08\,(100) + .06\,(0) + .05\,(0) + .04\,(100) + .05\,(50) + .04\,(0) = 35.40$$

By itself this number has no particular meaning, but the scores of several issues can be compared to find their relative ripeness.

The above issue ripeness model has several advantages. It forces the judges to agree on the relative importance of each factor in the model. Also, the fact that all judges use the same list of criteria assures consistency across judges (they all pass judgment on the same basis). If judges are explicitly forced to accept a set of prioritizing criteria and weights for them, arguments are seen more easily from a balanced perspective.

A second advantage is that the prioritizing model requires each judge to use each criterion in assessing an issue's ripeness, ensuring that no important criteria are overlooked, as is commonly the case. Often advocates of single issues use one criterion to promote favorite issues at the expense of others. The model thus gives us consistency of judgment within each judge as they evaluate several issues, and across judges as they evaluate the same issue.

A third advantage is that the model informs issue advocates of the reasons their issues received the scores they did. Too often, the process of prioritization is a mystery to those who did not participate. Using a formal prioritizing model should help clear up some of that mystery.

Intrinsic Importance

Issues are intrinsically important if they have a wide-ranging impact on individuals and society. For example, in U.S. politics, the state of the economy and the condition of the national defense may be addressed by policy analysis groups at any time. Nevertheless, intrinsic importance is usually modulated by ripeness; that is, until an issue becomes politically ripe the results of policy analysis are commonly set aside.

Table 13-4 describes a model to measure intrinsic importance. Each factor is described and weighted. Note that the panel of planners, providers, and change theorists that developed this model felt the direct impact of the problem on the health of the individual was the key factor. In other words, the more likely a problem is to kill or severely disable its victim, the worse it is. Health problems with mild or transient effects rate lower.

The first factor is *seriousness of problem,* a measure of the expected impact of the problem on an individual.

The second factor is *size of problem,* measured in terms of number of people directly affected. This factor can counterbalance factor 1. By assigning a high number to this factor, we can give a high priority to a problem that has a moderate impact on a large number of people.

The third factor addresses avoidable *costs,* the amount of savings available from an effective resolution of the problem. Interestingly, the experts rated avoidable cost the least important factor in establishing priorities.

The fourth factor is *trends,* which addresses the future of the problem if intervention remains at the current level. For example, if heart disease is now decreasing, other issues may deserve more attention even though heart disease is the nation's leading killer.

The fifth factor is the problem's impact on *community concern.* The rationale for this is that many health and social service problems have indirect and subtle impacts. For instance, perhaps the average quality of patient care would improve if a community hospital were closed and its patients treated elsewhere. But if the community hospital is the largest employer in town and a source of local pride, the damage that closing it will cause to community stability should be considered.

The sixth factor, *equity,* addresses the rather controversial notion that health care is a right, not a privilege. The judges assigned

Table 13-4 Criteria for Judging Intrinsic Importance

Weight	*Factor*	*Description*
24%	*Seriousness:* How badly does the problem impair health?	Consider both the severity and duration of the health problems. The more likely a problem is to kill or permanently disable its victims, the worse it is. Health problems with relatively temporary or mild effects rate lower.
20%	*Size:* How many people are being hurt by this problem?	Consider primarily the number of people whose health is impaired. Secondarily consider the number of people who are at risk and the number who are indirectly affected.
17%	*Costs:* How much does this problem cost the community?	What is the avoidable dollar cost for health, lost productivity and wages, wasted resources, etc?
15%	*Trends:* What will happen to this problem in the foreseeable future if no action is taken?	Consider first whether the problem is getting worse or better. Then consider whether it is now being addressed. A problem that is getting worse and/or for which no corrective actions are planned rates high; a problem that is getting better and/or is already being attacked on many fronts rates low.
14%	*Community concern:* How upset is the community about the problem, and how willing is it to support an effort to ameliorate it?	Consider primarily the attitudes of the public and consumers; to a lesser extent, consider attitudes of health care professionals and local officials.
10%	*Equity:* How much does this problem limit equal access to health care? How much does the problem prevent people from obtaining care?	Consider primarily financial barriers, secondarily geographic and attitudinal barriers.

this factor a weight of 10 percent, a value that might need to be reevaluated as citizens express greater concern for uninsured populations.

Unlike the ripeness model presented in Table 13-2, the intrinsic importance model does not present a series of alternative levels. Rather, it advises regarding what to consider in evaluating each factor that contributes to the issue's intrinsic importance. Each issue should be evaluated on all factors and scored between 0 and 100 for each one. Then calculate the intrinsic importance by multiplying the score of each factor by its importance weight and adding across all factors. Table 13-5 presents instructions for setting priorities. Table 13-6 presents sample problems and score sheets.

Use in Priority Setting

The process of setting priorities has three steps: drawing up a list of issues, scoring these issues on the intrinsic importance and ripeness scales, and choosing a subset of those issues for further analysis and close monitoring. Issues that are intrinsically important and ripe require in-depth analysis. Issues that are intrinsically important but not ripe can be actively and regularly monitored to see if their ripeness status has changed. Issues that are ripe but intrinsically unimportant may require instant attention but not necessarily in-depth analysis. Finally, unripe issues with low intrinsic importance can be passively monitored from time to time to determine changes in status. By using the priority-setting model in this way, a policymaker can rationally decide how to allocate the resources of a policy analysis team.

Another use of the model is to identify factors that contribute to an issue's ripeness and/or intrinsic importance. By monitoring changes in those factors, the analysis team can predict changes in the status of issues.

Table 13-5 Instructions for Setting Priorities on Intrinsic Importance

You will need the score sheets. Table 13-6 is a score sheet for the factor "equity"; score sheets for other factors (seriousness, size, costs, trends, and community concern) are similar. You will also need the list of issues generated in the issue identification phase.

1. Read the descriptions of a few of the issues to get a feel for how they are written. You will note that some facts and judgments are listed for each issue. You may accept or reject this information when you complete the worksheets requesting your judgments.

2. Begin with the score sheet for equity. Read the definition of the criterion and the examples at the top of the page.

3. Compare a problem statement to the extreme examples described on the score sheets. Be careful to consider only the one criterion! For instance, on working score sheet F (equity), don't consider how many people have the problem. Consider only the criterion of equity.

4. Assess the importance of the problem relative to the extremes, and assign an appropriate score between 0 and 100. A rating of 100 indicates that the problem is as extreme as the example; a problem that is half as serious gets 50. A rating of 0 means that the criterion is irrelevant to this problem.

5. Record the rating in the space provided on the score sheet. Asterisks on the sheet mean that the issue statement contains relevant information to help you; if there is no asterisk, your answer is strictly a matter of judgment.

6. Complete the score sheet by rating the rest of the problems against the criterion.

7. Complete the other score sheets using the same procedure.

Table 13-6 Score Sheet F "Equity"

How much is equal access to health care limited by the issue you are rating? Rate the problems on the extent to which financial, geographic, and attitudinal barriers limit equal access by all to health care. Consider primarily the financial barriers, secondarily geographic and attitudinal barriers. An issue marked by a highly inequitable distribution of access to health resources (many people are excluded) rates 100. An issue with no inequities in access rates 0. Rate other issues in between depending on the degree of inequity. Consider *only* equity, not seriousness, size, cost, trends, or community concern.

Issues

____ 1. Communicable disease
____ 2. Hypertension
____ 3. Obesity
____ 4. Sexually transmitted diseases
____ 5. Physical disabilities
____ 6. Alcohol abuse
____ 7. Mental illness
____ 8. Developmental disabilities
____ 9. Occupational illness/injury
____ 10. Unhealthy life-styles
____ 11. Premature births
____ 12. Perinatal mortality
____ 13. Infant mortality
____ 14. Preventable deaths
____ 15. Cardiovascular
____ 16. Cancer
____ 17. Accidents
____ 18. Nursing home alternatives
____ 19. Surgery costs
____ 20. Unavailable kidney transplants
____ 21. Dialysis costs
____ 22. Excess hospital education
____ 23. School health education
____ 24. Home health care*
____ 25. School nutrition education
____ 26. Inadequate health information
____ 27. Environmental health
____ 28. Basic services, rural*
____ 29. Medical emergencies*
____ 30. Nutrition assistance programs
____ 31. Financial incentives to maintain health
____ 32. Underserved areas*
____ 33. Teenage pregnancies

*Issue statement contains relevant information to help you.

Appendix 13-A. Determining Issue Ripeness

Answers to these questions help indicate when a potential issue will "ripen," i.e., develop to the stage where it must be addressed.

1. Has legislation been introduced on the floor to deal with the potential issue?
   a. Yes, its passage is almost certain this session.
   b. Yes, it has a reasonable chance of passing this session.
   c. Yes, it has a reasonable chance in subsequent sessions but not this session.
   d. Yes, but it has no chance of passage.
   e. No.
2. Is there any individual or group of legislators interested in the potential issue?
   a. Yes, legislators are committed to taking action on the potential issue.
   b. Yes, legislators are interested but have not yet committed themselves to an action.
   c. No.
3. Has the legislature requested a study of the potential issue?
   a. Yes, the legislature initiated and funded a study.
   b. Yes, the legislature requested a study from the executive branch.
   c. No.
4. Does it appear likely that the potential issue will surface as a campaign issue? Have any of the candidates taken a position on this potential issue or are they likely to do so?
   a. Yes.
   b. No.
5. Based on your personal contacts, reports of activities or records of meetings, what priority do bureau level policymakers assign to this potential issue?
   a. High priority level and urgent to deal with.
   b. Low priority level and not urgent to deal with.

6. Based on your personal contacts, reports of activities or records of meetings, what priority do county level policymakers assign to this potential issue?

   a. High priority level and urgent to deal with.

   b. Low priority level and not urgent to deal with.

7. Based on your personal contacts, reports of activities or records of meetings, what priority do bureau level program people assign to this potential issue?

   a. High priority level and urgent to deal with.

   b. Low priority level and not urgent to deal with.

8. Based on your personal contacts, reports of activities or records and agendas of meetings, what priority do county level people assign to this potential issue?

   a. High priority level and urgent to deal with.

   b. Low priority level and not urgent to deal with

9. Based on your personal contacts and discussions, what importance have public and private providers assigned to this potential issue?

   a. High priority level and urgent to deal with.

   b. Low priority level and not urgent to deal with

10. Is an alternative, which has been already implemented by other states, being advocated or pushed?

    a. Yes.

    b. No.

11. Based on your personal contacts, discussions and readings, has there been criticism related to this potential issue from professionals outside the mental health system?

    a. Yes.

    b. No.

12. Currently, are other systems (other than the mental health system) dealing with issues which will have an impact on the potential issue being considered?

    a. Yes, issues similar to the one being considered here have been or are being faced in other systems besides the mental health system.

b. Yes, other issues related to this potential issue have been observed in situations outside the mental health system. However, they have not developed to a stage in which they are addressed and debated substantially.

c. No.

13. Has there been a significant technological development (in the medical field or other fields) that has had an impact on the potential issue?

a. Yes.

b. No.

14. Currently, is the mental health system addressing issues that are related to or have an impact on this potential issue?

a. Yes.

b. No.

15. Is the potential issue being repeatedly identified through the formalized grievance mechanisms that exist?

a. Yes.

b. No.

16. Has the potential issue been addressed by the following federal government actions:

a. Either federal regulations, policy or funding changes have occurred.

b. A study has been initiated regarding the potential issue.

c. No action has been initiated.

17. Has the Governor or Secretary addressed this potential issue?

a. Yes, and made a specific written or verbal proposal related to this potential issue.

b. Yes, s/he has shown concern over the issue and raised it in a speech or a communique but has proposed no specific changes.

c. No.

18. At the state level, has a funding change been made that makes the potential issue more likely to occur?

a. Yes, there has been a change in ongoing support levels.

b. Yes, there has been a change in funding available for innovation.

c. Yes, a proposal for change in ongoing support has been made.

d. Yes, a proposal has been made to change funding available for innovation.

e. No.

19. Have there been or are there currently cases in the courts that will have some impact on the potential issue?

 a. There are important cases that deal with fundamental aspects of the existing system.

 b. There are an increasing number of less significant cases being tried that deal with the potential issue.

 c. No court cases, or only a few minor court cases, are dealing with the issue.

20. Has the potential issue drawn attention from special interest groups (including client groups, groups inside the system, and others)?

 a. Special interest groups have identified the potential issue as a high priority issue and have committed resources to the issue.

 b. Special interest groups have shown some concern by posing the issue and asking for an official position, but they have not yet committed resources and time to this issue.

 c. No.

21. Has the general public been exposed to the potential issue through mass media (e.g., major state-wide newspapers, television)?

 a. Yes, the media has often addressed this potential issue, or they have recently paid special attention to this issue by giving it more exposure.

 b. No.

22. Based on your personal contacts, written reports of activities or records of meetings, do first line workers (e.g.,

county social workers, police) identify this potential issue as important?

a. Yes.

b. No.

23. Has the issue accidentally or intentionally generated informal discussion among laymen or policymakers (e.g., letters to editors, discussion in internal newsletter, discussion in off-hours, conversation on the bus going home, etc.)?

    a. Yes.

    b. No.

24. Based on your personal contacts or written memos and reports, do policy analysts assign high priority to this potential issue?

    a. Yes.

    b. No.

25. Have critical incidents and events that are likely to focus the public attention on this issue occurred (e.g., the suicide of patient due to abuse)?

    a. Yes.

    b. No.

14

Conflict Analysis

Attitudes toward conflict have shifted. Conflict, once considered just a problem, is now seen as a potentially positive and creative force. This understanding has led us to prefer to manage, rather than eliminate, conflict. Effective management of conflict depends on an analysis that points the way to constructive outcomes—to win-win solutions.

This chapter describes a conflict analysis methodology (CAM) that is designed to resolve conflicts and build consensus among competing constituencies. The steps in CAM are identifying the constituencies, their assumptions, their goals, and the appropriate spokespeople to represent them; identifying the issues underlying the conflict and their possible levels of resolution (possible solutions); developing and analyzing "treaties" (resolutions with a separate agreement on every contested issue); and following a structured process of negotiation and consensus building to agree on a final treaty.

CAM can be used in two situations: if several constituencies recognize that a conflict must be resolved and are willing to take the necessary steps, and if one party wants a deeper understanding of the conflict.

Note the use of the term *constituencies* to describe parties involved in, or affected by, the conflict under study. We expect the spokespeople who represent each constituency to explain the group's goals, values, and assumptions during the early phases of the process.

If a constituency cannot or will not participate in the analysis, a group of "objective outsiders" should be asked to role-play it. While such refusal or inability to participate may reduce the chances that a chosen treaty will be implemented, having proxies is better than omitting pertinent viewpoints from the analysis. Note that it is helpful if the objective outsiders are highly regarded by the missing constituency.

A related point of discussion is our preference to meet alone with each constituency. To those who question the value of analyzing conflict with some parties absent, we answer that conflict analysis is a process to increase the understanding of each constituency's position, increase the consistency of its choices in subsequent negotiations, and identify different ways of solving the conflict by finding areas of possible agreement with the opponents. Many of these goals can be realized better in private meetings of constituencies. Plenary meetings too early in the process can interfere with CAM by stressing dispute instead of agreement and forcing the parties into a corner.

The premises underlying our methodology are:

1. People have cognitive biases that become more acute under conflict or crisis.
2. Preconceptions and false assumptions impair the ability to make the trade-offs that can lead to a solution.
3. A conflict is easier to grasp if it is broken into components (some of which may have little or no conflict).
4. Individuals can specify their values and prioritize them by using a structured process.
5. The analyst can learn about every viewpoint by gaining access to people who represent all sides of the problem and/or informed observers. These observers can role-play the values, preferences, and priorities of parties that cannot or will not take part in the analysis; they also serve as reality checks.

Our process classifies issues according to whether the disagreements behind them are ideological or technical. It also classifies levels of resolution according to their acceptability and effectiveness (as perceived by the constituencies).

When thinking about managing conflict, we must always consider escalation (the movement of conflict to a higher level of tension where potential losses are increased), a debilitating syndrome with many deleterious effects. The main sources of escalation are a lack of understanding among the parties, a lack of information about the opponent's position, and an emphasis on bargaining behavior. CAM counteracts these problems by increasing communication, emphasizing a problem-solving attitude, offering a joint definition of the problem, reducing the influence of ideology, and pointing the parties toward win-win solutions.

Conflicts can become so heated and the constituencies so stubborn that rational approaches cannot manage them. Often, however, all sides realize a decision must be made and that a better understanding of one another's position is essential. In that case, it becomes helpful for the parties to know the goals, attitudes, values, motivations, and levels of aspiration of the parties; the nature of the vital issues in the conflict and the options available to resolve them; and the consequences of taking each possible action. Unfortunately, research on policy formation suggests that people are inconsistent in their judgment, unaware of their values, and unable to explain accurately not only their opponents' positions but their own as well (Balke et al. 1973; Janis and Mann 1977).

The idea of a decision analysis framework for examining and resolving conflicts is not new. Theoretical models with specific prescriptions have long been discussed in the literature (Raiffa 1982; Von Neumann and Morgenstern 1944). This chapter, however, presents not a specific model but an entire methodology to increase all parties' understanding of the conflict. It is a methodology to *support* the process of conflict resolution by addressing the basic sources of conflict: lack of understanding, lack of information, and distorted communication. After these roadblocks are disposed of, we attempt to break down the conflict into component issues that can be traded off. If we succeed, each party achieves a victory on the issues it deems important.

An important aspect of conflict analysis is timing. Each issue has a life cycle and, depending upon the issue's current stage and the policymakers' intentions, policymakers may try to cause the issue to move to a different stage of its life cycle. Because conflict analysis is likely to draw attention to the problem, you need to know whether the constituencies want to add momentum to the issue—whether they wish to create a crisis or defuse a looming crisis.

The players in the analysis should reflect the purpose of the analysis. If a policymaker wants to explore how the other side might react to various treaties without committing to an action, the analysis must be performed with a group of objective outsiders. But if the purpose is to raise awareness about the problem or reach an agreement on it, the analysis should involve as many constituencies as possible.

Conflict analysis must represent each constituency's value structure without distortion. Difficult though it may be, a successful outcome depends on all constituencies' regarding the analyst as neutral and objective. By objectivity, we do not mean a cold and aloof tone but an ability to feel the concerns and values of all constituencies. Distorting or failing to integrate into the model every viewpoint might persuade the excluded constituency to refuse to participate or to sabotage the efforts at resolution. Distorting a constituency's values will also undermine the model's value as a description of the conflict and a tool to generate treaties. Thus, practitioners must develop not only model-building skills but also interpersonal abilities to interview, assimilate, and role-play other people.

We feel the methodology designed for conflict analysis (CAM) has several payoffs:

- It clarifies which issues are *really* causing conflict, by classifying the basic disagreements as pertaining to ideology or knowledge.
- It increases understanding by getting all the parties to agree on a conflict model and presenting the other side's values on the issues and levels of resolution. This also reduces the number of preconceptions and increases the likelihood of implementing trade-offs. Optimally, conflict analysis actually brings a resolution by helping devise a treaty acceptable to all parties.

Time pressures can impair the quality of the settlement. However, with the help of appropriate computer support, we can perform sensitivity analyses on the importance weights and preferences in the proposed conflict model very quickly. These sensitivity analyses allow us to explore quickly and with minimum effort the acceptability of several treaties by simulating a wide range of constituencies' responses.

The Methodology

So what is this methodology that promises to do so much? CAM consists of the three major phases (shown in Table 14-1), which can be broken down into 13 steps.

The first phase helps the analyst understand the underlying issues and gain a perspective on the background of the conflict. In the second phase we explore the problem by refining and decompos-

Table 14-1 Conflict Analysis Methodology

Phase	*Goals and Basic Actions*
Understanding the problem	Gain general understanding of problem through brief, informal interviews
	Identify constituencies
	Select spokespeople
	Assess importance and certainty
	Analyze catchwords
Exploring the problem	In-depth interview with one or more objective outsiders
	Identify goals, issues, and levels of resolution with stakeholders
	Obtain quantitative estimates for goals, issues, and levels of resolution
	Develop and score treaties
	Develop set of feasible treaties
Exploring the solution	Classify sources of disagreement
	Develop strategy of resolution
	Recommend course of action

ing the preliminary goals into specific components, which we call issues. In the third phase, we explore solutions to the entire conflict by analyzing treaties. The third phase analysis begins by asking the constituencies to weight the relative importance of the various issues and to state preferences for possible levels of resolution. We then package one level of resolution for each issue into a "treaty" and score it to see how well it meets each constituency's needs. In the next step we search for the optimum treaty for all parties (or for a specific party if that is our purpose). Then we convene all parties to search for a consensus resolution. Because the environment can be explosive at this stage, we determine the sequence in which to debate the issues on the basis of criteria such as the sources of disagreement and the levels of resolution.

The third phase concludes when we assemble the spokespeople to discuss the conflict model and suggested treaties and decide on further action. Note that even if the representatives agree on a treaty, they still must "sell" it to their constituencies, so agreement among spokespeople does not guarantee the conflict will end.

It may be useful to remember that the CAM can be applied in different situations:

- To assemble conflicting parties to find general areas of agreement
- To help a neutral observer or mediator understand the issues and priorities of the parties to a conflict
- To help one party clarify its position and perhaps roleplay the opposing positions (This roleplay clarifies the opposition's values and perceptions, enabling one side to understand the opposition's viewpoint and to develop a negotiating strategy.)

We will illustrate the application of CAM by discussing the issue of whether parental consent should be required before family planning services are provided to adolescents. We assume that a member of Congress has instructed a policy analysis team to develop a politically tenable position—one that does not alienate any major interest groups—on an upcoming family planning bill.

Phase 1. Understanding the problem

During phase 1 we will obtain an overview of the conflict, identify the players, and decide who will represent them. In our terminology, we will identify the constituencies, spokespeople, and goals.

Identifying constituencies. Constituencies—people or groups that stand to gain or lose as a result of a conflict—can be identified in several ways, such as by examining lists provided by lobbying organizations, by reviewing testimony on past legislation, and by interviewing key individuals by telephone. Commonly, many groups with various persuasions must be taken into account. However, the value systems of these differing groups toward family planning may be similar enough to be characterized by just two or three models. Such a simplification is a significant help in devising a useful solution. In our example, parents, ministers, physicians, health educators, social workers, and women's rights advocates might be grouped in ways they and we are comfortable with. We can decide whether groups can be lumped into one constituency by asking several prominent organizations to assign priorities to a set of goals about the conflict. This empirical approach works well if all relevant groups are surveyed.

A common mistake is to canvass only those groups that have lined up against each other. Although the first glance might suggest that only two groups are in opposition, further analysis may identify other important players. Using our example, at first it appears that just two constituencies are involved in the conflict: the "antiabortion forces" and the "family planning advocates." However, further examination might unearth a third group, which we could call the "concerned parents"—people who believe their teenage children are mature enough to make correct decisions if given balanced, unbiased information. These concerned parents care mostly about their children's well-being, and if they have an ideology on abortion, they do not want it to influence their children's actions.

Issues and issue weights. Once the constituencies have been identified, we can begin to examine their attitudes toward the parental notification issue. Specifically, we want to understand their goals and the perceived relative importance of each goal. Such information on goals can be obtained through interviews with informed individuals (who may or may not be interviewed in greater depth during later stages of the analysis). Interviews at this stage should be informal and focused on acquainting the analyst with the conflict.

The emphasis at this stage is on getting a comprehensive list of goals so we can obtain a realistic perspective of the problem and its sources. Later, goals may be dropped if close analysis shows them to be redundant, irrelevant, or trivial.

After having listed the goals on the basis of the first interviews, we can give them a preliminary rating. The most important goals (4 is a good number, but try not to list more than 10) are weighed for preliminary analysis. Because we only need to estimate the importance of each goal, we simply divide 100 points among the goals (Huber 1980).

Once we have identified preliminary goals and weights, the next step is to structure the information into a workable format. An example of such a format is seen in Table 14-2, addressing our parental notification problem. Suppose constituency I opposes giving family planning services to adolescents without parental notification, while constituency II prefers allowing adolescents to have access to these services. To the right of the goals in Table 14-2 are columns listing the relative importance weights assigned to each constituency's four top goals. Note that both constituencies agree on the importance of goal b but disagree on the importance of goals a, c, and d.

Table 14-2 Goals and Importance Weights for the Two Constituencies for the Issue of Parental Notification

	Importance Weights	
Goals	*Constituency I*	*Constituency II*
a. Reduce unwanted pregnancies	.15	.40
b. Teach children to be responsible	.25	.25
c. Preserve the family	.40	.10
d. Reduce number of abortions	.20	.15

Table 14-2 gets the formal conflict analysis started by listing a wealth of useful preliminary data. At this point in the analysis, we know which goals are important to each side, and we may start noticing opportunities for trade-offs or compromises. For instance, goal a is crucial to constituency II but relatively unimportant to constituency I, while the reverse is true of goal c. This means that constituency II may be willing to give a lot on goal c to reach goal a, and vice versa for constituency I.

Some people are unimpressed with this type of information and protest that any good politician notices these things intuitively. While this is true for some, unfortunately not all of us are so adept at politics. Furthermore, even good politicians are so busy that they may lack the time to examine the real positions of parties to the many conflicts they face. The result is that their intuition can be inaccurate because they depend on poor information. This step is one of many ensuring workable data for use in the process of conflict analysis.

Identifying spokespeople. Now that the preliminary analysis is finished, we must identify a spokesperson to represent each constituency participating in the conflict analysis. We need at least three types of spokespeople for our analysis: the proponents and opponents of the conflict and objective outsiders. The third group is valued for having a perspective that differs from those of the constituencies. In contrast, if policymakers do not want to negotiate with a constituency but merely wish to increase their understanding of the conflict, the analysis can be performed using only objective outsiders instead of a spokesperson.

Spokespeople should be good at identifying issues and solutions and comfortable with the task of quantifying preferences. While later in the process we may want to include individuals with institutional power to implement a compromise, during the analysis phase we need people with analytical skills, who are also sensitive, insightful, and articulate.

A nomination process to identify spokespeople begins by identifying five or six nominators—people who know the leaders and insightful people in the field and can identify individuals who might adequately represent a constituency. It is good practice to select only nominees who are suggested by several nominators because this indicates the person is widely respected. While in general having many people consider somebody is an expert will signify that person's credibility, even the most widely respected will have opponents and the most knowledgeable his or her cynics. Remember, what we seek is not respect across all constituencies as much as simply the ability to represent how a specific constituency perceives a given situation, and to offer some indication of how it will act. Universal acceptance is not essential.

When talking with the nominees, you can motivate them to participate by mentioning:

- Who nominated them (make sure to get permission to use the nominator's name)
- What the project is about
- Why their participation is important
- What will and will not be done with the results
- How their names will be used
- What tasks will expected of them
- How long each task will take and when it will occur
- What payment (if any) they will receive

In general, three spokespeople from each constituency are sufficient, although there may be good reasons, like attrition, to identify more nominees. During a long process, some spokespeople may drop out because of other commitments or a loss of interest. You must beware of running out of spokespeople before you conclude the analysis. A second reason for more than one representative is that in a complex analysis, especially one involving a series of technical and ethical issues, a single spokespeople may be unable to convey the full spectrum of a constituency's position.

Identifying assumptions. After the preliminary analysis is finished and the spokespeople have agreed to participate, you, as the analyst, must do some homework. This is because conflict about the definition or solution of a problem often arises from sharp disagreement over a fundamental factor that is not obvious to anybody, even the stakeholders themselves. We are referring to conscious or unconscious assumptions.

Now you must bring these assumptions to the surface for examination. This step has three purposes: to compare and systematically evaluate the assumptions, to examine the relationship between assumptions and each constituency's policies, and to formulate proposals that are acceptable in light of the assumptions.

To elicit the assumptions, you must gain as much information as possible about each constituency's views. Leaflets, brochures, advertisements, position papers, legislative hearings, and data used by a constituency to buttress its position are all valuable clues. Interviews may also be useful, but more often than not assumptions are so ingrained that people accept them unconsciously and have difficulty articulating them. Commonly, individuals are surprised by their assumptions once they are made explicit. This can lead to reconsideration.

Throughout this process, look for recurring catchwords or slogans in a constituency's statements. Slogans are chosen for their emotional content; they contain a wealth of information about the constituency's values. In the example of parental notification about family planning services for adolescents, constituency I said, "I want my children to have the courage to say no." Constituency II said, "Let's stop children from having children." While such statements simplify the conflict, they certainly give the flavor of the competing positions. This tone allows you to divine whether the debate is emotional or technical, and whether it concerns ethics or money. Again, remember that at this stage we are trying to *understand* where the different constituencies are coming from and to summarize their positions.

Table 14-3 shows some assumptions behind the worldviews of constituencies I and II. A close reading shows that assumptions can be classified as contradictory and noncontradictory. Constituency I's assumption that access to contraceptives lures teenagers into sex directly contradicts constituency II's assumption that access to contraceptives does not increase sexual activities. Clearly, the two assumptions cannot both be true at the same time. A noncontradictory pair of assumptions is also shown. While constituency I assumes that ad-

Table 14-3 Catchwords and Assumptions of Two Constituencies on the Parental Notification Issue

Constituency I	*Constituency II*
Catchwords	
Too much government.	Let's stop children from having children.
I want my children to have the courage to say no.	Contraception is better than unwanted pregnancies.
Assumptions	
Administration and red tape will eat as much as 90 percent of the funds.	The cost-benefit ratio in family planning programs is excellent
Contraception is dangerous, and people are misinformed about its effects.	Lack of pregnancy allows minors to take advantage of other possibilities (education, employment, etc.).
Parents do a better job providing sexual education.	Parents do not provide adequate sexual education.
Morality is the best contraceptive.	Counselors in family planning agencies provide the most persuasive influence against premarital sex.
Access to contraceptives lures teenagers into sexual activities.	Access to contraceptives does not increase sexual activities.
The decision to have sex is a good opportunity for establishing communication between parents and children.	Confidentiality is crucial in obtaining family planning services.

ministration and red tape will soak up as much as 90 percent of the funds, constituency II assumes that the cost-benefit ratio of family planning programs is excellent. These assumptions seem to be in opposition to each other, but both can be true at once.

At this stage, assumptions can be classified according to relative importance and relative certainty (Rowe et al. 1982) in order to explore critical assumptions behind a given constituency's position. Some assumptions crumble under such examination.

In analyzing assumptions, the analyst should:

- Determine which assumptions are fundamental to a constituency's position.
- Determine which assumptions the constituency feels most certain about, how great is this confidence, and what type of data, if any, the constituencies have to support the assumptions.
- Study the uncertain, but important, assumptions (these may lead to a breakthrough toward a resolution).

It is crucial, then, that you understand the assumptions involved in the conflict before interacting with the constituencies. Indeed, the constituencies must perceive you, the analyst, as a fair individual who is sensitive to their values. Thus, before you conduct sessions to identify goals, issues, and levels of resolution, make sure that you can roleplay the different value systems. Analysts are not actually called upon to roleplay, but they must understand the situation well enough to be able to do so.

Phase 2. Exploring the problem

After you have identified the constituencies and their general goals and assumptions, and the spokespeople have agreed to participate, it is time to model the conflict. In this phase we break down the conflict into its components and then quantify each one. These components are: the goals of each constituency, the issues that form the heart of the conflict and that must be addressed to resolve it, and the possible levels of resolution for each issue.

Goals, issues, and levels of resolution can be identified in several ways, including a questionnaire, a survey, Delphi technique, and/or individual interviews. A private meeting of each constituency is held to obviate the risky and possibly dysfunctional step of assembling the opponents before you know the depth and nature of the conflict. Later, after the conflict analysis model has been formed and some trade-off resolutions have been mapped out, the constituencies may be assembled to search for solutions, or treaties.

Identifying goals. Identifying goals of each constituency is important not only because they help us understand the different parties but also because they help explain why the conflict exists. Only when we understand why the conflict exists can we best define ways to resolve it. The list of goals obtained in the preliminary analysis is a logical starting point for in-depth discussions of goals with the spokespeople.

Identifying issues. Issues are the basic building blocks of conflict; they are fundamental factors that must be understood and addressed in order to reach resolution. Once the issues are identified, we can classify them on a continuum from agreement to disagreement. This will allow us to concentrate on finding acceptable compromises or trade-offs on intensely disputed issues. The classification may reveal, for example, that the constituencies agree on several issues, that they are mildly opposed on others, and that the conflict really concerns just a couple of issues. This makes the conflict appear more manageable and focused and thus simplifies its resolution.

If disputed issues are not ripe for resolution because their political time has not arrived, or if they are too thorny to be resolved, you can try to develop a partial solution by concentrating on more tractable issues. Either way, separating the conflict into component issues allows you to localize the problem to specific areas and use resources to solve problems more effectively.

Identifying levels of resolution. Finally, the levels of resolution—the specific actions, laws, services, etc., that can resolve an issue—must be identified. Typically, constituencies identify levels of resolution that range from optimal to unacceptable. This step tells you what each constituency is considering or fearing in terms of suitable and unsuitable solutions, and it serves as a foundation for generating new resolutions.

Preliminary interviews. During this phase, you will have two sets of meetings, one with objective outsiders and one with the spokespeople for each constituency. Performing in-depth interviews with a few objective outsiders is an excellent way of preparing and rehearsing before the actual sessions with the spokespeople. Objective outsiders need not be experts in the subject matter—they can be trusted associates who are acquainted with the problem and not intimately involved with its solution. These people are chosen for their analytical skills, candor, and willingness to participate. The sessions with the objective outsiders give you a preliminary conflict model that serves as a starting point for the constituency sessions.

During both the proxy interviews and the spokespeople group sessions, the analyst:

1. Discusses the problem in general terms and notes examples of goals, issues, and levels of resolution. The analyst asks

questions to help define the problem and ensure that its key elements are understood.

2. Asks for descriptions of the parties: Who are the proponents? Who are the opponents? How do they view the conflict? What would each side like to see in terms of a resolution? Why?
3. Asks for a list of the goals that drive the parties, and gives examples of goals so participants know the meaning of each. Brainstorming is a good way to draw up a comprehensive list.
4. Asks for a list of the issues that are dividing the sides, to make sure they are as independent of each other as possible (i.e., there is no overlap). Possible questions include: What is the underlying conflict? On which issues do the opponents agree or disagree? Which issues must be resolved for the sides to reach agreement? The list of issues must be exhaustive.
5. Asks for a list of all levels of resolution for each issue. Some levels will be preferred by one side, and other levels by the opponents. Some levels will be compromises that are not preferred by any side but may be acceptable to all. The analyst tries to identify as many levels of resolution as possible on each issue even though only a few may be considered in the final analysis. The reason for seeking so many levels is to promote the development of creative levels. Ask questions like: What is this side's position on the issue? What resolutions might it accept? What resolutions are totally unacceptable? Why are the levels considered in this way? Try to keep the levels as independent of each other as possible, so that no overlap can occur.

Let's return to our example of the issue of parental notification.

Suppose that after performing these interviews with two objective outsiders and with meetings of the spokespeople, the list of goals and weights shown in Table 14-2 was found. Those goals were to reduce unwanted pregnancies, teach children to be responsible, preserve the family, and reduce the number of abortions. Furthermore, suppose the interviews revealed that the conflict can be distilled into two issues: which components of family planning should be available to minors and under what conditions. Suppose also that both constituencies agree that any family planning program must have at least three components:

- *Education* includes topics such as values, morals, biological processes, birth control, decision making, goal setting, sex roles, pregnancy, and parenting skills.
- *Counseling* focuses on some of the same issues as education but involves more interaction between provider and client. The focuses of counseling might include crises, pregnancy, abortion, elective nonparenthood, and preparation for childbirth.
- *Services* might include birth control devices, adoption services, abortions, prenatal care, sexually transmitted disease testing and treatment, and financial assistance.

With this information, you are now in a position to explore the issues at the heart of the conflict. In this example, the crucial issues are:

- Should values and morals be taught when delivering family planning services?
- Should counseling of adolescents start from the position that premarital sex among adolescents is bad?
- Which is more important, allowing easy access to services or having services controlled by organizations with what are considered high morals?
- What are the optimum technical qualifications of the personnel?
- Who, if anybody, should regulate the provision of family planning services to adolescents?

Once the issues have been identified, the objective outsiders and spokespeople are asked, separately, to define a set of feasible resolutions to each issue. A set of issues and their possible levels of resolution are listed in Table 14-4, which shows the format we will use in performing the next step of the analysis.

Once the issues and levels of resolution have been formulated by the full group of spokespeople, they must be checked, detailed, and rephrased in separate sessions with spokespeople from each constituency. It is critical to obtain a consensus on the phrasing and substance of all issues and all levels of resolution from all parties

Table 14-4 Issues and Levels of Resolution

Issues	*Levels of Resolution*
A. To what extent should family planning programs try to *convince* clients that *adolescent sex is bad*?	A1 Should not do so A2 Should be available to clients A3 Should be required of all clients A4 Should be a fundamental part of every service of a family planning program
B. To what extent should family planning *programming* be oriented *toward strengthening the family*?	B1 Not at all B2 Depends on the client B3 Always, with all clients
C. What limitations to *access* to family planning programs should exist for adolescents?	C1 No parental notification C2 Parental notification before counseling or services C3 Parental permission before counseling or services
D. What *type of supervision* should be required for people who provide family planning services to adolescents?	D1 Social work supervision D2 Physician supervision D3 Theologist supervision D4 Experience as parent of adolescent
E. What *organization* (with what *moral qualifications*) should be allowed to deliver family planning services?	E1 Nonprofit E2 Educational E3 Governmental E4 Health care (doctor's office or hospital) E5 Religious
F. Who should *regulate* the provision of family planning services for adolescents?	F1 Peer review F2 Local government F3 State government F4 Federal government F5 Community

before proceeding. Note that at this time the goal is to have all the constituencies agree on the *components* of the conflict, not on a *solution* to it.

Phase 3. Exploring the solution

A treaty is a set of levels of resolution, one level of resolution per issue. Each level of resolution represents a level of satisfaction for a given constituency; each issue is assigned a given priority by each constituency. We analyze the treaties by examining the "satisfaction score" of each treaty for each constituency. The goal is to find a treaty—a combination of resolutions—that has a high satisfaction score for all constituencies.

Eliciting weights and preferences. Treaty development should be the most time-consuming aspect of conflict analysis. Developing quantitative estimates is certainly an important part of the methodology, but this should take less time than developing the model's structure—the goals, issues, and levels of resolution. This is true because model development requires a thorough understanding of the conflict.

To compute scores of various treaties, we must have quantitative estimates for each constituency's preference for all goals, issues, and levels of resolution. You can obtain the weights and preferences for these items by using structured group processes in separate sessions with the objective outsiders and the spokespeople. Methods to elicit weights and preferences are found in Chapter 7 and in Huber (1980) and Keeney and Raiffa (1976).

Once levels of resolution for all issues have been identified and the weights and preferences assessed, it is useful to do an initial review of the situation. For instance, we know that preserving the family is important to constituency I and of relatively little importance to constituency II (see bottom of Table 14-5). This suggests that resolutions that enhance the family might be included in the model as long as they do not impair other goals. For example, family planning organizations could teach parenting skills to parents of adolescents or develop educational and counseling programs to teach adolescents to get along with their families.

The importance weights (in percentages) and utilities of various issues and levels of resolution in our parental notification model are listed in Table 14-5, which shows a complete consensus model of the

Table 14-5 Importance and Preference for Issues, and Levels of Resolution by Each Constituency

Constituency I Preference for Issue Level		*Constituency II Preference for Issue Level*			*Issue/Level of Resolution*
.14		.07		A	*Is adolescent sex bad?*
	0		100	A1	Not part of program
	30		90	A2	Programs available
	80		20	A3	Programs required
	100		0	A4	Built into all components
.20		.04		B	*Family*
	0		100	B1	Not part of program
	30		90	B2	Programs available
	90		20	B3	Programs required
	100		0	B4	Built into all components
.25		.48		C	*Notification*
	0		100	C1	No notification
	80		10	C2	Notification
	100		0	C3	Permission
.11		.14		D	*Qualifications*
	0		100	D1	Social work
	60		50	D2	Medical
	100		30	D3	Theological
	50		0	D4	Parent of adolescent
.23		.16		E	*Organizations*
	0		100	E1	Nonprofit
	60		30	E2	Educational
	30		15	E3	Governmental
	80		10	E4	Health care
	100		0	E5	Religious
.07		.11		F	*Regulation*
	0		100	F1	Peer review
	90		20	F2	Local government
	40		30	F3	State government
	10		40	F4	Federal government
	100		0	F5	Community

Table 14-5 Importance of Goals for Each Constituency

Importance Weight		
Constituency I	*Constituency II*	*Goals*
.13	.36	Reduce unwanted pregnancy
.32	.33	Teach responsibility
.43	.10	Preserve the family
.12	.21	Reduce number of abortions

conflict for each constituency. The importance weights attached to the issues were obtained by asking spokespeople in the meetings to divide 100 points among the issues. This assessment was checked by asking spokespeople to assign 10 points to the least important issue and use it as a benchmark to weight the other issues. The weights were then normalized.

The levels of resolution were scored by asking spokespeople to assign a preference value or "utility" between 0 and 100 to each resolution level. A 100 was given to the optimum resolution level according to their constituency, 0 for the worst acceptable resolution. The levels of resolution in between were scored directly (e.g., if level of resolution x was the best possible resolution for issue i, it was assigned a utility of 100; and level of resolution y was the worst acceptable resolution, it thus was assigned a utility of 0. We asked a spokesperson who considered level of resolution x the best possible and level of resolution y the worst possible to assign a utility value between 0 and 100 to all "in-between" levels of resolution by comparison to levels x and y.

When the scores from different spokespeople for the same constituency varied about a goal, issue, or level of resolution, we discussed the discrepancy. If it remained, we averaged the weights and values to obtain a value to represent the constituency. (Averaging should be a last resort used only if differences are relatively small.) Note that during the discussion of differences you might find certain factions of a constituency to be more amenable to compromise. This is signaled by separate factions having similar scores on issues and levels of resolution. In the extreme case, you might find that the spokespeople on one side actually represent several positions, not just one. If so, you may need to develop a distinct model for each position.

We then had a consensus among all constituencies in our parental notification example about which issues form the heart of the conflict, and about how important each goal and issue was to each group. While we also had preferences for each level of resolution, we did not have, and might never achieve, consensus on the relative importance of the issues and acceptable levels of resolution. Finally, we sought the trade-offs to resolve the entire conflict.

Creating and scoring treaties. The point of analyzing treaties is to identify those treaties acceptable to all parties. A large number of treaties is typically available—in our example on parental notification, 4,800 treaties could be formed. In general, the number of possible treaties N is equal to:

$$N = \prod_{i=1}^{n} x_i \qquad (2)$$

where x_i is the number of levels of resolution attached to issue i, and n is the number of issues.

Fortunately, only a few of these conceivable treaties are conducive to an agreement. Assuming we have correctly measured the values of the constituencies, the best treaty can be identified by the following scoring system.

We can calculate the value of a treaty by multiplying the value of one level of resolution by its issue importance weight, doing this for all issues, and summing the results. The values reached in step 1 could be called "weighted satisfaction scores" because they represent how happy a constituency would be with a given level of resolution weighted by the importance it attaches to the issue.

Suppose family planning legislation passed with these levels of resolution (taken from Table 14-5):

- A_2 Programs stress the negative aspects of adolescent sex.
- B_2 Programs to strengthen the family would be required in any adolescent family planning service.
- C_2 Parents must be notified when an adolescent uses a family planning program.
- D_2 All providers of family planning services must have medical qualifications.
- E_2 Educational institutions will carry out family planning programs.

F_2 State governments must regulate family planning programs.

The value of this treaty is calculated by multiplying the relative importance weight of an issue by the utility assigned to the level of resolution, then adding across all issues. (The scores of all issues are added using the processes discussed in Chapter 7.)

That is, the score of treaty k for constituency c is equal to,

$$\text{Score of treaty } k_c = \sum_{i=1}^{n} W_{ci} Uc_{ij} \qquad (3)$$

Where k is a specific treaty consisting of a set of levels of resolution for all n issues
W_i is the importance weight of issue i
U_{ij} is the utility of level of resolution j to issue i
n is the number of issues underlying the conflict

For constituency I, the value of the treaty described above is (values taken from Table 14-5):

$$(.14 \times 30) + (.20 \times 30) + (.25 \times 80) + (.11 \times 60) + (.23 \times 60) + (.07 \times 40) = 53.4$$

For constituency II, the value of the treaty is:

$$(.07 \times 90) + (.04 \times 90) + (.48 \times 10) + (.14 \times 50) + (.16 \times 30) + (.11 \times 30) = 29.8$$

How good is this treaty? Can we satisfy both constituencies better? These questions are addressed in the following section.

Performing Pareto optimality analysis. In common-sense terms, we realize that a treaty should allow one party to get as much as possible without damaging the opponent's position, and vice versa. This concept of mutual improvement is called the Pareto optimality criterion (Raiffa 1982). Pareto optimality is reached when one side cannot improve its position without degrading the position of its opponents.

The purpose of analyzing treaties is to devise a few that are acceptable to all constituencies and that can recast the conflict as a win-win situation. The analyst and constituencies should explore all trade-offs and compromises in a cooperative manner.

The analysis of treaties may be a time-consuming process, requiring calculations, figures, and computer programs, so it should be performed by the team of analysts. After the analysis is done, convene meetings with spokespeople from each constituency separately to review the promising treaties and do more analysis if necessary. Again, you are wise to consult a few objective outsiders on the preliminary results before conducting these meetings.

The examination of alternative treaties is an iterative process that can be simplified with computer programs or examined intuitively by using Table 14-5 to ponder improvements in the negotiations. The important point is that understanding the relative importance of issues and levels of resolution often allows us to find treaties that trade off and improve the outcome for everybody.

It is common to approach conflict resolution by seeking a compromise on each issue in turn. However, we suggest that a more unified approach, called logrolling, has significant advantages. Logrolling is a trading of issues so each party can win the ones it considers most important. Researchers have found that conflict resolution processes that allow logrolling lead to more Pareto optimal settlements and higher satisfaction than processes based on issue-by-issue compromise (Froman and Cohen 1970).

Logrolling's advantages over compromising are that it

- Defuses ideological disputes
- Increases Pareto optimality
- Satisfies both parties better
- Encourages looking at the big picture
- Does not divide the original conflict into several conflicts

On the other hand, the one-by-one compromise approach tends to exacerbate the total conflict. If one side wins issue *x*, the losers may try all the harder to win issue *y*. The very nature of the procedure encourages the parties not to search for the best overall solution.

Let's see how compromising on issues one by one can lead to inferior solutions. Suppose that a conflict has only two issues: the amount of funding allocated to family planning services and the age at which adolescents can use those services without parental consent. Figure 14-1 shows that constituency I's preference toward funding holds constant up to about $400,000 and then decreases as funding increases. On the other hand, constituency II's preference increases with increases in funding. The two constituencies might compromise on this issue by deciding that $700,000 (i.e., the intersection of both preference curves) would make both sides equally happy.

Figure 14-2 reflects the constituencies' preferences for age of eligibility. Constituency I's preference increases markedly after 14 years of age, while constituency II's preference is to give eligibility at age 13. Thus, this constituency's satisfaction decreases as age of eligibility rises above 13 years. Again, the constituencies could compromise by selecting age 16 (the intersection of the preference curves) as the age of eligibility, because this is the age where both sides have the same preference value.

If constituency I feels that the issue of age is twice as important as funding and constituency II feels funding is three times as important as age, we can create Table 14-6 to contain our data.

Table 14-6 shows that the maximum score constituency I can get on age is 66 (.66 multiplied by the utility of the best resolution level, 100), and the maximum score constituency II can get is 25. The maximum score for constituency I on funding is 34, while constituency II can get 75. The curves in the figures are redrawn to reflect these weights.

Figure 14-1 Preferences for the Amount of Funding for Family Planning Services

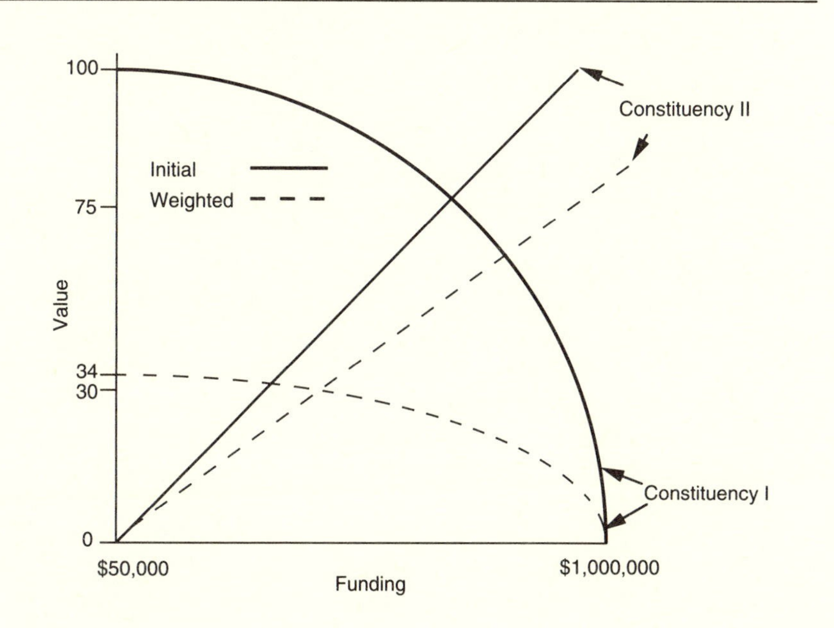

Figure 14-2 Preferences for the Age of Eligibility to Have Access to Family Planning Services Without Parental Consent

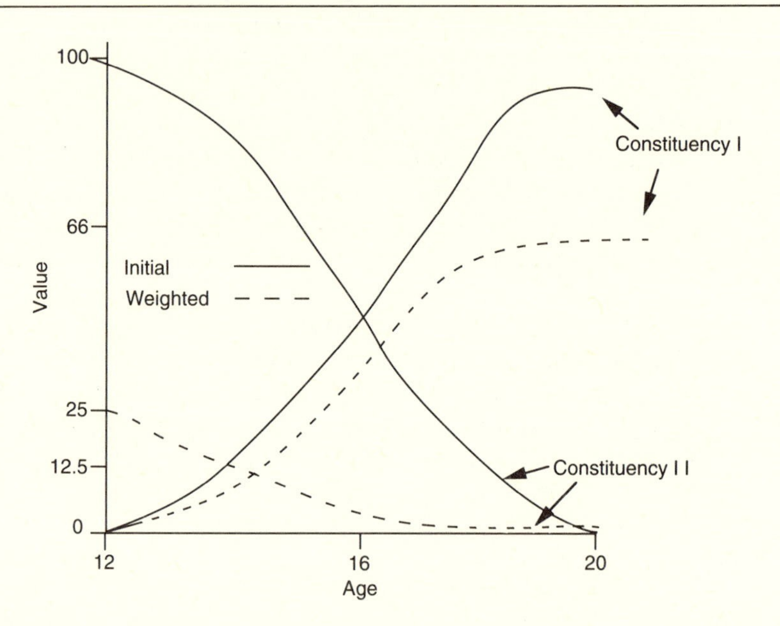

Table 14-6 Importance Weights for Age of Eligibility and Funding

	Age	*Funding*
Constituency I	.66	.34
Constituency II	.25	.75

The value of this treaty, with age at 16 and funding at $700,000, is 42.5 (30 + 12.5) for both constituencies. We can show that many treaties would be superior. One would let constituency II win completely on funding, the issue it considers more important, and would let Constituency I win its issue, age. The values for this treaty would be (66, 75). Thus, we believe this type of logrolling should be encouraged at the expense of compromising on issues one by one.

Logrolling fits nicely with the Pareto optimality analysis, which teaches that any treaty that increases the value for one constituency without hurting the other is an acceptable improvement. The cross-hatched area of acceptable improvements shown in Figure 14-3 in-

Figure 14-3 Pareto Optimal Analysis of Treaties

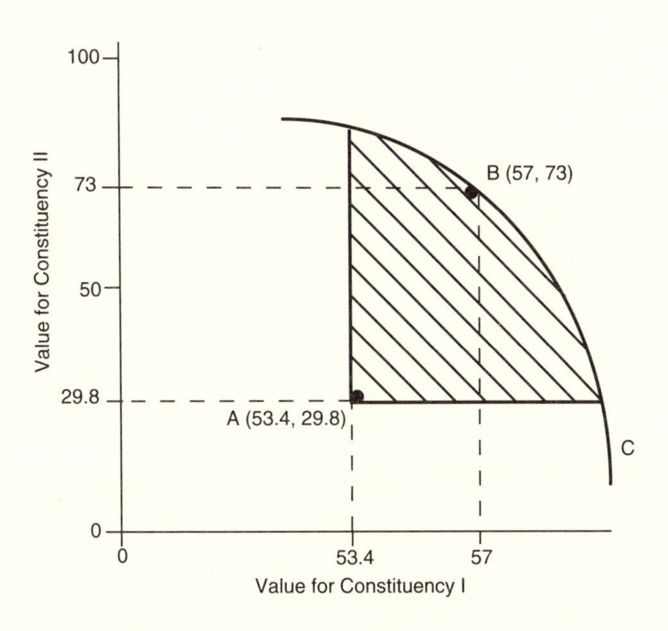

cludes point B (57, 73). Point B represents a treaty made up of levels of resolution that let each constituency win its important issues.

The treaty analyzed above is a middle-ground solution because each of its levels of resolution grew out of a compromise. This treaty was worth 53.4 to constituency I and 29.8 to constituency II (point A in Figure 14-3).

Now let's see what happens if we do some logrolling and each constituency is allowed to win its most important issues. This would mean constituency I wins on issues A, B, and E, and constituency II wins on C, D, and F. So the resolution treaty would be (from Table 14-5): A_4, B_4, C_1, D_1, E_5, and F_1.

The scores for the constituencies would be, using equation (3) above:

Constituency I: $(.14 \times 100) + (.20 \times 100) + (.25 \times 0) + (.11 \times 0) + (.23 \times 100) + (.07 \times 0) = 57$

Constituency II: $(.07 \times 0) + (.04 \times 0) + (.48 \times 100) + (.14 \times 100) + (.16 \times 0) + (.11 \times 100) = 73$

Both constituencies improve their position, although II gains much more than I. So, if both constituencies want to improve their respective positions and not to block or defeat the other, this treaty would be better for both. In other words, it would satisfy the requirements of the Pareto optimality criterion.

There is, however, a set of treaties beyond which neither side can improve its position without hurting the opponents. Those treaties fall on the Pareto optimal curve (C). The goal is to find acceptable treaties that come as close to the Pareto optimal curve as possible.

It is crucial that before the next step—in which the analyst and all constituencies jointly generate the resolution treaty—each constituency understands the other side's view and values. This understanding can be increased and clarified through the analysis of treaties presented here.

The process of negotiation. Now we have a model of conflict acceptable to all parties, a set of weights representing the issue importance, and preferences for the various levels of resolution. Each set of weights represents the value structure of one constituency. With our model of conflict and value structures, we have created, scored, and analyzed treaties and discovered, after discussions with spokespeople for each constituency, a few treaties that might be acceptable to all parties.

Coming up with an agreed-upon conflict model is a big achievement in itself. The opposing constituencies have been delineated, the conflict has been specified, the structure of the model (goals, issues, and levels of resolution) has been defined, and the value structures of the constituencies have been articulated.

Now we assemble the constituencies for the first time. The delay emphasizes the need for each constituency to understand its own position before discussing it with its opponents. To minimize conflict caused by lack of understanding of one's own position, the CAM spends a great deal of time clarifying what those positions are and what feasible treaties might look like.

Now it is time to choose a course of action. Note that at this point the parties have agreed on the issues and feasible levels of resolution but not on the relative importance of the issues or on their preferences for levels of resolution. It is unlikely that agreement will be reached on all these matters, because some issues are purely ideological, but it is realistic to aim for agreement on certain issues and levels of resolution and to hope to compromise on the rest. The following discussion offers suggestions and guidelines to increase the probabilities of success at this stage.

Because the issues and level of resolution have been agreed upon, the emphasis of the meeting is to reduce differences of perception and judgment on the relative importance of disputed issues and levels. But which issue should be addressed first? The answer arises from an examination of the source of the disagreements, which the analyst performs before the plenary meeting. In our view, the source of disagreements can be classified along a continuum ranging from purely *ideological* to purely *technical.*

Ideological or technical issues. Ideological issues (e.g., the suggestion that abortion is a form of murder) are largely immune to assaults by data alone. Ideological issues are based on values and beliefs; they are not subject to proof or disproof because data either do not exist or they can be interpreted in widely different ways. Technical issues, on the other hand, can be settled by reference to hard data (such as the results of controlled experiments). Some issues blend the two characteristics.

Examining the nature of the disagreements has two benefits: it helps you decide how to approach individual issues, and it helps you decide in which order to present issues. The process of negotiation should depend on the nature of the disagreement. For example, you can break an ideological issue down into components to localize the value difference, or you can logroll it. But a technical difference may be resolved by bringing in an expert or searching for relevant information.

If you work first on technical issues, you will gain a sense of agreement and progress that will help when you must approach the more contentious ideological issues. People tend to be feel much more strongly about ideological differences; they often deal with technical differences in a cooler, more rational fashion. And since ideological disputes are seen by the parties as zero-sum games, it is important to address them carefully.

A similar problem can arise when assessing the effectiveness of a level of resolution. For example, abortion can be rejected as a way of controlling unwanted pregnancies because of a dispute over a technical matter of *effectiveness* (is abortion safe and nontraumatic?) or an ideological dispute about *acceptability* (is abortion murder?).

If the disagreement is based on a constituency's rejection of a level of resolution, you can investigate whether the resolution can be packaged in another way, whether substitutes exist for this level of resolution, and under what conditions and for what issue the opponents would trade it off. On the other hand, a disagreement over perceptions about the effectiveness of the resolution might lead you to identify sources of expertise and information to determine its effectiveness.

Once the sources of disagreement on issues and levels of resolution have been classified, you must determine an order in which to present the issues to the constituencies. Here it might help to classify the issues as issues on which the constituencies:

1. Agree about the relative importance and the level of resolution to implement
2. Disagree about relative importance but agree about the level of resolution to implement
3. Agree about relative importance but disagree about the level of resolution
4. Disagree about both relative importance and level of resolution

The order in which the cluster of issues should be presented to all spokespeople is shown in Figure 14-4. Figure 14-4 shows that the process should start with those issues on which the constituencies agree (number 1). It is important to quickly establish a mood of goodwill and cooperation. By showing the areas of agreement, you can enhance the mood for the later negotiations.

You should then discuss issues where both constituencies have agreed about a preferred level of resolution. Again, this is to prove ground for agreement exists on some issues.

Discuss third the issues on which the constituencies agree about relative importance but disagree about the level of resolution to implement. Start by discussing issues having a low importance rating.

Finally, negotiate the differences on issues about which the constituencies disagree on both levels of resolution and importance. These are the crux of the matter and will take the most negotiating skill and likely the most time as well.

A further refinement would reflect the level of disagreement on specific issues. Issues within each type can be presented in a sequence reflecting the range of the weights (or variance if more than two sets of weights are involved). For example, assume constituency

Figure 14-4 Classification of Issues to Determine Order of Presentation for Analysis and Consensus Building (numbers indicate sequential order of presentation)

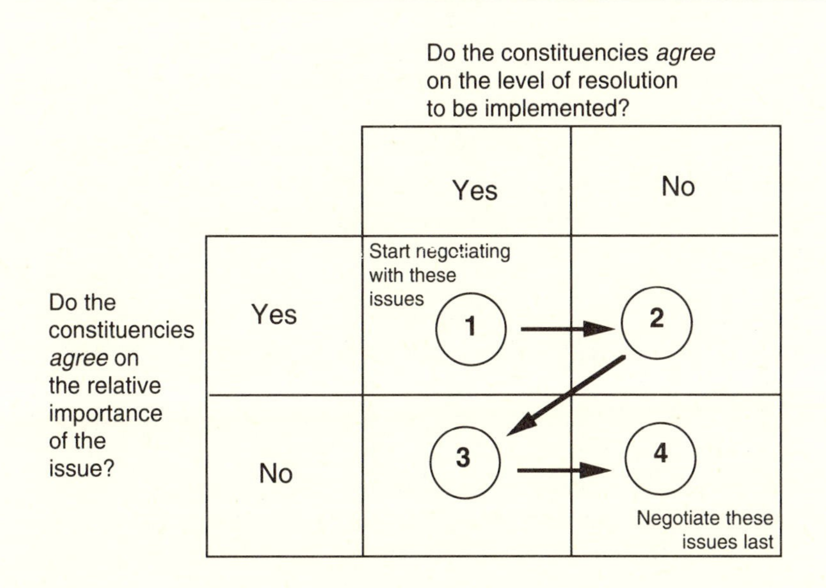

I assigned issue *x* a value of 10 and a value of 25 to issue *y*, and that constituency II rated these issues 60 and 30, respectively. The range of ratings on issue *x* is 50 (60 – 10); the range on issue *y* is 5 (30 –25). The goal is to discuss issues with low ratings before moving to those with larger ratings. (If more than two constituencies were involved, you would use the standard deviance or variance instead to measure range on the issues.)

The more disagreement and polarization there is about an issue, the larger the range (or variance) of the weights on it and the later in the discussion it should be presented. As before, this ordering typically implies dealing with purely technical issues first, and then moving gradually to the more ideological issues.

Why spend so much time determining the order to present and discuss issues? Primarily because when the parties to a hot conflict assemble, they can create such an explosive atmosphere that it's good to get some agreement on some issues as quickly as possible. If a feeling of accomplishment and understanding can be instilled, the constituencies may suspend or water down their negative preconceptions of the other side and adopt a constructive, problem-solving attitude. Also, if a controversial issue cannot be resolved, the fact that some issues *were* settled diminishes the frustration of deadlocked negotiations and assures the constituencies that their efforts were at least partially successful.

Throughout the negotiations of differences, the analyst must keep in mind the Pareto optimal treaties that were developed earlier with the objective outsiders. At the beginning, when relatively noncontroversial issues are being discussed, it is better not to bring up these "fair" treaties, so the constituencies cannot focus on the final outcome and shortcut the early stages, which help establish a positive attitude. Later, if negotiations are deadlocked, use the fair treaties to generate a breakthrough. These Pareto optimal treaties can prove that one constituency can improve its satisfaction without damaging the satisfaction of its opponents.

Finally, it is important to end the meeting with a concrete plan for action. If you succeeded in bringing about an overall resolution and drafting a final treaty, then actions to implement this treaty should be agreed upon. On the other hand, if no treaty or a partial treaty was obtained, you should list actions that will continue the conflict analysis and resolution. Remember that conflict analysis is being done with spokespeople who must sell the resulting agreement to their constituencies.

Deescalating the conflict. We have presented this conflict analysis methodology as a strategy to increase the understanding of the underlying sources of conflict. In this section, we show how CAM can deescalate a conflict and increase the probability of a negotiated settlement to conflicts that seem intractable.

CAM checks this escalation by:

- Preventing the parties from negotiating on the overall treaty until some agreements have been reached
- Discussing issues in a sequence that minimizes frustration
- Dividing the conflict into component issues
- Increasing the probability of finding a few areas of agreement

Parties to escalating conflicts frequently forget their initial concerns and turn to trying to beat each other. The parties may think about "saving face," "getting even," "teaching the others a lesson," or "showing them they can't get away with this." At this point, they more closely resemble a battered prizefighter than a reasonable participant in a public policy debate. When this stage has been reached, the conflict is likely to expand to other areas where it should not logically exist. These ancillary conflicts may be created by a party to make sure it has some ground to retaliate if it loses the main battle. Examples can be found in parties that try to block Pareto optimal improvements—situations in which one constituency could gain satisfaction without another losing satisfaction.

Escalation is likely to increase the number and size of the issues under dispute and the hostile and competitive relations among the parties. The opponents may pursue increasingly extreme demands or objectives, using more and more coercive tactics. At the same time, whatever trust existed between the parties is likely to corrode. The ultimate stage in escalation is reached when the parties think differences exist across many issues (even some that were created solely for the sake of bargaining). A feeling of frustration settles in, along with the impression that the parties are incompatible, that compromise is impossible, and a fight for total victory is the wisest course.

Researchers have noted that where substantial differences exist, parties must ventilate their feelings toward each other and talk about the issues dividing them before they can seek a solution that integrates the positions of all important parties (Thomas 1975). That is, a ventilation phase generally needs to precede an integration phase. CAM encourages, and actually demands, that parties state their positions, their perception of the sources of conflict, and their assumptions. The model-building phase usually has a cathartic effect, alleviating hostility and creating an atmosphere conducive to such an integration phase.

Thomas (1975), in a comprehensive review of the literature, identified several causes of escalation, among them lack of reevaluation, self-fulfilling prophecies and biases of perception, distortions in communication, increase in distrust and hostility, losing sight of the original issue and expanding the conflict to other issues, feeling of general incompatibility, and lack of a ventilating phase.

How can CAM address these causes of escalation? The reevaluation process may happen when a party, hearing the other side's arguments, reevaluates its definition of the issue or its choice of preferred alternative. CAM allows for reevaluation at several points in the process, by facilitating communication, explaining positions, and emphasizing problem-solving behavior. The methodology insists throughout that parties develop an understanding not only of their position but of their opponent's positions as well.

Self-fulfilling prophecies and biases in others' perception directly affect the ability to make trade-offs. An interpretation of another's behavior is largely a response to the image each party has of the other. A constituency that sees the other as cold and calculating will be suspicious of compromises its opponent suggests. If it sees the opponent as compassionate and respectful, the resulting goodwill can simplify the negotiations. CAM's main thrust is to elicit rational, analytical behavior and to clarify, through an iterative process of questioning, the perceptions of all parties.

In conflicts, parties tend to use information to manipulate and coerce their opponents and/or the public. Communication becomes distorted, trust is diminished, and messages between parties lose credibility or are ignored. At this stage, a party can further distort its view of the opponents without contradiction. CAM tries to dispel misunderstandings by allowing an explicit presentation of the parties' positions, and tries, by developing treaties, to serve as catalyst for discussion of issues and potential resolutions.

Summary

This chapter describes a methodology to reduce conflict and build consensus called the conflict analysis methodology. CAM consists of identifying the constituencies, their assumptions, their goals, and the appropriate spokespeople to represent their position; identifying the issues underlying the conflict and possible levels of resolution; developing and analyzing treaties; and following a structured process of negotiation and consensus building to agree on a final treaty. A summary of the methodology is shown in Table 14-7.

Our process of negotiation classifies conflicts by the nature of the disagreements behind them (ideological or technical) and classifies levels of resolution—or acceptable agreements on component issues—by their acceptability and effectiveness as perceived by constituents. Issues should be negotiated in a logical sequence so ideological issues are not approached until the easier technical issues have been disposed of.

The main sources of conflict escalation are a lack of understanding among the parties, a lack of information about the oppo-

Table 14-7 Summary of the Conflict Analysis Methodology

Step	*Actions*	*Purpose*
1	Hold informal interviews with people informed about the conflict.	To get an overview of the conflict and write a preliminary list of goals and their importance.
2	Identify constituencies and their spokespeople.	To define the coalitions and identify individuals who will later develop the conflict model and choose treaties.
3	Analyze assumptions.	To obtain a general understanding of the problem, identify ideological and technical sources of conflict.
4	Perform an in-depth interview with one or more objective outsiders.	To refine the goals identified in step 1, check understanding of issues and different set of values involved, and begin exploring grounds for resolution.
5	Have spokespeople and a few objective outsiders identify goals, issues, and level of resolution. Have separate sessions with each constituency.	To develop the conflict model and identify key levels of resolution that lead to an overall compromise.
6	Obtain quantitative estimates for issues and levels of resolution from spokespeople. Have a separate session for each constituency.	To quantify the importance and preference of the component issues and the levels of resolution.
7	Form and score treaties.	To generate a set of feasible solutions and explore Pareto optimal solutions.
8	Analyze the treaties with spokespeople during a separate session with each constituency.	To generate treaties that are likely to resolve the conflict.

Continued

Table 14-7 Continued

Step	Actions	Purpose
9	Classify unresolved issues to develop a strategy of negotiation.	To increase the likelihood of a positive outcome.
10	Present results of all spokespeople.	To generate an acceptable treaty and develop guidelines to implement the treaty or agree on further actions in the resolution process.

nent's position, and an emphasis on bargaining behavior. CAM addresses these problems by using a joint definition of the problem, emphasizing win-win solutions, and helping to minimize the impact of ideological differences by identifying areas of agreement.

CAM promotes consensus building by underscoring the importance of a clear and structured resolution process, by eliciting an understanding of the positions held by the constituencies, and by finding potential trade-offs among them.

References

Balke, W. M., K. R. Hammond, and G. D. Meyer. 1973. "An Alternative Approach to Labor Management Relations." *Administrative Science Quarterly* 18: 311–27.

Froman, L. A., and M. D. Cohen. 1970. "Compromise and Logroll: Comparing the Efficiency of Two Bargaining Processes." *Behavioral Science* 15: 180–86.

Huber, G. P. 1980. *Managerial Decision Making.* Glenview, IL: Scott, Foresman.

Janis, I. L., and L. Mann. 1977. *Decision Making: A Psychological Analysis of Conflict, Choice, and Commitment.* New York: Free Press.

Keeney, R. L., and H. Raiffa. 1976. *Decisions with Multiple Objectives: Preferences and Value Trade-Offs.* New York: John Wiley.

Raiffa, H. 1982. *The Art and Science of Negotiation.* Cambridge, MA: Harvard University Press.

Rowe, A. J., R. O. Mason, and K. Dickel. 1982. *Strategic Management and Business Policy: A Methodological Approach.* Reading, MA: Addison-Wesley.

Thomas, K. W. 1975. "Conflict and Conflict Management." In *The Handbook of Industrial and Organizational Psychology,* Vol. II, edited by M. D. Dunnette. Chicago: Rand McNally.

Von Neumann, J., and O. Morgenstern. 1944. *Theory of Games and Economic Behavior.* New York: John Wiley.

15

Using Bayesian and MAV Models to Analyze Implementation

This chapter discusses the design and construction of two decision aids to estimate the probability that an attempt to change will succeed. These decision aids can also be used to explain that prediction. This chapter has four parts. First, we review the literature on the major factors affecting successful change. Second, we describe and compare Bayesian and multiattribute value models, both of which comprise 24 attributes grouped in five categories. Third, we describe field tests on the validity and reliability of the models. Fourth, we propose these modeling processes as paradigms with wider application in social science research.

Most literature about decision aids sees them as supporting the choice of a best solution from a set of feasible options. Recently, some have suggested that these aids should begin to operate earlier in the process, to help structure and define problems and generate options, not just choose from them. We believe that support is needed after a decision has already been made, so that decision aids can be used to help plan the implementation.

Indeed, quality decisions are not always carried out, and the real concern is how to do so effectively. The many reasons for implementation failure have been extensively reported: Ackoff (1960); Burns and Stalker (1961); Marquis and Marquis (1969); Schultz and Slevin (1975); Watson and Marett (1978); Cooper (1980); Lee and Steinberg (1980); and Ginzberg (1981). A policymaker who knows which factors affect success may have a hard time combining those factors to assess the success of a specific implementation plan (Anderson, Chervany, and Narasimham 1979).

We believe it would be useful to have a predictive and explanatory model so policymakers can examine their implementation plans. Predictions from this model could help policymakers gauge overall chances of success. The model could indicate the greatest threats to the plan and suggest changes to improve its prospects.

Previous Research on Implementation

The literature on implementation, innovation, and change suggests that a multitude of factors influences the likelihood of successfully implementing a policy. Those factors can be classified into five groups according to the characteristics of (1) the target group, (2) the change agent, (3) the change itself, (4) the process used to develop and introduce the change, and (5) the environment for the change. The evidence for each category is described below.

Characteristics of the target group

The target group is the person, group, department, or community that will be affected most directly by the change. Research indicates that certain characteristics of the target group influence subsequent success. Several investigators (Greiner 1967; Dalton 1969; Brereton 1972; Huse 1975; Anderson and Narasimham 1979; Alter 1980; and Alavi and Henderson 1981) suggest that a sense of tension or felt need for change is a key condition for a change to occur. Lippitt, Watson, and Westly (1958) postulate that tension for change may result from dissatisfaction with the status quo because of decreased performance, lowered self-esteem, increased aspirations, or external pressures from the environment and/or power or peer groups.

The organization's history of successful change affects the target group's attitude toward change (Rogers and Schoemaker 1971; Vertinsky et al. 1975; Powell 1976; Alter 1980; Lippitt, Watson, and Westly 1958; Wilson 1966; Greiner 1967; Reisman and de Kluyver 1975). Incentives for change also affect the target group's attitude (Wilson 1966).

The target group must be confident it can deal with change. Lippitt, Watson, and Westly (1958) and Bandura (1977) state that organizations and individuals will resist change unless they are confident that their attempts to change will be successful. Thus, resistance to change can arise if the target group fears that it lacks the strength, understanding, skill, and/or economic ability needed to succeed (Zaltman, Duncan, and Holbeck 1973; Guimaraes 1981).

Characteristics of the proposed change

Characteristics of a proposed change influence the likelihood of successful implementation. The change must have good technical quality (Bennis 1965; Maher and Rubenstein 1974; Delbecq 1975; Schultz and Slevin 1975) and must be likely to meet the needs of its customers (Dickson 1976; Cooper 1980). The proposal must be simple to understand and control; modifications should be easily made to retain its effectiveness. A policy requiring uniform reporting of falls by nursing home residents would face greater resistance if reports must be presented on 3½-inch disks using a specific word-processing program, rather than any legible medium (Ackoff 1967; Bolan 1969; Harvey 1970; Huysmans 1970; Little 1970; Maher 1973; Vazsonyi 1973; Lee and Steinberg 1980).

The degree of uncertainty regarding a policy's irrevocability is inversely proportional to the likelihood of success (Rogers 1983; Radnor and Neal 1973; Lee and Steinberg 1980; Alavi and Henderson 1981; Guimaraes 1981). For instance, some policies may be tested in a small pilot test before full-scale implementation. If a major flaw is found, the policy can be modified or scrapped. During the pilot test, the policy is not irrevocable.

The less radical (i.e., the less deviation from prior practice) the proposed change, the greater the probability that it will be accomplished with minimal conflict (Stevens and Galanter 1957; Fliegel and Kivlin 1966; Hage and Aiken 1970; Stimson and Stimson 1972; Delbecq 1975; Kotter and Schlesinger 1979).

The proposed change should be considered new and clearly distinguishable from the status quo (Rogers 1962; Maher 1973; Manley 1975; Schultz and Slevin 1975; Vertinsky, Barth, and Mitchell 1975; Kotter and Schlesinger 1979; Cooper 1980).

Characteristics of the change agent

The change agent is the person or group trying to help the target group implement the proposed change. Like the previous categories, the change agent influences the climate for change (Greiner 1967;

Argyris 1971; Vertinsky, Barth, and Mitchell 1975; Powell 1976; and Gustafson, Rose, and Howes 1975). The change agent's position within the organization is important to ultimate success (Wilson 1966; Dane, Gray, and Woodworth 1979; Kimberly and Evanisko 1981). Characteristics improving the change agent's prospects include prestige (Duncan and Zaltman 1975; Freeman 1982); respectability and worthiness (Dalton 1969); sharing values and attitudes with the target group (Duncan and Zaltman 1975; Delbecq and Mills 1985); and acting professionally (Lippitt 1973).

Duncan and Zaltman (1975) and Harvey (1970) emphasize the importance of the change agent's values, inasmuch as they shape the definition and diagnosis of the problem and the selection of strategies and solutions.

The change agent's strategy and style of implementation are also important (Lippitt 1973). The change team should assess resistance to change, psychosocial forces, the consequences of change, and the power, organizational, and communication structures (Bennis 1965; Huysmans 1970; Stimson and Stimson 1972; Zaltman, Duncan, and Holbeck 1973; Lippitt 1973; and Lockett and Polding 1978).

Psychological support by the change agent for the target group during implementation is another important variable (Bennis 1965). Ginzberg (1978) reports that involving the analyst as the project terminates helps guarantee that the proposed change will be solidified.

Characteristics of the planning and implementation process

Although they precede the period of carrying out the change, the processes of defining the problem and developing the solution can profoundly affect success. The target group's level of involvement is crucial to success (Leavitt 1964; Greiner 1967; Evans and Black 1967; Lawrence 1969; Hage and Aiken 1970; Lucas 1975; Manley 1975; Schultz and Slevin 1975; Ginzberg 1981). Specifically, Lonnstedt (1985) reports that the target group collaborated in the problem exploration in 80 percent of projects considered successful. This success rate was double that of projects in which no early collaboration occurred.

Bean et al. (1975) and Alter (1980) found that management support correlates positively with successful implementation. On the other hand, the lower the number of people who must carry out or be affected by change, the better the chances for implementation (Lee and Steinberg 1980).

The degree of line manager responsibility for staffing and managing the change process and the extent of operating manager accountability for the impact of the change both correlate with success (McKinsey 1968; Wynne 1979; Lee and Steinberg 1980).

Finally, the literature suggests that the likelihood of implementation depends on the stated purpose (Dane, Gray, and Woodworth 1979); the clarity and justification of the need for and objectives of the proposal; and the clarity and definition of the activities and the tasks of the change process (Lonnstedt 1985; Lee and Steinberg 1980).

Characteristics of the environment

Environmental conditions can increase tension for change or hamper an organization from attaining its goals. Environmental characteristics include the social system of which the target group is a member and the interrelated systems over which the target group has no direct control. For instance, a state policymaker wishing to alter nursing home regulations may face more difficulty if the modifications conflict with federal regulations.

Research indicates that the relative power of the target group and other constituencies will affect success. Greiner (1967) and Geisler and Rubenstein (1987) found that strong support for change from all organizational levels was characteristic of success. Anderson and Narasimham (1979), Cain (1979), Lee and Steinberg (1980), and Maher (1973) propose that top management not only support the target group but also help establish the climate for change. Huysmans (1970) states that top management must be interested in the change and understand it, while others (e.g., Meyer and Goes 1988) emphasize the importance of a project champion.

Organizational factors, such as environmental turbulence, complexity, formality, centralization (Robey and Zeller 1978), and size (Kimberly and Evanisko 1981) affect the likelihood of implementation. Carter and Schlesinger (1979) suggest three key situational variables: (1) the position of the initiators vis-à-vis the resistors in terms of power and trust, (2) the locus of relevant data needed in the implementation effort, and (3) the stakes.

These studies provide important insights about the impact of specific factors, but the multidimensional nature of the implementation problem requires that the analysis extend beyond an item-by-item specification of the factors. We must be able to aggregate and trade off the influence of individual factors before we can estimate the overall probability of success. Such an aggregation model must address these points: What is the relative importance of each factor? How can we measure each factor's degree of achievement? How can we combine the various factors to yield an overall prediction of success or failure?

We know of no quantitative model specifically designed to predict implementation success in the field of health and social policy. But research and development project selection models have been developed. Adaptations to his simple model have included rating (rather than dichotomous scoring), weighting of factor importance (Roberto and Pinson 1972), and cutoff criteria, a means of combining judgments of several factors (Hamilton 1974).

Cooper's (1980) research advanced the field of project selection by using factor analysis to reduce 48 variables to an eight-factor model. Using a regression analysis on 195 projects, the model correctly predicted the success or failure in 84 percent of cases. Because this model was not developed for the social policy arena, in which politics plays a major role in acceptance and dissemination, many of the factors mentioned above were not considered. Still, this model offers reason to hope that a similar model might be developed for social policy.

Model Development

The literature provided a foundation for creating a model to predict and explain implementation success, but four other steps were also needed to complete the model. We completed the model using a variant of the integrative group process. First, we selected the most relevant factors for predicting implementation success; second, we developed a means to measure each factor; third, we determined how to aggregate these factors into a mathematical prediction; and fourth, we estimated the relative importance weight of each factor.

Selecting factors

The literature suggests that a plethora of factors may be important in predicting success of implementation, but for a prediction model the number must be reduced to a manageable size, so we convened a panel of experts to select the vital factors.

The panel gave us theoretical and practical experience in organizational change and program planning. The members were three nationally recognized academics; the director of the division of health in a midwestern state; the director, assistant director, and senior planner for a technical assistance organization to health planners; and the director, associate director, and senior planner for the lead health policy agency in a state government.

From the list of factors in the literature review, the panel chose ones they considered particularly diagnostic for distinguishing successful implementations from unsuccessful ones. Later the planners and change theorists reviewed the refined list of factors to identify strong dependencies. Where the majority of these experts deemed two factors to be strongly dependent, we either created two new factors that reflected the unique aspects of the original factors or, failing that, we eliminated one of the factors. The panelists identified five categories as shown in Table 15-1.

Measures

Most of the variables selected for the model could not be measured objectively. For instance, performance on factor 2 (problem exploration) is very much a matter of judgment, and the panelists could not find any objective measure for it. The panel felt these aspects of problem exploration were important: target group involvement, the use of empirical data to document the problem's existence, and the use of observation to promote firsthand understanding.

We decided to measure each factor by writing a description of excellent, mediocre, and poor performance (according to whether such a performance would increase or decrease the likelihood of implementation). Then a real implementation effort could be compared to those benchmarks. The benchmarks for problem exploration were:

Highest rating: The planner makes a strong effort to involve the target group, to observe the system, and to use the data in problem exploration.

Middle rating: The planner has firsthand experience with the system and feels he or she knows its problems. No attempt is made to explore data or draw upon target group opinion.

Table 15-1 Categories of Factors for Predicting Implementation Success

Target group
Tension for change
Perceived chance of success
Previous history with change
Proposed change
Radicalness of design
Evidence of effectiveness
Relative advantage
Change agent
Reputation
Commitment
Role
Consistency of values
Planning and implementation process
Planning mandate
Problem exploration
External expertise
Alternative solutions
Complexity of implementation process
Feedback
Implementation materials
Environment
Endorsement of key power groups
Impact on supporters/opposition
Staff qualification
Power groups involvement
Endorsement of middle managers
Funding availability

Lowest rating: No empirical data are used to set priorities. No attempt is made to draw on the target group's opinions or gain firsthand experience with the system.

The description of factors and at least two benchmarks for each are found in Appendix 15-A.

Aggregation methods

Once the factors were identified and measures specified, a mathematical model was developed to combine assessments of each variable and estimate the potential for success. A mathematical model not only can quantify the variables in the analysis but also can produce a score. We can use this score to compare several plans, which helps us explore several approaches to implementing a policy. Such simulations allow us to test-drive policies from the safety of the office before moving into the field—where mistakes have real consequences.

Mathematical models also force a rigor in the analysis that might be lacking in less formal approaches. As we specify each factor in quantitative terms, we must examine and quantify the relationships among factors, leading us to examine in depth the concepts underlying the model. Two different types of model are reasonable candidates for the task of predicting implementation success.

Bayesian model

Bayesian statistical models (see Chapter 8) estimate the probability of an event based on information about factors that influence it. A Bayesian model could, for instance, use information about the extent of problem exploration to predict the likelihood of successful implementation. A second reason for using a Bayesian model is that it uses primarily judgmental, rather than empirical, evidence. Bayesian models using subjective estimates have been used (Gustafson et al. 1969) to aid in problem areas where empirical data are limited but expertise is plentiful, such as suicide detection (Gustafson et al. 1977).

Bayesian models depend on certain definitions. In this case:

- In predicting implementation, there will be two hypotheses: successful implementation (H) and failure of implementation (H*)
- In our model, 24 factors were used to predict success, and each factor has several levels. We define F_{ij} to be the *j*th level of the *i*th factor.
- The prior probability of success, or failure, will be designated by p(H) and p(H*), respectively.

- The *likelihood* of finding the *j*th level of the *i*th factor in a situation where the implementation was successful is $p(H/F_{ij})$.

As usual, we employed the odds form of Bayes' theorem, where

$$\frac{p(H/F_{ij})}{p(H^*/F_{ij})} = \text{Posterior odds}$$

$$\frac{p(F_{ij}/H)}{p(F_{ij}/H^*)} = \text{Likelihood ratios}$$

$$\frac{p(H)}{p(H^*)} = \text{Prior odds}$$

Assuming conditional independence of the variable (we have tested this already), Bayes' theorem is written:

$$\frac{p(H/F_{il}, \ldots, F_{i24})}{p(H^*/F_{il}, \ldots, F_{i24})} = \frac{p(F_{il}/H) \ldots p(F_{i24}/H)}{p(F_{il}/H^*) \ldots p(F_{i24}/H^*)} \; \frac{p(H^*)}{p(H^*)} \quad (3)$$

Because Bayesian models are multiplicative, they can yield a probability estimate using as many or as few variables as are known. While greater information yields a better estimate, missing data are not a technical problem.

The primary disadvantage of the Bayesian model is that its multiplicative nature causes errors to multiply. Significant mistakes can result from erroneously estimating likelihood ratios. As a result, the model may yield extremely large (10,000 to 1) or small probabilities. Still, the model has been quite effective in several settings (Gustafson and Huber 1976), such as estimating the probability that a person complaining of suicidal thoughts will attempt suicide (Gustafson et al. 1977) or measuring quality of care (Gustafson et al. 1989).

MAV model

The volatility of Bayesian models led us to choose another approach: the multiattribute value (MAV) model, introduced in Chapter 7. This model treats strength of an implementation effort as a utility or "goodness" score. Unlike in a Bayesian model, the score is not directly interpreted. An implementation score of 50 does not mean a 50-percent chance for success—it is simply twice as good as a score of 25. The results are interpreted by comparing them to alternative implementations of the same policy and of other policies.

MAV models tend to be more stable than Bayesian ones, and they have successfully been used in more situations (see von Winterfeldt and Edwards (1986) for a review of many applications). The disadvantage of MAV models is that they require complete data, so we must make assumptions whenever information on a factor is missing. (If we do not make an assumption, the factor must be dropped from the model and the weights of the remaining factors normalized.)

Formally, the MAV model is:

$$U = \Sigma W_i * U(X_i)$$

where W_i = the relative importance of the *i*th factor in predicting success

$U(X_i)$ = the score assigned to the implementation effort in terms of its status on a given level of factor *i*

U = the overall utility score assigned to the implementation effort

Estimating likelihood ratios and weights

In developing the implementation model, we need estimates for likelihood ratios and prior odds (see Chapter 8 for definitions) in the Bayesian model and for the weights and utilities in the MAV model. These estimates were obtained from the panel of experts, using estimation methods reported in the decision theory literature (Slovic and Lichtenstein 1971; Hogarth 1975, 1980).

Each expert was asked to estimate the likelihood ratio for each benchmark on each variable in the model. When the estimates were similar, they were averaged. When they differed substantially, panelists were asked to discuss the rationale for their estimates and then reestimate. These discussions enhanced understanding of the topic and reduced differences. The estimators were given logarithmically calibrated scales on which to make the estimates. This amounted to approximately 80 estimates since several variables had high, low, and medium benchmarks. An example of the Bayesian likelihood ratio estimation task is shown in Table 15-2.

For the MAV model, two types of estimates were needed: the relative importance of the factor and the utility scores of different

Table 15-2 The Format for Rating Projects in Terms of Their Implementation Potential with a Comparison of MAV and Bayesian Estimates for Problem Exploration

Problem Exploration (MAV Weight = .04)

	MAV Utility	*Benchmark*	*Bayes Likelihood Ratio*
How thoroughly did the planners explore the problem before establishing the goals for the plan?	100	Planners made a concerted effort to involve the community, observe the system, and use data in problem exploration.	6/1
FIRST Describe the actual situation here: __________ __________ __________ __________	25	Planners had firsthand experience with the system and feel they know its problems. No attempt to explore data or draw on community opinion.	1/3.5
THEN Enter a score in the space below indicating how well the project performs according to the benchmarks: _____	0	Empirical data were used to establish priorities. No attempt was made to draw on the opinions of the community or to observe the system.	1/9

levels of each factor. To obtain these estimates, the panel of experts grouped the model's 24 factors in rough order of importance and assigned a score of 10 to the least important. Finally, the experts estimated scores for the relative importance of all other factors. (A detailed discussion of this method can be found in von Winterfeldt and Edwards [1986].) For example, if "evidence of policy effectiveness" was the least important predictive factor, it would receive a 10. If "complexity of the implementation process" was considered twice as important, it would be assigned 20.

After all scores were assigned, the panelists' estimates were again compared. If a large difference of opinion was noted, those panelists were asked to reconsider their estimates. Scores from each panelist were averaged and normalized, and these normalized scores became the factor weights shown in Table 15-2.

The second type of estimate needed for the MAV models was the relative utility of each benchmark for each factor. We could have estimated these utilities independently of the Bayesian estimates by asking the experts to assign each factor a utility between 0 and 100. However, we converted the likelihood ratio scores to utilities to save panel time. A score of 0 was assigned to the descriptor with the worst implication for implementation. A score of 100 was assigned to be most encouraging description. The likelihood estimates indicate where intermediate descriptions stood between these extremes. The relative position on the 0-to-100 scale was determined by interpolation. Suppose there were three likelihood ratios in a model, 1/4, 2/1, 5/1. The 1/4 would be assigned a utility of 0, and the 5/1 would be assigned a score of 100. Interpolation occurs by taking the ln(1/4), ln(2), ln(5) to get scores of –1.386, .693, and 1.609. The range 1.609 to (–1.386) is 2.995. The distance from –1.386 to .693 is 2.079, which is 69 percent of the range (2.079/2.995). So, on a scale of 0 to 1.00, the 2/1 would be converted to 0.69.

The MAV utilities and Bayesian likelihood ratio estimates for all factors are shown in Appendix 15-A. As an example, factor 2 ("problem exploration") is reproduced in Table 15-2.

Model Validation

As with any innovation, a model needs validation to be sure it works and to justify acceptance. Evaluations of these MAV and Bayesian models included: a reliability test, with several people rating identical cases; a reliability test of the process of translating interview data into MAV utilities and Bayesian likelihood ratios; a comparison of individuals on the implementation potential of a specific policy; and a field test comparing MAV model predictions with a real case.

Reliability tests

Twenty graduate students in a health administration master's degree program (with an average of 3.25 years of experience in the field) used the models to evaluate descriptions of 10 actual health plans. The descriptions were obtained through interviews conducted at four health planning agencies in two midwestern states. The subjects were given information about the plans in terms of the model's factors. For instance, the plans included implementation of a new budgeting system for a mental health council and the addition of beds to a community hospital. Subjects rated each plan on each factor in our implementation model.

After rating the factors, the subjects were asked to estimate the probability of success. At the same time, we used each model to calculate the implementation potential and compared the scores to the subjects' estimates. Both models correlated with expert judgment quite well (Bayes = .77 and MAV = .78) (see Table 15-3). There are no firm standards against which these performance figures can be compared. But one benchmark would be how well these correlations compare with performance figures for other models. It turns out that our models predict judgments as well as other Bayesian and MAV models developed for other health and social service problems (Gustafson et al. 1986; Gustafson et al. 1977).

One way to validate a model is to use it to predict the success of a project while asking a panel of experts to make the same prediction. A comparison of the two predictions yields one measure of validity. Since we would expect that the model would be best at predicting the judgments of the panel that developed it, we prefer to use separate panels for development and validation.

The model should be able to predict a subject's overall rating of a project if it is given that person's component scores. However, the results may underestimate the model's validity. Low correlations can be caused not only by a poor model but also by the subject's inconsistent rating of the overall project. Alternatively, correlations can be misleading when we compare the model's prediction of success to an expert panel's predictions, not to reality. The evaluation shown in Table 15-3 is weaker still because it used subjects' component scores to predict their own overall ratings, not those of someone else.

The second reliability test was designed to remove the effects of (1) individual subject's inconsistencies in factor scores and overall ratings and (2) using the model to predict one subject's overall ratings based on another person's factor scores. The test consisted of using the factor scores assigned by one set of subjects to predict overall project ratings assigned by another set of subjects. The overall project ratings of six subjects randomly selected from the twenty were averaged (arithmetic means for the MAV model, geometric means for the Bayesian) for each health plan. The average ratings

Table 15-3 Correlation between the Calculated Implementation Scores Using Individual Subject Ratings on the Implementation Success Factors and the Individual Subject's Overall Rating of Each Plan: Probabilities of Implementation Success

	Bayes	*MAV*
Correlation averaged across subjects	.77	.78
Average standard deviation	.13	.09

are shown in column 2 of Table 15-4 along with the standard deviation (in parentheses).

The factor scores from each of the remaining 14 subjects were then averaged and used in the MAV and the Bayesian models to predict success. The estimates assigned by the MAV and Bayesian models are shown in columns 3 and 4 of Table 15-4. The correlation between overall ratings by 6 subjects and the model scores from 14 others was .95 with either model. This suggests that models using average scores from one set of judges can predict average overall ratings of implementation success by another set of judges.

Field test

The MAV implementation model was field-tested as part of an evaluation of nursing home regulatory models in Wisconsin. Two approaches to regulating quality of care in nursing homes were being compared. In both approaches a team of state surveyors identified deficiencies, issued citations, and consulted with the nursing homes on ways to improve care.

The first approach, the Quality Assurance Project (QAP), was an experimental process that relied on an in-depth review of a statistical sample of residents and a screening visit to the facility. If problems were found in that sample and/or during the visit, a more extensive examination was initiated. The second approach has been standard procedure for monitoring nursing home quality and was labeled the traditional (TR) process. It consisted of briefly interviewing all nursing home residents and evaluating all homes on 1,547 regulations. The TR process was the same regardless of how good the home appeared.

Table 15-4 Means and Standard Deviations for Subjective Probability Estimates of a Subgroup of 6 Judges compared to Means and Standard Deviations of MAV and Bayesian Scores Obtained from a Subgroup of the Other 14 Judges

Case	*Overall Ratings*	*MAV Model*	*Bayes Model*
1	53 (.18)	.53 (.08)	.61 (.10)
2	34 (.09)	.40 (.08)	.28 (.15)
3	77 (.08)	.70 (.08)	.87 (.05)
4	49 (.15)	.47 (.07)	.47 (.21)
5	79 (.08)	.76 (.08)	.76 (.26)
6	29 (.12)	.41 (.08)	.35 (.23)
7	21 (.11)	.38 (.09)	.29 (.17)
8	35 (.15)	.41 (.09)	.36 (.22)
9	69 (.12)	.60 (.08)	.80 (.11)
10	39 (.09)	.62 (.06)	.36 (.21)

Note: Correlation between models and overall ratings: MAV = .95; Bayes = .95; standard deviations are in parentheses.

To evaluate the two approaches, a research team, consisting of a nurse and a recreation therapist, observed their application in 13 homes and rated the proceedings using the MAV implementation model factors. Ideally, more nursing homes would have been involved, but each review took several days of costly professional time. It should be noted that the implementation model was adapted in two ways for this study. First, seven factors (1 and 19–24) were eliminated as inappropriate to nursing homes. Second, the factors were rewritten to apply specifically to nursing homes. For instance, factor 3 ("use of outside experts") was adapted to recognize that the state surveyors were to some extent outside experts.

Instead of rating a particular deficiency, the research team used the implementation model factors to rate the significance of all deficiencies as a whole, the tenor of the target group (nursing home personnel), the general problem-solving and implementation planning effort, and the change agents (state surveyors). Four months later, the research team revisited each home and judged what percentage of deficiencies had been corrected. Scores from the first visits were compared to the implementation estimated by the two survey methods, QAP and TR.

Predictions of success (using the MAV model) were compared to the percentage of corrections found during the follow-up inspection to give a correlation of .80. The results provide further evidence of the model's usefulness for predicting implementation success. The field test also suggests how the model might be used to improve implementation. We present this use of the models as an example of how the technique might help a change agency work better.

If a change agent were interested in maximizing the chance of successful implementation, he or she could describe the situation in terms of the implementation model factors and use the model to predict the chance of success. By weighting the factors in terms of relative importance, we can identify which changes would most improve the implementation scores.

The TR and QAP survey methods can also be compared by looking at the average scores each method receives on each factor. An example from Figure 15-1 might help. Factor 16 in that figure is "endorsement of middle managers." Change theory suggests that a process intended to promote change will be more successful if it induces middle management to support the change. The QAP method was substantially more effective than TR in securing such support (a value score of 61 vs. 39). In fact, the ability to secure that endorsement was one of the big differences between the methods. Similarly, the TR method received a score of .62 on "problem exploration," while the QAP method received .42. This suggests that problem exploration was carried out more effectively in the TR model.

In this way, the analysis of factor scores help us understand the strengths and weaknesses of each regulatory system. For instance, the changes proposed by the TR method tended to have a better "history of success" (15) while those from QAP tended to create a better "long-term monitoring process" to ensure that change actually took place (8). Neither method was particularly effective at creating "tension for change" (12) in the organization. QAP seemed to have little ability to create several "alternative solutions" (4) while TR tended to create more "complex solutions" (6). To determine the most efficient means of improving the regulatory system, we would need to multiply these difference scores (e.g., .62–42 for problem exploration) by the weight of each factor and sum these products.

Figure 15-1 Average Intervention Scores for the TR and QAP Methods

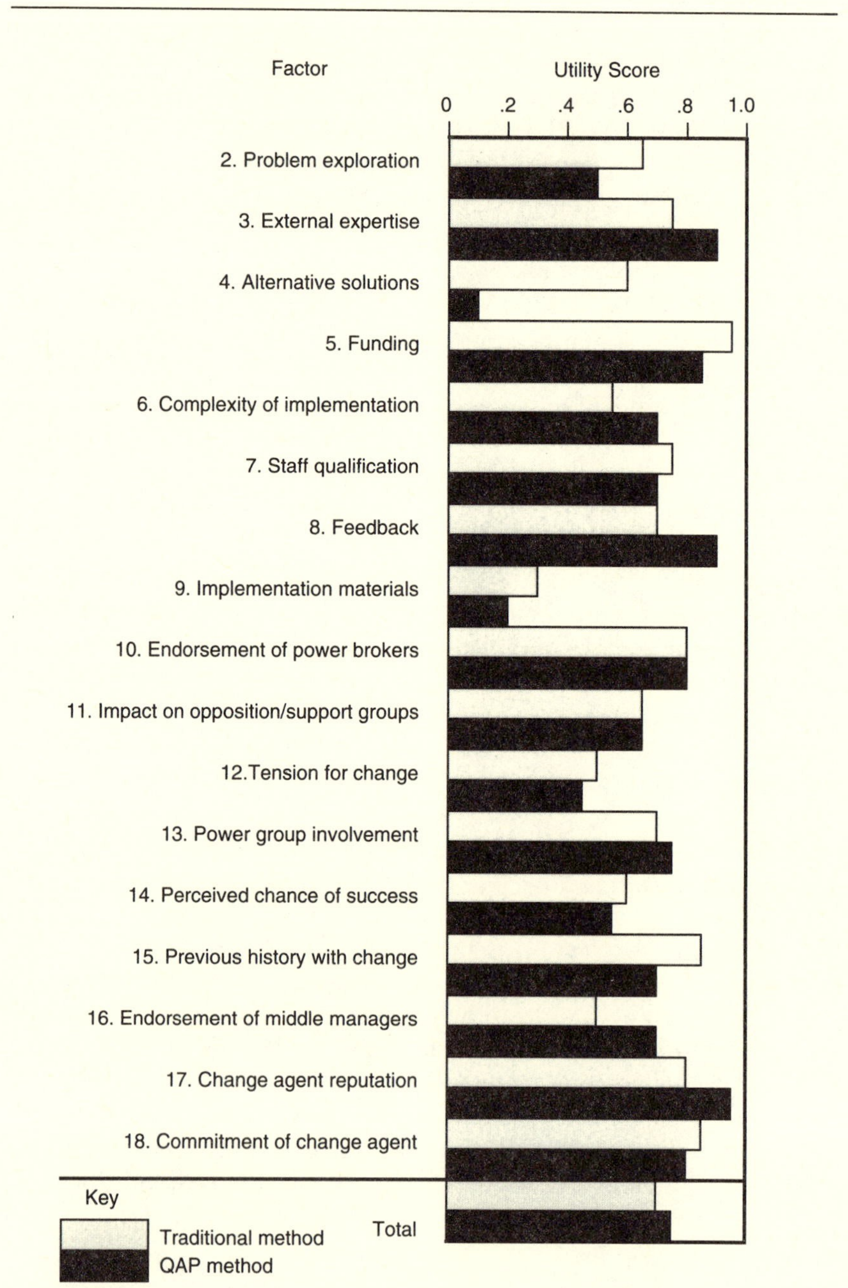

Summary

This application of Bayesian and MAV models focused on developing and testing a model for predicting the potential for successfully implementing a proposed change. Results are encouraging. The predictions of the Bayesian and MAV models were similar to those of several judges who reviewed descriptions of an innovative program. The models also effectively predicted the actual outcomes of change programs in several nursing homes. The models may also be useful in pinpointing the strengths and weaknesses of an implementation, and in suggesting where adjustments can be made during all stages of the process.

Still, several questions remain unanswered. Will the predictive success suggested by a study of these nursing homes be demonstrated in studies of larger samples? Will the model's predictive capability generalize to other settings? (The model seems generalizable as long as irrelevant variables can be deleted.) Will the models actually be useful for planning further interventions if a project requires?

An interactive computer-based implementation analysis tool has been developed using the MAV model. Using as a framework the factors and benchmarks shown in Appendix 15-A, the computer interviews the user to collect data on implementation. The computer then processes the data to predict the likelihood of success and gives feedback on the effort's strengths and weaknesses. The manager can examine "what if" questions about possible alterations in the implementation plan and can modify inappropriate weights in the model. Preliminary trials suggest the system may be quite helpful in thinking through implementation strategies.

Appendix 15-A MAV Utilities and Bayesian Likelihood Ratio Estimates

	Benchmarks and Value Measures		
Factor/Description	*Benchmark 1*	*Benchmark 2*	*Benchmark 3*
1. *Planning mandate.* How strong was the planning mandate from external power brokers when this project began?	A group with substantial influence and with support of the community studied the goals carefully and gave this project a high priority before planning began.	A group of substantial influence did not seriously study the goals. They merely adopted the priorities recommended by the staff.	The planning was carried even though the board was clearly opposed to addressing this goal.
MAV weight = .05	Utility = 100 Likelihood ratio = 6/1	Utility = 40 Likelihood ratio = 1.5/1	Utility = 0 Likelihood ratio = 1/4
2. *Problem exploration.* How thoroughly did the planners explore the problem before beginning to develop the proposed design?	Planners made a concerted effort to involve the target group, to observe the system, and to use the data in problem exploration.	Planners had first hand experience with the system and feel they know its problems. No attempt was made to explore data or draw on target group opinion.	Empirical data was used to establish priorities. No attempt was made to draw on the opinions of the target group or to observe the system.
MAV weight = .04	Utility = 100 Likelihood ratio = 6/1	Utility = 45 Likelihood ratio = 1/3.5	Utility = 0 Likelihood ratio = 1/9

3. *External expertise.* Were outside experts and literature reviews used in designing the solution to the problems?	People who have studied or dealt with this problem strongly influenced the design team. There is evidence that latest advances in the field have been considered. But the design has been tailored to the unique needs of this setting.	Action was patterned after successful application elsewhere.	Almost no attempt was made to draw on the experience of people who have studied or dealt with this problem before.
MAV weight = .02	Utility = 100 Likelihood ratio = 3/1	Utility = 30 Likelihood ratio = 1/1/	Utility = 0 Likelihood ratio = 1/2
4. *Alternative solutions.* To what extent were significantly different alternatives to the recommended action developed during the planning process?	Several different alternatives were developed, advantages and disadvantages documented and used to select the best alternative.	Several different alternatives were developed but no formal documentation of advantages and disadvantages was done.	No serious attempt was made to explore the alternatives that were really different from the one chose.
MAV weight = .02	Utility = 100 Likelihood ratio = 3/1	Utility = 60 Likelihood ratio = 1.5/1	Utility = 0 Likelihood ratio = 1/3

Appendix 15-A Continued

Factor/Description	Benchmarks and Value Measures		
	Benchmark 1	*Benchmark 2*	*Benchmark 3*
5. *Funding.* Are funds available to support implementation of this project?	Funds needed to implement have been committed.	No money is needed. A funding source has been identified who appears sincerely interested in the project.	Money is needed. Initial contact suggest that potential funders have many other options and have expressed pessimism about funding for this project.
MAV weight = .05	Utility = 100 Likelihood ratio = 4.5/1	Utility = 65 Likelihood ratio = 1.5/1	Utility = 0 Likelihood ratio = 1/8
6. *Complexity of implementation.* How complex is the process for implementing this recommended action?	The implementation process should be very easy to accomplish. Little system reorganization, red tape, or target group reorientation is involved.	The change will involve some reorganization, red tape, and reorientation, but this should not be drastic.	Extensive system reorganization, red tape, and target group reorientation is involved.
MAV weight = .03	Utility = 100 Likelihood ratio = 2.5	Utility = 60 Likelihood ratio = 1.5/1	Utility = 0 Likelihood ratio = 1/2.5
7. *Staff qualification.* Is staff provided with sufficient background information and training to carry out proposed change?	Personnel are available who are properly trained or have the experience/education necessary to learn the change.	Personnel are not currently qualified and training will be provided but it will not cover the critical issues thoroughly.	Appropriately trained personnel will not be available.

MAV weight = .02	Utility = 100 Likelihood ratio = 1.5/1	Utility = 40 Likelihood ratio = 1/2	Utility = 0 Likelihood ratio = 1/4
8. *Feedback.* How will the change agent obtain feedback from the target group during implementation?	A system has been developed to obtain and respond to feedback. Communications are open and frequent.	No system has been designed to obtain and respond to feedback but information communications are fairly good.	No system has been designed to obtain and respond to feedback and communication with the target group is strained.
MAV weight = .02	Utility = 100 Likelihood ratio = 2/1	Utility = 60 Likelihood ratio = 1.3/1	Utility = 0 Likelihood ratio = 1/2
9. *Implementation materials.* To what extent do written materials exist to explain the new project and guide implementation?	Educational materials and guidelines have been developed to help this implementation	Parties affected have no written materials describing how the new system should work.	
MAV weight = .02	Utilty = 100 Likelihood ratio = 2/1	Utility = 0 Likelihood ratio = 1/4	
10. *Endorsement of key power groups.* How strong were the endorsements of key power brokers regarding the proposal before implementation begins?	Key influentials not only endorse the proposal but committed some resources (time, money, etc.) to support its implementation.	The key influentials have all endorsed this proposal. Utilty = 60 Likelihood ratio = 2/1 This is a controversial issue. Key influentials are about evenly split about this proposal.	Key influentials have expressed uniform opposition to this proposal.

Appendix 15-A Continued

	Benchmarks and Value Measures		
Factor/Description	*Benchmark 1*	*Benchmark 2*	*Benchmark 3*
MAV weight = .22	Utility = 100 Likelihood ratio = 32/1	Utility = 40 Likelihood ratio = 1.5/1	Utility = 0 Likelihood ratio = 1/24
11. *Impact on supporters opposition.* To what extent will implementation (or failure to implement) affect those who support and oppose the proposal?	The outcome is much more important to supporters of the proposal than to opponents.	Outcome is equally important to both sides.	The outcome is much more important to opponents than to supporters of the proposal.
MAV weight = .10	Utility = 100 Likelihood ratio = 10/1	Utility = 50 Likelihood ratio = 1/2	Utility = 0 Likelihood ratio = 1/20
12. *Tension for change.* How dissatisfied are those people most directly affected (e.g., rank and file), with the current situation that initiated this proposal?	The people affected are very dissatisfied with their current situation. They believe change is essential.	While most of the parties affected are satisfied with their current situation, a small number of key influentials feel change is essential in this area.	The people affected feel they can live with the existing system. They see no need for recommended action.
MAV weight = .06	Utility = 100 Likelihood ratio = 6/1	Utility = 55 Likelihood ratio = 2/1	Utility = 0 Likelihood ratio = 1/9

13. *Power groups involvement.* Were the key groups (who are important to succesful implementation) actively involved in problem exploration and program design?	An intensive effort was made to involve all affected parties.	Most of the groups were involved in planning.	The most important groups were not involved in planning.
MAV weight = .05	Utility = 100 Likelihood ratio = 4/1	Utility = 70 Likelihood ratio = 2/1	Utility = 0 Likelihood ratio = 1/8
14. *Perceived change of success.* What are the perceptions of parties affected of the change of successful implementation?	Parties affected believe they are capable of implementing the change.	Parties affected generally feel that there are several obstacles to overcome but they feel they may be able to implement the change.	Parties affected generally feel that it is unlikely they will be capable of implementing the change.
MAV weight = .045	Utility = 100 Likelihood ratio = 3/1	Utility = 80 Likelihood ratio = 1.5/1	Utility = 0 Likelihood ratio = 1/8

Appendix 15-A Continued

	Benchmarks and Value Measures		
Factor/Description	*Benchmark 1*	*Benchmark 2*	*Benchmark 3*
15. *Previous history with change.* To what extent has the target group resisted attempts to implement projects of the same general character as this recommended action?	Similar projects have been successfully implemented.	There has been a mixed record of success and failure.	Implementation of similar projects has failed in the past.
MAV weight = .045	Utility = 100 Likelihood ratio = 4/1	Utility = 30 Likelihood ratio = 1/1.1	Utility = 0 Likelihood ratio = 1/6
16. *Endorsement of middle managers.* How strong were the endorsements of middle level internal managers regarding the proposed design before the implementation phase began?	Key middle managers not only endorse this action but committed their support to implementation.	Key middle managers are keeping a hands-off attitude regarding the project.	Key middle managers have expressed opposition to this action.
MAV weight = .02	Utility = 100 Likelihood ratio = 3/1	Utility = 40 Likelihood ratio = 1/2	Utility = 0 Likelihood ratio = 1/4
17. *Change agent reputation.* What is the change agent reputation?	Change agent is trusted and respected by the parties involved.	Change agent has not established a reputation.	Change agent is distrusted and disrespected by parties involved in the change.

MAV weight = .04	Utility = 100 Likelihood ratio = 4/1	Utility = 75 Likelihood ratio = 1.5/1	Utility = 0 Likelihood ratio = 1/6
18. *Commitment of change agent.* How committed is the change agent to this project?	The change agent believes in and is excited about the project. He is committed to make it work.		The principal change agent considers this project just part of the job. He is not particularly excited about it.
MAV weight = 0.26	Utility = 100 Likelihood ratio = 6/1		Utility = 0 Likelihood ratio = 1/4
19. *Role of change agent.* What power to force implementation does the change agent have?	The planning group has power to force implementation.	The planning group is supporting another party with enforcement power who has assumed principal responsibility for implementation.	The planning group has no enforcement power and is the change agent.
MAV weight = .024	Utility = 100 Likelihood ratio = 6/1	Utility = 80 Likelihood ratio = 4/1	Utility = 0 Likelihood ratio = 1/1
20. *Consistency of values.* To what extent do the change agent and target group share similar values?	The change agent and the target group share the same values.	The change agent and the target group have quite different values.	
MAV weight = .02	Utility = 100 Likelihood ratio = 2/1	Utility = 0 Likelihood ratio = 1/4	

Appendix 15-A Continued

	Benchmarks and Value Measures		
Factor/Description	*Benchmark 1*	*Benchmark 2*	*Benchmark 3*
21. *Radicalness of design.* How radical is the proposed design?	The proposed design is consistent with the prevailing philosophy of the community or organization.	The proposed design is a significant departure for this community/organization. The concept is new to most of the people affected.	
MAV weight = .02	Utility = 100 Likelihood ratio = 3/1	Utility = 0 Likelihood ratio = 1/2	
22. *Evidence of effectiveness.* What kind of concrete evidence exists to suggest this proposed design will work?	Concrete evidence exists that documents the success of this change in similar settings.	Concrete evidence exists that documents that this program has almost always failed when tried elsewhere.	
MAV weight = .02	Utility = 100 Likelihood ratio = 1/5	Utility = 0 Likelihood ratio = 1/4	
23. *Relative advantages of design.* How do parties involved perceive the advantages and disadvantages of the recommended action?	Parties involved feel the proposed design has many more advantages than disadvantages.	Parties involved do not have a clear perception of advantages and disadvantages.	Parties involved feel the proposed design has many more disadvantages than advantages.
MAV weight = .02	Utility = 100 Likelihood ratio = 3/1	Utility = 60 Likelihood ratio = 1.5/1	Utility = 0 Likelihood ratio = 1/3

24. *Flexibility of design.* How hard is it to modify (without changing the critical elements) the proposed design?	The proposed design easily incorporates many of the suggestions made for its modification.	The proposed design would be very difficult to modify.
MAV weight = .02	Utility = 100 Likelihood ratio = 3/1	Utility = 0 Likelihood ratio = 1/1

References

Ackoff, R. L. 1967. "Management Misinformation Systems." *Management Science* 14 (4): B147–56.

———. 1960. "Unsuccessful Case Studies and Why." *Operations Research* 8 (4): 259–63.

Alavi, M., and J. C. Henderson. 1981. "An Evolutionary Strategy for Implementing a Decision Support System." *Management Science* 27 (11): 1309–23.

Alter, S. A. 1980. *Decision Support Systems: Current Practice and Continuing Challenges.* Reading, MA:. Addison-Wesley.

Anderson, J. C., and R. Narasimhan. 1979. "Assessing Project Implementation Risk: A Methodological Approach." *Management Science* 25 (6): 512–21.

Anderson, J. C., N. L. Chervany, and R. Narasimham. 1979. "Is Implementation Research Relevant for the OR/MS Practitioner?" *Interfaces* 9 (3): 52–56.

Argyris, C. 1971. "Management Information Systems: The Challenge to Rationality and Emotionality." *Management Science* 17 (6): B275–92.

Bandura, A. 1977. *Social Learning Theory.* Englewood Cliffs, NJ: Prentice-Hall.

Bean, A., M. Radnor, R. Neal, and P. Tansik. 1975. "Structural and Behavioral Correlations of Implementation in U.S. Business Organizations." In *Implementing Operations Research/Management Science,* edited by R. L. Schultz and D. P. Slevin. New York: Elsevier.

Bennis, W. G. 1965. "Theory and Method in Applying Behavioral Science to Planned Organizational Change." *Applied Behavioral Science* 1 (4): 337–60.

Bolan, R. S. 1969. "Community Decision Behavior: The Culture of Planning." *American Institute of Planners Journal* 35: 301–10.

Brereton, P. R. 1972. "New Model for Effecting Change." *Journal of Extension* (Spring): 19–27.

Cain, H. 1979. "The Intangibles of Implementation." *Interfaces* 9 (5): 144–47.

Cooper, R. G. 1980. *Project NEWPROD: What Makes a New Product a Winner?* Montreal: Quebec Industrial Innovation Center.

Dalton, G.W. 1969. "Influence and Organizational Change." Paper presented at the Conference on Organizational Behavior Models, Kent State University.

Dane, C. W., C. F. Gray, and B. M. Woodworth. 1979. "Factors Affecting the Successful Application of PERT/CPM Systems in a Government Organization." *Interfaces* 9 (5): 94–98.

Delbecq, A. L. 1975. "Contextual Variables Affecting Decision Making in Program Planning." Discussion paper, School of Social Work, University of Wisconsin.

Delbecq, A., and P. Mills. 1985. "Managerial Practices that Enhance Innovation." *Organizational Dynamics* (Summer): 24–34.

Dickson, J. W. 1976. "The Adoption of Innovative Proposals as Risky Choice: A Model and Some Results." *Academy of Management Journal* 19 (2): 291–303.

Duncan, R. B., and G. Zaltman. 1975. "Ethical and Value Dilemmas in Implementation." In *Implementing Operations Research/Management Science*, edited by R. L. Schultz and D. P. Slevin. New York: Elsevier.

Evans, W. M., and G. Black. 1967. "Innovation in Business Operations: Some Factors Associated with Success or Failure of Staff Proposals." *Journal of Business* 40.

Fliegel, F. C., and J. E. Kivlin. 1966. "Attributes of Innovations as Factors in Diffusion." *American Journal of Sociology* 72: 235–48.

Freeman, C. 1982. *The Economics of Industrial Innovation.* Cambridge, MA: MIT Press.

Geisler, E., and A. Rubenstein. 1987. "The Successful Implementation of Application Software in New Production Systems." *Interfaces* 17 (3): 18–24.

Ginzberg, M. J. 1981. "Early Diagnosis of MIS Implementation Failure: Promising Results and Unanswered Questions." *Management Science* 27 (4): 459–78.

———. 1978. "Steps Toward More Effective Implementation of MS and MIS." *Interfaces* 8 (3): 57–63.

Greiner, L. E. 1967. "Patterns of Organization Change." *Harvard Business Review* (May-June): 119–30.

Guimaraes, T. 1981. "Understanding Implementation Failure." *Journal of Systems Management* 32 (3): 12–17.

Gustafson, D. H., and G. Huber. 1977. "Behavioral Decision Theory and the Health Delivery Systems." In *Human Judgment and Decision Processes*, edited by Martin F. Kaplan and Steven Schwartz. San Diego: Academic Press.

Gustafson, D. H., W. Edwards, and L. Phillips. 1969. "Subjective Probability in Medical Diagnosis." *IEEE Transactions in Man-Machine Systems* 10, no. 3.

Gustafson D. H., D. Fryback, J. Rose, V. Yick, C. Prokop, D. Detmer, and J. Llewelyn. "A Decision Theoretic Methodolgy for Severity Index Develompent." *Medical Decision Making* 6 (1): 27–35.

Gustafson, D. H., J. H. Greist, F. F. Strauss, H. Erdman, and T. Laughren. 1977. "A Probabilistic System for Identifying Suicide Attempts." *Computers and Biomedical Research* 10, no. 2.

Gustafson, D. H., S. W. Johnson, M. J . Sateia, and F. Sainfort. 1989. "Measuring Quality of Care in Psychiatric Emergencies: The Construction and Validation of a Bayesian Index." Maine Health Information Center, Augusta.

Gustafson, D. H., G. L. Rose, and N. J. Howes. 1979. "Roles and Training for Future Health Systems Engineers." In *The Commission on Education for Health Administration*, Vol. II.

Hage, J., and M. Aiken. 1970. *Social Change in Complex Organizations.* New York: Random House.

Hamilton, H. R. 1974. "Screening Business Development Opportunities." *Business Horizons* (August).

Harvey, A. 1970. "Factors Making for Implementation Success and Failure." *Management Science* 16, no. 6.

Hogarth, R. M. 1975. "Cognitive Processes and the Assessment of Subjective Probability Distributions." *Journal of American Statistical Association* 70, no. 350.

Huse, E. 1975. *Organization Development and Change*. St. Paul: West Publishing.

Huysmans, J. H. B. M. 1970. *Implementation of Operations Research*. New York: John Wiley.

Kimberly, J. R., and M. J. Evanisko. 1981. "Organizational Innovation: The Influence of Individual, Organizational, and Contextual Factors on Hospital Adoption of Technological and Administrative Innovations." *Academy of Management Journal* 24 (4): 689–713.

Kotter, J. P., and L. A. Schlesinger. 1979. "Choosing Strategies for Change." *Harvard Business Review* 57 (2): 106–14.

Lawrence, P. R. 1969. "How to Deal with Resistance to Change." *Harvard Business Review* (January-February).

Leavitt, H. J. 1964. "Applied Organizational Change in Industry: Structural, Technological, and Humanistic Approaches." In *Organizations*, edited by J. G. March and H. A. Simon. New York: John Wiley.

Lee, W. B., and E. Steinberg. 1980. "Making Implementation a Success or Failure." *Journal of Systems Management* 3 (4): 19–25.

Lippitt, R., J. Watson, and B. Westly. 1958. *The Dynamics of Planned Change*. New York: Harcourt, Brace and World.

Lippitt, S. L. 1973. *Visualizing Change: Model Building and the Change Process*. Fairfax, VA: Learning Resources Corporation.

Little, J. D. C. 1970. "Models and Managers: The Concept of a Decision Calculus." *Management Science* 16 (8): B466-85.

Lockett, A. G., and Polding, E. 1978. "OR/MS Implementation-A Variety of Processes." *Interfaces* 9 (1): 45–50.

Lonnstedt, L. 1985. "Factors Related to the Implementation of Operations Research Solutions." *Interfaces* (February).

Lucas, H. C. 1975. *Why Information Systems Fail*. New York: Columbia University Press.

McKinsey & Co. 1968. "Unlocking the Computer's Profit Potential." New York.

Maher, P. M. 1973. "Attitudes and Conclusions Resulting from an Experiment with a Computer-Based R&D Project Selection Techniques." *R&D Management* 4 (1): 1–7.

Maher, P. M., and A. H. Rubenstein. 1974. "Factors Affecting Adoption of a Quantitative Method for R&D Project Selection." *Management Science* 21 (2): 119–29.

Manley, J. H. 1975. "Implementation Attitudes: A Model and a Measurement Methodology." In *Implementing Operations Research/Management Science,* edited by R. L. Schultz and D. P. Slevin. New York: Elsevier.

Marquis, S., and D. Marquis. 1969. *Successful Industrial Innovations: A Study of Factors Underlying Innovation in Selected Firms.* National Science Foundation (NSF 19-71).

Meyer, A., and J. Goes. 1988. "Organizational Assimilation of Innovations: A Multilevel Contextual Analysis." *Academy of Management Journal* 31 (4): 897–923.

Powell, G. N. 1976. "Implementation of OR/MS in Government and Industry: A Behavioral Science Perspective." *Interfaces* 6 (4): 83–89.

Powers, R. F., and G. W. Dickson. 1973. "MIS Project Management: Myths, Opinion, and Reality." *California Management Review* 15, no. 3.

Radnor, M., and R. Neal. 1973."The Relation between Formal Procedures for Pursuing OR/MS Activities and OR/MS Group Success." *Operation Research* 21, no. 2.

Reisman, A., and C. A. de Kluyver 1975. "Strategies for Implementing Systems Studies." In *Implementing Operations Research/Management Science,* edited by R. L. Schultz and D. P. Slevin. New York: Elsevier.

Roberto, E., and C. Pinson. 1972. "Compatability Analysis for Screening New Products." *European Journal of Marketing* 6, no. 3.

Robey, D., and R. L. Zeller. 1978. "Factors Affecting the Success and Failure of an Information System for Product Quality." *Interfaces* 8 (2): 70–75.

Rogers, E. M., and F. F. Shoemaker. 1971. *Communication of Innovations: A Cross-Cultural Approach.* New York: Free Press.

Schultz, F. L., and D. P. Slevin (Eds.). 1975. *Implementing Operations Research/ Management Science.* New York: American Elsevier.

Slovic, P., and S. Lichtenstein. 1971. "Comparison of Bayesian and Regression Approaches to the Study of Information Processing and Judgment." *Organizational Behavior and Human Performance* (5).

Stevens, S. S., and E. H. Galanter. 1957. "Ratio Scales and Category Scales for a Dozen Perceptual Continua." *Journal of Experimental Psychology* 54: 377–409.

Stimson, D. H., and R. H. Stimson. 1972. *Operations Research in Hospitals: Diagnosis and Prognosis.* Chicago: Hospital Research and Educational Trust.

Vazsonyi, A. 1973. "Why Should the Management Scientist Grapple with MIS." *Interfaces* 5, no. 2.

Vertinsky, J., R. Barth, and V. Mitchell. 1975. "A Study of OR/MS Implementation as a Social Change Process." In *Implementing Operations Research/Management Science,* edited by R. L. Schultz and D. P. Slevin. New York: Elsevier.

Von Winterfeldt, D., and W. Edwards. 1986. *Decision Analysis and Behavioral Research.* New York: Cambridge University Press.

Watson, H. J., and P. G. Marett. 1979. "A Survey of Management Science Implementation Problems." *Interfaces* 9 (4): 124–27.

Wilson, J. Q. 1966. "Innovation in Organization: Notes Toward a Theory."

In *Approaches to Organizational Design,* edited by J. D. Thompson. Pittsburgh: University of Pittsburgh Press.

Wynne, B. E. 1979. "Keys to Successful Management Science Modeling." *Interfaces* 9 (4): 69–74.

Zaltman, G., R. Duncan, and J. Holbeck. 1973. *Innovations and Organizations.* New York: John Wiley.

Zand, D. C., and R. E. Sorenson. 1975. "Theory of Change and the Effective Use of Management Science." *Administrative Science Quarterly* 20 (4): 532–45.

16

Using MAV Models to Test the Effectiveness of Hospital Categorization

This chapter shows how MAV models, which were introduced in Chapter 7, can be applied as severity indexes for the evaluation of health policy. Hospital categorization is a system of classifying a hospital's ability to deliver certain services; the procedure is promoted as a vehicle for ensuring that patients receive care in the most appropriate setting, which should, at least theoretically, improve the quality of hospital care. Categorization of emergency care is mandatory in some states and has been voluntarily initiated by hospitals in others. Here we discuss a process for evaluating the effectiveness of a scheme for categorizing hospital burn care facilities. The chapter also demonstrates how MAV-based severity indexes can strengthen program evaluation and raises certain methodological issues about evaluation with MAV models.

Hospital categorization became a cornerstone of the emergency medical service system and of such treatment areas as neonatal care, open heart surgery, and burn care with the passage of the EMS Systems Act in 1973. Categorization would improve efficiency, said proponents of the technique, because it would indicate which patients (usually the more severe cases) should be transferred to hospitals in a higher category. In short, the idea was to match hospital capability to patient need.

Like most policies, categorization had negative aspects that

Thanks to Spencer Graves, Ph.D., and Joseph Rossmeissl for their contributions.

caused it to be resisted or diluted. Some states, such as Wisconsin, used a voluntary categorization system in which level was determined through a process of bargaining and negotiation. Other states avoided categorization entirely. Some reasons for this resistance are obvious: hospitals low on the categorization ladder fear losing patients to hospitals in higher categories. Moreover, the loss in prestige from a low categorization could have spillover effects, such as hindering staff recruitment or convincing the public that a low rating on one area proves the hospital inferior in general.

Resistance to categorization has taken several forms. Internists have complained that protocols for triaging trauma patients routinely and unfairly bypass private physicians and prevent them from exercising control of their patients' care. Private practitioners have accused the government of removing accident victims from their care.

But there is another reason for resisting categorization: we really do not know if it works. Perhaps more severe cases really do as well in standard care settings; perhaps the only effect of categorization is to boost health care costs by forcing additional transportation, technology, and referral costs. Perhaps the concept works but is hobbled by inappropriate categorizing criteria. Further, the process may work in one treatment area but not others. Attempts to answer these questions have been hampered by a number of technical problems.

These problems point up the need for a methodology to examine the effectiveness of categorization systems. In this chapter, we show an application of that methodology to a case study of burn care categorization, an area that has used the process for many years.

The National Burn Information Exchange (NBIE) was our primary source of data for this effort. The NBIE is a computer-based system that collects, stores, and retrieves abstracts of burn patient medical records. At the time of this study, 18 hospitals that cared for burn patients were active subscribers to NBIE, and our data set consisted of 6,223 medical record abstracts. The data included information on both direct admissions and transfer admissions.

At the time of this study, the NBIE had four categories of hospital burn care: burn centers, burn units, burn programs, and standard care programs (see Feller and Crane 1971):

- A *burn center* has a burn program with a physical facility housing only burn patients, a single doctor directing care for these patients, and a research and educational function.

- A *burn unit* is a burn center with no teaching or research components.
- A *burn program* is an organized plan of care directed by a single physician, but without a unit housing only burn patients.
- A *standard care program* has no formal program of care; burn care is supervised by several physicians, and burn patients are distributed throughout the hospital.

Methodology

Because we believe the methodology leading to the decision to construct a MAV model is an important part of using this tool, we summarize it here. We considered a number of factors for evaluating the categorization scheme. Some of these (selection of outcome measures and hospitals for the study) are important but straightforward. We will emphasize the use of the less straightforward concept of severity indexes to stratify cases, and how statistical analysis can examine differences in and among categories.

Selection of outcome measures

Although a categorization scheme should ideally be tested against both the costs it incurs and the benefits it offers, existing cost and benefit criteria left much to be desired. Mortality rate (the most common benefit criterion) did not address residual disability for patients discharged alive. But since morbidity indexes were in an early stage of development and not yet usable, we selected mortality rates as our outcome statistic. With a large number of cases, we felt it would be possible to detect differences in mortality rates for different categories. We also chose mortality because it was a universally understood outcome measure that would likely be accepted by policymakers interested in the results.

Because data on hospital charges did not represent actual hospital costs for the care of specific patients, we used length of stay as a surrogate for actual costs. Nevertheless, we recognize that both our cost and benefit measures have weaknesses.

Hospital selection. We evaluated hospital categorization by examining the performance (mortality rates and length of stay) of several institutions in a single category. In doing this, we needed enough data on the selected hospitals to help decide if differences in other criteria might explain our results better than the categorization. For instance, bed size, location, and type of patient are potentially important stratification variables.

We used the full population of hospitals participating in the NBIE as the "subject" population. Not all burn care services participated in the NBIE. This was a particular problem for standard care hospitals, because any hospital interested enough to participate in the NBIE could have quite different capability from typical standard care hospitals. This self-selection limited our attempts to generalize our findings.

Stratification of the patient population. Different quality of care is not the only factor that can account for differences in patient outcome. One of the more influential alternative explanations is that different kinds of patients are being treated in various hospitals. It is entirely plausible that the severity of conditions among patients varies widely between various hospitals. Nevertheless, if categorization works, we would expect to find greater outcome differences between hospitals in separate categories than between hospitals in the same category.

The concept of severity has received substantial attention in the literature (see Levy et al. 1978; Champion et al. 1980; Gustafson and Holloway 1975; and Gustafson et al. 1980). The basic intent of much of the literature is to develop an index that would stratify patients according to their severity of illness into different risk groups. Linn's controversial study of burn care programs is an interesting example of an attempt to account for patient severity. Linn et al. (1977) researched hospital effectiveness in Florida to determine if more burn care facilities or educational programs on burn care were needed in a region. Their study compared outcomes of randomly selected pairs of patients, one in a hospital with a specialized burn unit and another in a hospital without such a unit. In an attempt to control for severity, patients were matched for percentage of body surface burned. The results showed that patients treated in special burn care facilities studied "did not do significantly better statistically in mortality or morbidity than patients in hospitals without such facilities" but that "costs, loss in revenue, and length of stay were markedly greater in those hospitals with burn units or programs."

One problem with the Linn study (and with any effort to measure severity) is that matching patients solely on one dimension (such as amount of burned surface) may not control for other factors that could explain differences in cost or effectiveness. To put it another way, different hospitals within the same category could be treating different kinds of patients. Some hospitals in the NBIE categorization system, for instance, specialize in children, while others are devoted to veterans. Factors such as age, type of burn, location of burn, and depth of burn also help determine treatment cost or effectiveness. In fact, we performed an analysis of covariance on patients from NBIE files and found that both age and number of past medical problems were more significant predictors of mortality and length of stay than percentage of the body covered by full-thickness burn. This experience indicates that multidimensional severity indexes generally should be developed to control for confounding factors.

The severity index

The development and testing of the severity index used in the burn study was described by Gustafson and Holloway (1975). The index used the MAV modeling process (described here in Chapter 7) to combine these factors to create a single severity value: area of full-thickness burn, area of partial-thickness burn, location of burn, patient age, and medical history. These factors were derived from two sources: a literature review and structured problem-identification sessions with four physicians and a nurse, all of whom had extensive experience in burn care. The burn experts then assigned weights to each factor, developed measures, and assigned utility (value) functions to them.

Table 16-1 presents an example of how the severity index would be applied to a 38-year-old patient. This patient had a history with one important preexisting medical problem, full-thickness burns covering 40 percent of his body, partial-thickness burns covering 20 percent, and burns on all of his body except face, back, and arms. These patient characteristics were converted to scores by multiplying the weighted value of the criterion by the patient's score on the criterion. The weighted scores were summed to yield a single index value, in this case 35.

A bias in severity indexes. Severity indexes are useful, but we do not consider them a panacea for program evaluation, primarily because

Table 16-1 Research and Analysis Burn Severity Index Applied to Sample Patient

Score	*Full-Thickness Burn Area*	*Score*	*Number of Past Medical Problems*
0	Less than 10% of body	22	Four or more
5	10–24%	20	Three
11	25–39%	16	Two
18*	40–54%	12*	One
28	55–64%	0	None
33	65–74%		
37	More than 74%		

Score	*Partial-Thickness Burn Area*	*Score (add all that apply)*	*Burn Site*
0*	Less than 30% of body	1.5	Face
1	30–45%	.6*	Chest and abdomen
3	45–65%	.6*	Perineum
5	66–74%	.4	Back of head
7	Over 75%	.4*	Neck
		.4	Back
		.2	Arms and hands
		.2*	Legs and thighs
		.2*	Buttocks

35 = Sum of scores on each of the five dimensions
(18 + 0 + 3 + 12 + .6 +.6 + .4 + .2 + .2 = 35)

Score	*Age*
24	0–1 year
8	2–6
3*	7–40
8	41–50
15	51–60
22	61–70
30	Over 70

they cannot be measured with certainty. Brian Joiner, a respected consultant in quality improvement, suggested an example that demonstrates how error in a measure of severity can bias results. First, assume (Figure 16-1A) that we can measure severity perfectly and that the only source of error in predicting outcome is the randomness in our outcome variable (for example, some patients die who shouldn't). Also assume that the real mortality rates for hospitals A and B are points A and B. When we averaged mortality rates for those hospitals at two times (1 and 2), the rates were A_1, A_2, B_1, and B_2 (located one unit above and below points A and B). Under least squares (regression) analysis, points A_1 and A_2 and points B_1 and B_2 are averaged to give A and B, and the regression line is a good predictor of outcome.

But if we use the reasonable assumption that severity indexes are imperfect measures, then not only do mortality rates vary but so do severity scores. Figure 16-1B shows what happens when both vary by pretending that patients with a real severity score of A sometimes are scored a little lower or higher than A. That is a reasonable consideration because sometimes we err in estimating percentage of burn area. But the problem now is that least squares regression analysis doesn't do quite as good a job. Using least squares, we average those points vertically to give A_3, A_4, B_4, and B_3 and try to fit a straight line through them. Since a straight line won't fit, we draw the straight line that fits approximately. But least squares causes the points farthest out to have the greatest influence on the slope of that line (the solid line in Figure 16-1B). The effect is to reduce the slope of that line, leading to a biased estimate of its slope.

Finally, assume hospitals A and B are being examined and that hospital A treats only low-severity patients while hospital B treats only high-severity ones. For simplicity, assume all patients from both hospitals are combined to yield the estimated line in Figure 16-1B. We might conclude that if the points for one hospital tend to fall below the estimated line, the probability of death for its patients is below expectation. Both hospitals have an equal number of points above and below the true line, but hospital A has more points below the estimated line (that is, its mortality rate is lower than predicted) while hospital B has more points above it. From this observation, one might reach the simplistic conclusion that hospital B does worse than expected while hospital A does better than expected. However, this type of evaluation places hospitals with more severe patients at a disadvantage. (It should be noted that the plot in Figure 16-1B is an exaggeration used to show error in severity measurement.) Although the assumptions used in this example tend to inflate actual biases, such biases are likely to be present in practice. The better the severity index estimates the true severity, the lower the error in the analysis. The point is that it pays to do a good job of MAV modeling.

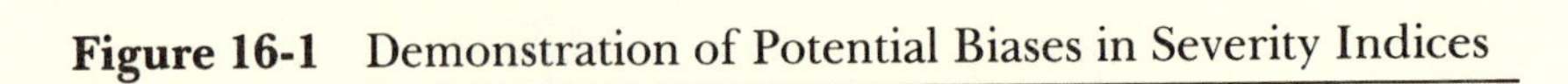

Figure 16-1 Demonstration of Potential Biases in Severity Indices

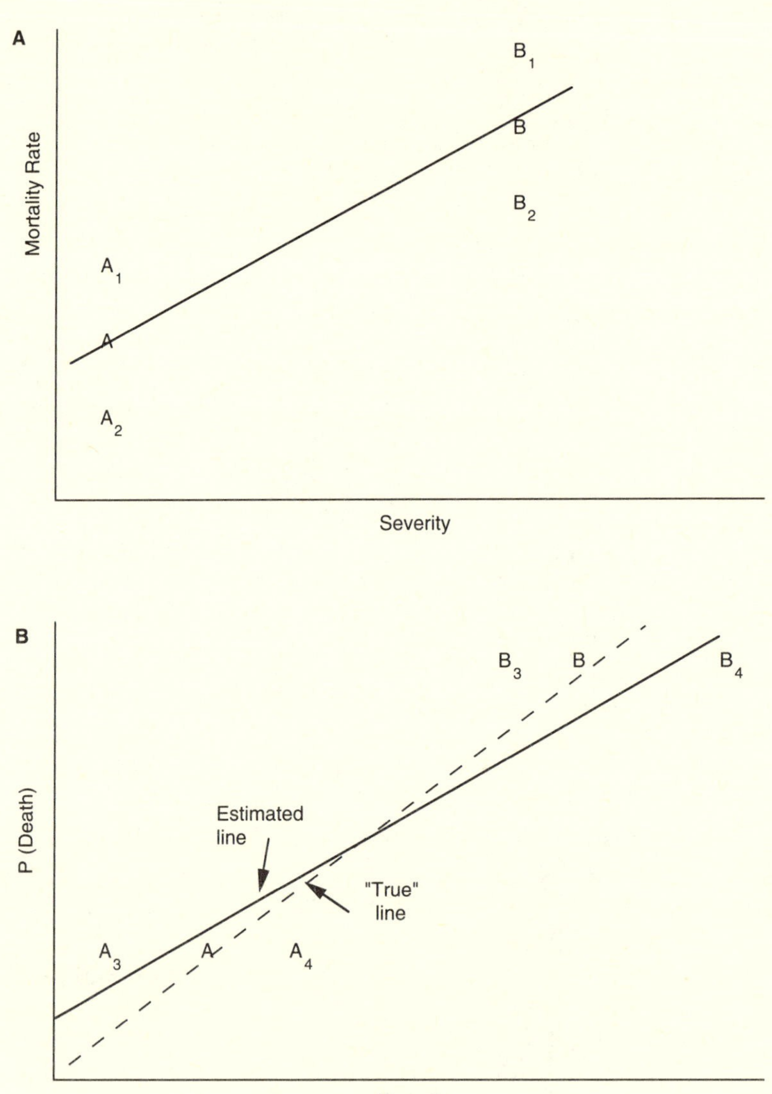

The severity index employed by Gustafson and Holloway (1975) seemed to perform quite well on the assessment criteria established for it. Index scores reflecting patient severity were correlated with severity judgments by burn specialists in three hospitals. The average Spearman's rank correlation was .83. As shown in Table 16-2, severity levels also were strongly correlated with survival rates.

Comparing outcome between categories

The severity index for the burn care evaluation was intended to answer the predictable criticism from certain hospitals that "we treat sicker patients so naturally our mortality rate is higher." But criticisms won't stop there. A good strategy for those who dislike or disbelieve the results of an evaluation is to criticize the method, so it is doubly important to ensure that the methodology is sound and accepted by all affected parties before it is used. An example might clarify this issue. Table 16-2 lists the survival rates from the categorized hospitals in the burn study, with severity shown on the left column. A score of 1.0–3.9 is least severe in this scheme.

Table 16-2 Comparison of Survival Rates between Burn Systems

Severity	*Center*	*Unit*	*Programs*	*Standard Care*	*Overall*
1–3.9	.984	.990	.995	.973	.986
4–7.9	.945	.987	.992	1.000	.957
8–11.9	.945	.989	.936	.972	.949
12–15.9	.875	.950	.900	.952	.888
16–19.9	.876	.934	.900	1.000	.894
20–27.9	.697	.969	.761	.823	.733
28–40	.481	.711	.526	.656	.517
Over 40	.096	.120	.180	.166	.115
Simple average survival rate	.827	.903	.844	.922	
Standardized survival rate	.827	.903	.851	.885	

A straightforward approach to analyzing these data would be to plot survival rates in different categories against severity, as in Figure 16-2. Apparently there is a wide difference in survival rates in the middle range of severity. Most interesting, the burn centers appear least effective. But are these real differences or just artifacts of small samples or other analytical failures? This is sure to be one of the first questions asked by people concerned that their hospitals seem to be doing a poor job of caring for patients.

In checking our results, an obvious tactic would be to test the differences between categories with a technique like analysis of covariance. However, analysis of covariance or even simple regression analysis is inappropriate for analyzing mortality data because the dependent variable (survival rate) is a probability score derived from binary data (live/die) and is limited to the range 0 to 1.0. If regression analysis is done with 0 to 1.0 data, the equation could predict scores

Figure 16-2 Survival Rates of Patients Treated in Hospitals Categorized into Different Levels, Survival Rates Compared Across Severity

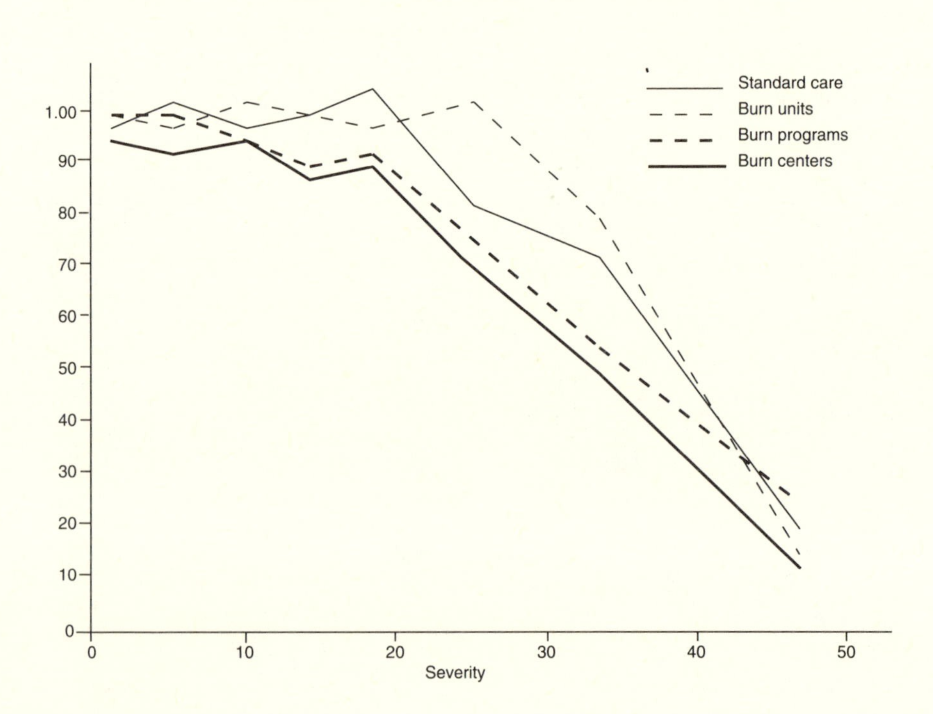

less than 0 or greater than 1.0. (See Figure 16-3.) To prevent such erroneous scores, you must perform a mathematical trick called a transformation. If mortality rate is a probability p, then survival rate is 1 – p. If you take the log of p(1 – p), you find the resulting scores remain between 0 and 1.0. This process is called a log odds transformation. As it turns out, logistic regression analysis (see Cox 1970; Graves 1980) is a form of regression analysis appropriate to binary outcome data.

The fact that the data in Figure 16-2 don't fall in a straight line causes another problem for statistical analysis because most structured procedures assume straight-line relationships. We know it would be nice to forget statistics, but the naysayers out there like to ask, "Are the lines really different?" Well, as it turns out, our transformation also took care of this "nonlinearity" problem. Figure 16-4 plots the log odds {log [p/(1 – p)]} of survival against severity. Now we have straight lines, but our outcome variable has lost its meaning, because it is the log odds of survival. We deal with that by placing a second score on the left to let people "reconvert" to the original survival rates.

The significance levels of our logistic regression analysis of differences between hospitals in the four category levels are shown in Table 16-3. The results suggest that each category level is signifi-

Figure 16-3 Characterization of Linear and Logistic Regression

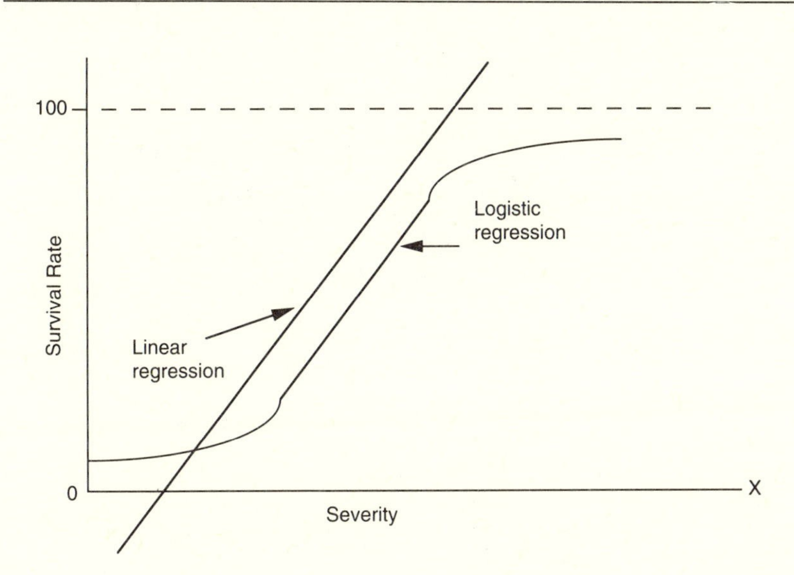

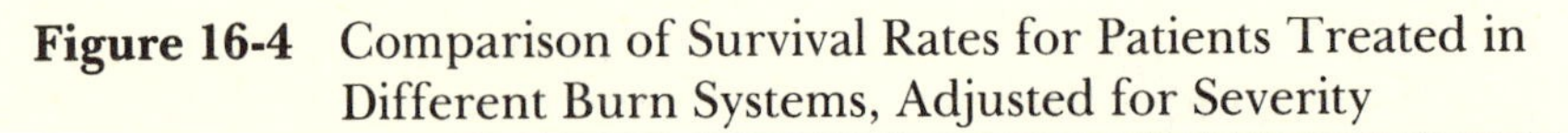

Figure 16-4 Comparison of Survival Rates for Patients Treated in Different Burn Systems, Adjusted for Severity

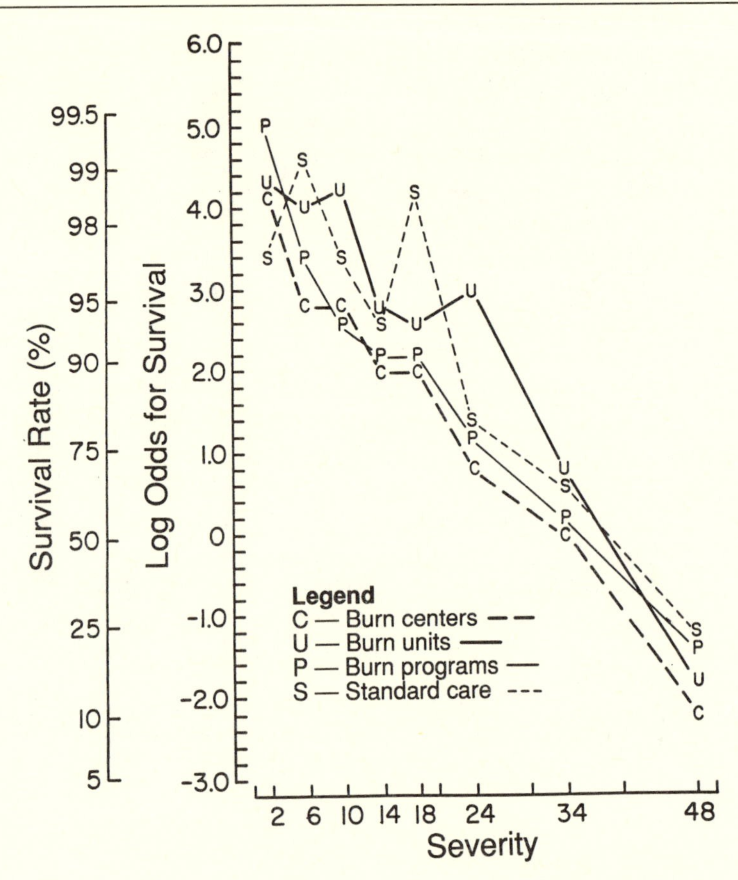

Table 16-3 Levels of Significance in the Difference between Categories of Burn Treatment (Two-Tailed Significance Probability)

	Centers	*Standard*	*Units*
Units	.0001	.66	.0003
Programs	.02	.02	
Standard care	.0003		

cantly different from the others except that burn units and standard care hospitals perform at about the same level. And the results are surprising: it seems burn centers are the least effective means of care.

The relationship between the outcome variable and severity need not be linear, even if we transform the data. If tests of significance are still important, we must use a nonparametric analysis such as Kendall's Coefficient of Concordance (Kendall 1962; Diaconis and Graham 1977). When Kendall's Coefficient of Concordance was used on the burn data in Table 16-2, differences between systems were found at the .0001 level of significance.

Comparing outcome within categories

Categorization systems must examine not only variation between categories but also within categories. If there is less variation within categories than across categories, we are probably clustering the hospitals correctly. However, if variation within categories is equal to or greater than variation between categories, then we wonder if the categories are realistic. The analytical methods used to examine between-category variation are also suitable to within-category variation. Logistic regression can be used if assumptions of linearity can be met. Otherwise, use nonparametric analyses, such as Kendall's Coefficient of Concordance.

The within-category variation among burn center hospitals is shown in Figure 16-5, plotting the log odds of survival against severity score. Because the plots appear reasonably linear, we performed logistic regression. Table 16-4 presents the levels of significance for differences found in the analysis between hospitals.

Table 16-5 plots survival rates (of burn center hospitals only) against severity. The results suggest that hospitals 1, 2, and 5 are not substantially different but that 3 and 4 are quite different from each other and from the other three. When Kendall's Coefficient of Concordance was used on the data in Table 16-5, differences between hospitals were found at a significance level of .0001. The differences between some hospitals in the burn center category were substantial. (Later we will address the question of whether differences between categories exceed differences within categories.)

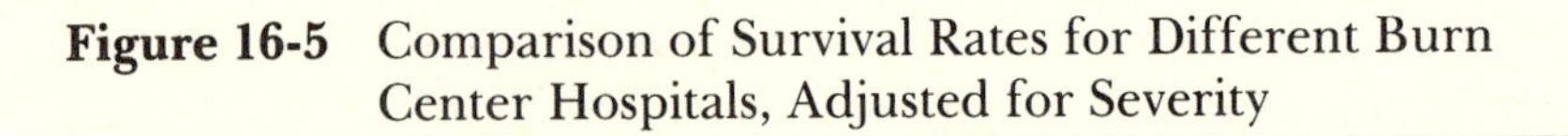

Figure 16-5 Comparison of Survival Rates for Different Burn Center Hospitals, Adjusted for Severity

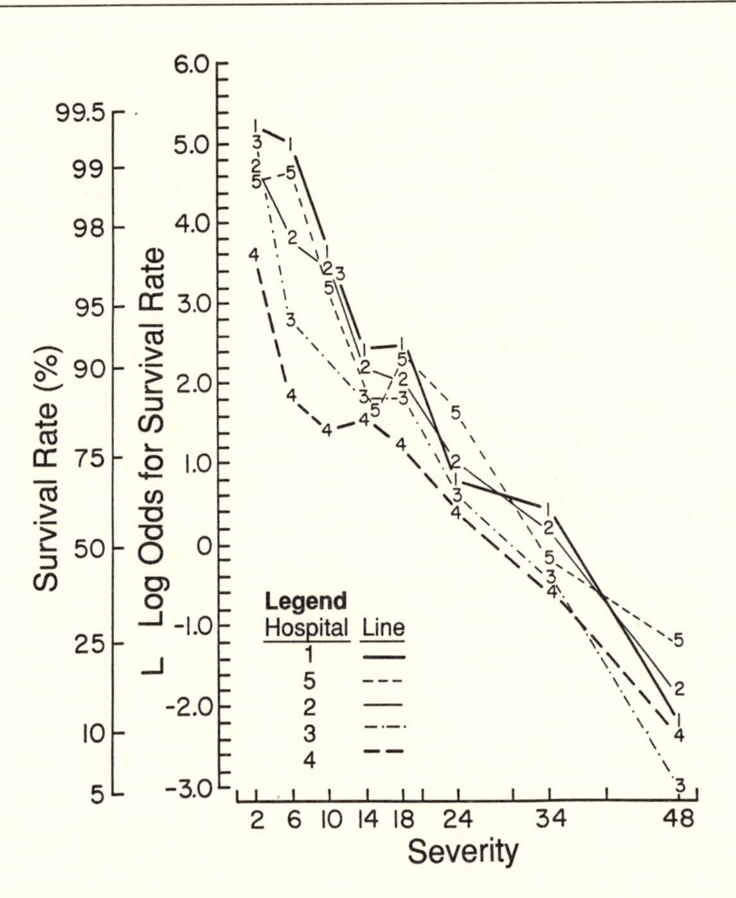

Table 16-4 Levels of Significance in Differences between Hospitals in the Burn Center Category

Hospitals	*2*	*3*	*4*	*5*
1	.44	.0003	.0001	.57
2		.0007	.0001	1.0
3			.0001	.025
4				.0001

Table 16-5 Comparison of Survival Rates among Hospitals in the Burn Center Category

	Hospital					
Severity	*1*	*2*	*3*	*4*	*5*	*Overall*
1–3.9	1.000	.994	.995	.975	1.000	.985
4–7.9	1.000	.979	.948	.856	1.000	.945
8–11.9	.978	.968	.972	.792	.970	.945
12–15.9	.919	.908	.871	.829	.838	.876
16–19.9	.929	.893	.866	.776	.923	.876
20–27.9	.707	.747	.663	.620	.833	.697
28–40	.609	.545	.399	.356	.459	.481
Over 40	.091	.143	.042	.062	.211	.097
Standardized survival rate	.868	.856	.817	.748	.857	

Can the outcome data be generalized?

The data presented so far provide convincing evidence that the hospitals in the study, when stratified by category, do have different survival rates. The data also suggest that survival rates differ substantially in hospitals categorized as burn centers. But what can we say about hospitals outside the study? The first question to ask when generalizing is whether we can assume that the hospitals selected for this study are representative of all hospitals in each category. At least regarding the standard care hospitals, this assumption is dubious because their choice to participate in the NBIE suggests that they have more interest—and perhaps more capability—in burn care than other hospitals in the category.

If the sample of hospitals was truly representative, we could test the generalizability of results by estimating the severity-adjusted log odds of survival for each hospital in the study. Then we could perform an analysis of variance on the results. Our analysis provides the severity-adjusted log odds of survival for each burn center hospital, and it appears that survival rates differ substantially between the hospitals. Unfortunately, the hospital-specific data for burn program, burn unit, and standard care hospitals were not available when we wrote this chapter, so these analyses were not included.

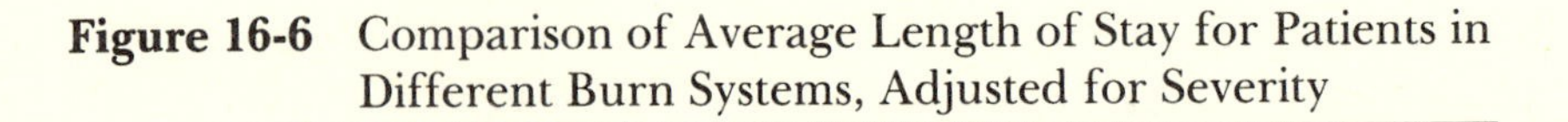

Figure 16-6 Comparison of Average Length of Stay for Patients in Different Burn Systems, Adjusted for Severity

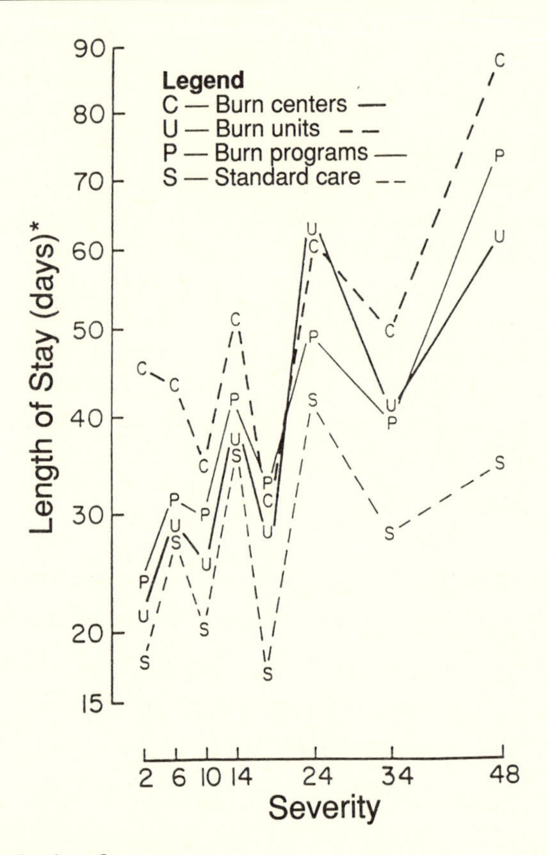

Statistical analysis of costs

As mentioned earlier, the difficulty of assembling good cost data often leads us to seek surrogate data, but there are weaknesses in most surrogates for cost. We used length of stay data as a surrogate, even though that may unfairly penalize less sophisticated forms of care. The daily cost at a standard care hospital may be substantially lower than the cost at a burn center, with its sophisticated equipment and specialized personnel. On the other hand, burn centers by definition are teaching units, so they may benefit from using low-cost student doctors. A second problem with using length of stay is deciding which patients to include. Patients who stay a very short time may be a very different population from those who stay longer. Furthermore, patients who die often leave sooner than survivors, so it might make sense to study only survivors. In our example, we elected to study only surviving patients who stayed longer than one day.

Despite its weaknesses, length of stay seemed the best available surrogate for cost. Length of stay is a continuous variable that can be a dependent variable in analysis of covariance. Figure 16-6 plots length of stay data for each category of hospital against severity. (To transform data and permit linear regression analysis, we actually plotted the square root of length of stay.) Despite wide variation in the data points, straight lines still appear to be reasonable approximations. Analysis of covariance on length of stay gives an $F_{3,6375}$ statistic of 45.2, a highly significant value. A nonparametric analysis also found differences between category levels at the .0001 significance level. The principal differences appear to be between burn centers and all other categories. Standardized lengths of stay from the analysis of covariance (see Table 16-6) were:

Burn center	45.6
Burn unit	31.5
Burn programs	34.1
Standard care	24.5

The burn center hospitals were compared to examine the differences within the burn center category. Again, the analysis of covariance found highly significant differences between hospitals ($F_{20,6358}$ = 28.6). We standardized the statistics for these comparisons (meaning that we applied the performance statistics of each type of hospital, i.e., mortality rates for different levels of severity) to an identical mix of patients. The standardized lengths of stay in the five burn centers (see Table 16-7) were

Center 1	41.8
Center 2	32.6
Center 3	34.4
Center 4	81.8
Center 5	34.1

As with survival rate, no data were available on individual hospitals in other categories. Generalization to hospitals outside NBIE, then, was impossible, even though this type of finding is usually important.

Table 16-6 Comparison of Length of Stay for Survivors between Burn Systems

	Burn Centers	*Burn Units*	*Burn Programs*	*Standard Care*
1–3.9	45.856	21.727	24.689	17.866
4–7.9	42.949	28.217	31.008	27.786
8–11.9	35.295	25.818	29.718	20.076
12–15.9	51.047	36.793	40.875	36.294
16–17.9	31.977	27.833	33.716	16.970
20–27.9	60.631	61.462	48.714	41.923
28–40	48.729	40.414	39.609	27.632
Over 40	88.368	62.000	72.900	34.000
Standardized length of stay	45.572	31.515	34.083	24.480

Table 16-7 Comparison of Length of Stay for Survivors among Hospitals in the Burn Center Category

	Burn Center				
	1	*2*	*3*	*4*	*5*
1–3.9	22.772	24.547	24.054	71.041	23.100
4–7.9	35.387	30.455	31.571	92.670	41.205
8–11.9	35.112	29.750	57.530	81.298	28.163
12–15.9	39.526	43.747	39.691	91.788	44.478
16–19.9	35.750	27.779	27.324	53.821	30.519
20–27.9	80.862	52.550	44.000	87.795	47.611
28–40	65.536	27.625	40.022	84.526	47.467
Over 40	141.330	76.000	15.000	142.000	50.750
Standardized length of stay	41.768	32.547	34.442	81.790	34.067

Joint analysis of costs and benefits

As concern about health care costs increases, so does the need to assess the trade-offs between costs and benefits. We do not argue that decisions on where to send patients should be made solely on the basis of a cost-benefit ratio, but such analysis can identify profitable topics for further investigation. One way of presenting cost-benefit data is to plot cost against outcome data. To do this, we must collapse curves of "severity to probability of death" and curves of "severity to length of stay" into a single number for each hospital or category level. This can be done using a standard mix of patients for all hospitals and calculating an expected survival rate and length of stay.

Figure 16-7 plots length of stay against severity for each hospital category and for the five burn center hospitals. Ideally we would choose hospitals that are closest to the lower right corner in Figure 16-7, because this is the region with highest survival and shortest stay. If only category levels were plotted, we might conclude that burn centers are the least cost-effective of all categories, because they seem to have high costs and low survival rates. However, when the individual burn centers are plotted, hospital 4 shows up as an outlier. The other burn centers are much closer to burn programs and units in terms of costs (but are still poorer than the average burn unit or standard care hospital).

Alternative Explanations

This chapter presented an example of using MAV models to measure severity of illness and how to use these severity scores to evaluate health care delivery systems. We discussed how severity indexes can be misleading unless they are developed with attention toward possible biases. The use of logistic regression analysis was shown along with an explanation of why such analysis is needed. We demonstrated several ways to portray data (standardized survival rates and lengths of stay, logistic curves, and cost-benefit plots) that may help to highlight differences. Finally, we considered some of our errors which limit utility and generalizability, such as hospital selection, lack of hospital specific data for burn programs, and so on.

Our findings can be subject to alternative explanations: they do not prove standard care hospitals are better than hospitals with specialty burn programs. If personnel in standard care facilities are relatively unfamiliar with burn victims, they might tend to overrate severity. This bias would shift the "standard care" line in Figure 16-2 to the right. Alternatively, perhaps the standard care hospitals participating in the NBIE are a special breed whose physicians and staff are particularly interested in burns and are thus unrepresentative of all standard care hospitals. In fact, because of their self-selection into

Figure 16-7 Predicted Survival Rates vs. Length of Stay for Different Burn Systems and Burn Center Hospitals Using a Common Severity Adjusted Case Mix

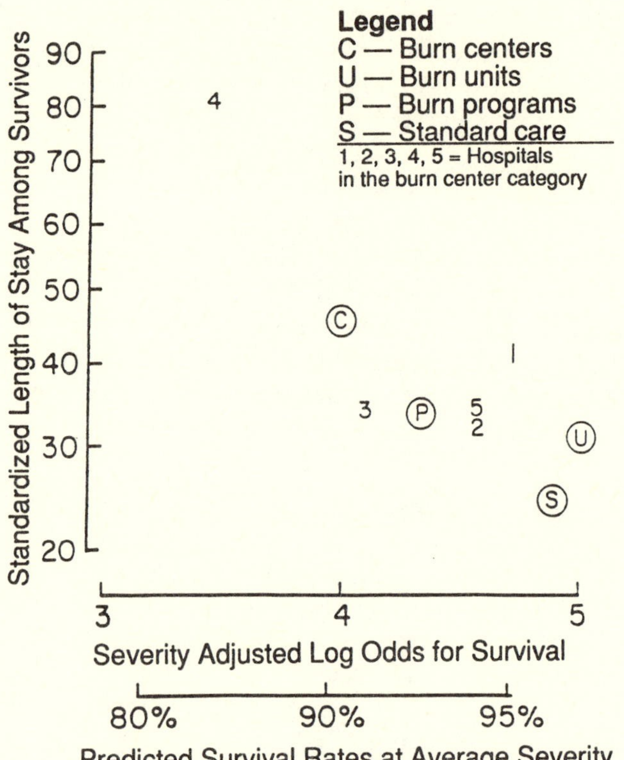

the NBIE, none of the results can be extended to hospitals that did not participate.

Another possible explanation for the surprising performance of standard hospitals lies in the selective nature of hospitalizations. Perhaps the standard care hospitals have more direct admissions while the burn specialists have more transfer patients. Although the severity index should account for differences in severity, transfers may differ from direct admits in dimensions not considered by the index.

Finally, despite everything that could have skewed our conclusion, the performance of standard care hospitals makes us wonder about the utility of technology-rich delivery systems. Considering the skyrocketing cost of health care, this analysis stresses the importance of carefully evaluating sophisticated approaches to care. They may not be cost effective.

References

Champion, H., W. Sacco, D. Hannan, R. Lepper, E. Atzinger, W. Copes, and R. Prall. 1980. "Assessment of Injury Severity: The Triage Index." *Critical Care Medicine* 8: 201–8.

Feller, I., and K. H. Crane. 1971. "Classification of Burn-Care Facilities in the United States." *JAMA* 215: 463–66.

Gustafson, D. H., and D. C. Holloway. 1975. "A Decision Theory Approach to Measuring Severity in Illness." *Health Services Research* 10: 97–106.

Gustafson, D. H., M. E. Hiles, and C. Taylor. 1980. "Report on the Trauma Severity Index Conference." Sponsored by the National Center for Health Services Research (Grant No. HSO4149) and the American Trauma Society. University of Wisconsin Center for Health Systems Research and Analysis.

Kendall, M. G. 1962. *Rank Correlations Methods,* 3rd ed. London: Griffin Co.

Levy, P., R. Mullner, J. Goldberg, and H. Gelfand. 1978. "The Estimated Survival Probability Index of Trauma Severity." *Health Services Research* 13: 28–35.

Linn, B. S., S. E. Stephenson, Jr., and J. Smith. 1977. "Evaluation of Burn Care in Florida." *New England Journal of Medicine* 296: 311–15.

17

Using Subjective Probabilities to Predict the Impact of National Health Insurance on Low-Income People

This chapter presents a case study on using subjective probabilities to predict the impact of national health insurance programs on five low-income populations. As noted in Chapter 8, subjective estimates can be used instead of empirical data if there are problems in acquiring or using data. The available data in this case were inappropriate to the population being studied, but often time or money constraints are the problem. In addition, data may be unreadable, outdated, incomplete, or misleading. The methodology described here requires no expensive data collection and can generate results relatively quickly.

The current spiral in health care costs is causing the U.S. government and major payers to again consider adopting national health insurance. The implementation of national health insurance (NHI) would introduce a new and uncertain variable to the complex equation that determines who will receive what care at what cost. While this social welfare plan would affect all segments of the population, it would have profound effects on people who currently rely on federal programs administered by the Bureau of Community Health Services (BCHS) in the Department of Health and Human Services. These groups include migrant farm workers, Native Americans, mothers and children needing preventive or special care, residents of medically underserved areas, people desiring family planning services, and those lacking adequate health insurance coverage.

The unique circumstances of migrant workers and Native Americans necessitated the creation of special services responsive to their needs. If NHI results in termination of such assistance, the result could be a financial burden on current beneficiaries—and a reduction in their access to care.

Many previous efforts to forecast the effects of NHI have been limited by time and data restrictions; in general, they overlooked repercussions on BCHS users. This chapter describes how subjective probabilities were used in an analysis to estimate changes in the out-of-pocket costs incurred by BCHS beneficiaries under three NHI schemes proposed in the mid-1970s. The study was done under severe time constraints so the results could be made available to Congress as it debated, and ultimately failed to pass, national health insurance. The model compared the current costs to BCHS families (for all health services obtained from any source) with the costs for the same services obtained through a national health insurance plan.

To accurately appraise changes that would occur under NHI, we had to find the utilization patterns of BCHS families and the unit cost of services consumed. We also had to ascertain the eligibility requirements and cost-sharing provisions of the three NHI proposals. Finally, we had to determine which currently used services were included in the various NHI benefit packages. Primarily because of time constraints, we could not collect data and were confined to using the best available information. These data were only marginally useful because they were based on middle-class families, not the populations served by BCHS. Therefore, we used subjective probabilities to improve both the reliability of utilization estimates and their applicability to the study population.

Assessing the impact of NHI on any socioeconomic group was complicated by the multiplicity of proposals that were considered by Congress. The bills varied in extent of federal participation, means of administration and payment, and relationship to existing programs and delivery mechanisms. All NHI plans, however, addressed four basic concerns: eligibility requirements, benefit coverage, cost-sharing stipulations, and initiation of catastrophic coverage.

The model focused on the effects of three NHI designs: the Ford administration bill (CHIP, HR-12684); the Kennedy-Corman bill (HR-22, supported by the AFL-CIO and United Auto Workers); and the Ullman bill (HR-1, supported by the American Hospital Association). The five BCHS client populations studied were users of community health centers, migrant health projects, maternal and infant projects, children and youth projects, and family planning services.

The NHI simulation was designed as a short-term study to help the planning and budgetary processes in the BCHS. The planning function necessitated that the study be completed within six months, and thus existing empirical studies had to suffice as sources of data.

Existing Studies

Many studies have examined the potential impact of NHI. The RAND Corporation Health Insurance Study (Newhouse 1974) sought to determine how various financing mechanisms (including variations in how consumers shared the costs) would affect demand for services. While this experimental study provided some basis for projecting demand changes under CHIP, the results were contaminated because the study guaranteed that participants would not pay more for health care than they would have under their previous insurance, so different cost-sharing provisions were really never implemented. Also, the subjects were drawn largely from the middle class, and some people have suggested that fluctuations in middle-class demand are different from those of disadvantaged groups, such as BCHS users. Reservations about the absence of supply variables in the determination of demand changes (Hester and Levenson 1974) also limited the RAND study's accuracy.

Little and Company (USDHEW 1971) in a study for the Department of Health, Education, and Welfare, extrapolated data from a federal employee health insurance program to the general population and predicted a considerable increase in demand after the addition of catastrophic health insurance. The researchers cautioned, however, against applying their conclusions about a relatively homogeneous sample to specific populations such as migrant workers.

These studies point out the limitations of traditional analyses. They can be costly, time-consuming and of limited value because they depend on extensive simplifying assumptions. Significantly, the populations on which they are based are not those affected by our policy analysis. We decided that, if we used the traditional approach, our study would not be available when the planners needed it, and it would have limited value anyway because of the limitations listed above. Thus, we needed a new approach: subjective probabilities.

Methodology

Several assumptions must be made in a study of this sort. A major assumption was that BCHS would cease funding projects such as family planning and migrant health after NHI was implemented. This "worst case" allowed us to determine whether NHI could sustain the present level of services to our target populations in the absence of BCHS resources.

The purpose of the NHI simulation was to estimate family utilization patterns. The simulation was accomplished by having a computer generate utilization statistics for each of several family profiles. Family profiles differed by family size, age, and sex and information on income, employment status, and employer characteristics.

Utilization patterns were created for individual family members (for each BCHS program) and stratified into appropriate age groups. Utilization patterns for individuals were then aggregated to achieve a family utilization description. The computer also determined extent of coverage under different NHI schemes and compared this figure to present costs.

Our NHI simulation study departed from previous work because it considered the effects of each of three major bills on a particular segment of the population. Earlier studies had either broadly examined the effects of NHI plans on the general population or narrowly looked at the effects of one bill on one group.

The model assumed that utilization of health services by one family member (such as a parent) was independent of utilization by others (such as children). The model did account for causal relationships between utilization patterns of some services. For example, to receive a prescription, a family member must have obtained a related service, perhaps from an ambulatory care physician.

The Data Required

Before directing the computer to simulate some sample user families, we needed information on *family characteristics,* particularly the population's socioeconomic and demographic status. These characteristics included such factors as number and size of families, age and sex of family members, employment status, whether employment was longer than 400 hours per year per employer, size of employing firm, income levels, and Medicaid status. This information was primarily collected from U.S. census data on BCHS populations. In addition, a Providence, Rhode Island, household survey (RHIHSR 1973) gave us frequency distribution data for the number of people in a family within a particular age group served by BCHS (for instance, what percentage of families had what number of children between the ages of 1 and 14 years).

The simulation also required us to input frequency distributions of utilization rates for each of a set of health services (such as hospitalizations or prenatal visits) for each existing BCHS project. Separate distributions were created for different levels of age and income. For each service listed in Table 17-1, initial utilization rates were obtained from existing data, census tables, household interviews, BCHS program data systems, research studies, and the National Ambulatory Medical Care Survey. In many cases, these preliminary estimates were national averages for use of the particular service. In some cases, we decided that the best estimates of utilization came from regional sources, such as the Community Health Survey for Rockford, Illinois (Community Health Survey 1973) or the Mental Health Registry of Monroe County, New York (Mental Health Registry 1973). The quality and reliability of these estimates were highly variable. Equally important, the available data reflected populations significantly different from BCHS users. Therefore, we employed subjective probability estimates (Ludke et al. 1977) to improve the accuracy and utility of our estimates of these distributions.

Sixteen experts were selected to form a judgment panel for each of the five BCHS programs, for a total of 80 people. Some of these authorities were project directors with experience caring for BCHS clients at organizations with reliable data systems; others were researchers who had studied utilization of BCHS programs. Experts were nominated by personnel from BCHS's central and regional offices and by advisors to our study.

We showed the panelists "clues" in the form of the initial utilization estimates from existing sources. Table 17-2 presents the clues given to the panelists predicting utilization rates for well-person examinations. Note that estimates are divided into five age groups and that the data were separated into two stages. First, panelists saw estimates of the chance of at least one exam, then they saw estimates of the number of exams for clients who received at least one. Panelists were told the source of the data and asked to revise those estimates in light of their experience, using a form similar to Table 17-2 but with the percent column blank. In cases for which no existing

Table 17-1 Services Considered in the NHI Simulation

CMHC mental health visit	Home health—acute
Private mental health visit	Home health—chronic
Medical/social service	Home health—midwife
Podiatry visit	Prescriptions
Orthodontia visit	Hospital—admissions
Eyeglasses	Hospital—days
Hearing aids	Hospital—physician visit
Orthopedic shoes	Ambulatory care—acute visit
Dental prosthetics	Ambulatory care—mental health
Artificial limbs	Mental health—admissions
Vision test—ophthalmologist	Mental health—days
Vision test—optometrist	Mental health—physician visit
Hearing test	Physiotherapy
Well-person exam	Immunizations
Transportation—outreach	Prenatal/postpartum
Transportation—emergency	Health education
Transportation—nonemergency	Initial health assessment*
Emergency medical service	Recall health assessment*
Nutrition service	Child development*
Skilled nursing facility—admissions	Continuity service*
Skilled nursing facility—days	Nursing service*
Skilled nursing facility—physician visit	Evaluation service*
Dental—preventive	
Dental—restorative	

*These services are only included in children and youth analysis. They represent the emphasis that the children and youth program places on continuity of care.

utilization data could be found, the experts were asked to give their own final estimates.

Each panel was then divided into groups of four and asked to discuss their estimates. Each group within each panel concentrated on a single user population. For example, at the community health center meeting, one table represented rural health centers, another small urban centers, and two others large urban health centers. Following their discussions, each panelist made final, independent estimates. This estimation process followed the estimate-talk-estimate guidelines established for probability estimation (Gustafson, Shukla, Delbecq, and Walter 1973).

The revised utilization rates were aggregated into one set of estimates for each service. The aggregation across experts was done

Table 17-2 A Sample of the Materials Panelists Received to Help Them Estimate Utilization Patterns for the Service (Well-Person Exams) and the Program (Community Health Centers)

Service: *Well-person exams*

Think only about the patients served by "your" program. Estimate the percentage of that population that received at least one *well-person exam* in the last year. *Five* estimates are required, one for each of the following ages: <1, 1–14, 15–44, 45–64, 65+.

Percent	*Age*
80%	<1
60%	1–14
40%	15–44
40%	45–64
30%	65+

(These are national data for all people.)

Now think about only those people who received at least one *well-person exam* in the last year. Estimate what percentage had one and only one, two and only two, etc. *Five* sets of estimates are required, one for each of the following age groups: <1, 1–14, 15–44, 45–64, 65+.

Age <1

Percent	*Visits*
20%	1
30%	2
50%	3+
100%	

Age 1–14

Percent	*Visits*
50%	1
20%	2
30%	3+
100%	

Age 15–44

Percent	*Visits*
75%	1
15%	2
10%	3+
100%	

Age 45–64

Percent	*Visits*
80%	1
10%	2
10%	3+
100%	

Continued

Table 17-2 Continued

Age 65+	
Percent	*Visits*
60%	1
20%	2
20%	3+
100%	

(These are national data for all people.)

by weighting the estimates according to the proportion of the total BCHS user population each estimate represented. For instance, the estimates of rural health center panelists received less weight than large urban health centers because fewer people participate in rural programs.

To simulate utilization under each NHI proposal, we examined NHI provisions of three bills, using studies of the 1974 legislative proposals (U.S. Congress 1974). Further analyses of differences between the 1974 and 1975 bills came from studies by the Office of Research and Statistics, Social Security Administration (1976), and the legislative services of BCHS (1975). Wherever the special studies were incomplete, we referred to the original legislation.

To simulate current costs to BCHS user families, we collected information on the existing regulations and payment schemes from Medicare and Medicaid legislation and regulations implementing BCHS programs.

To simulate the current cost per unit of service, we used statistics from the BCHS programs, professional associations, and the state department of health and social services.

Simulation Method

The computer simulation began by creating user families from information on family characteristics. The simulated family was described by size, age, employment, income, and sex. All factors were chosen from probability distributions as described above. Once the family profiles were developed, the model simulated utilization for each family member according to the utilization patterns derived from the expert's subjective probability estimates. The exact rate of utilization for each service was determined by looking at probability distributions for that service. These utilization patterns were stratified by age groups appropriate to the program being studied: generally 0–5, 6–14, 15–44 females, 15–44 males, 45–64, and 65 and over. For instance, suppose one member of the family was 61 years old. The computer would have a bag of 100 chips—80 blue and 20 white. The blue represent the percentage of people who were between 45 and 64 years old who had at least one well-person visit. The white chips represented people in the same age group with no visits. In essence, the computer reached into the bag and drew out a chip. Each time it was white, the person was defined to have no well-person visits. A blue chip would lead the computer to another bag of chips representing number of visits. We then aggregated utilization rates for individuals to form family utilization patterns, a step that was necessary because most NHI plans determine copayments and deductibles on the basis of family utilization patterns.

We developed eligibility patterns for BCHS families with regard to NHI provisions. For each NHI bill, eligibility criteria were identified and the computer calculated the number of families meeting each criterion. From this, we estimated the number of families at each level of eligibility to determine copayment, deductible, and premium levels for the simulation. Next we computed the number of services provided to each eligible family. Then, depending on the cost-sharing stipulations of the NHI bill under study, we calculated costs to the user family, to the national health insurance fund, and to third-party payers.

We started determining the extent of cost sharing under current payment provisions by counting how many families met the various standards of BCHS programs. Then we allocated each eligibility category a percentage of the total BCHS user population. Having identified the number of services provided, the number of services used by the family, and the eligibility of those families, the computer calculated the number of services provided at each eligibility level. Costs to users, to BCHS, and to third-party payers were tallied using this information.

The simulation was run over a total of 500 families. A different set of families was, of course, generated for each BCHS program. We could economically and effectively use 500 families in each simulation because earlier versions, run on 2,500 hypothetical community health center families, showed that results stabilized after 500 families were simulated.

Results

Tables 17-3 through 17-7 depict our results, with each table describing one BCHS program. The tables list the estimated cost per BCHS family under each NHI bill and under the current system. The tables also itemize the costs of the various components of the cost-sharing mechanism and the percentage of total costs paid by families and other parties. The results suggest that the national health insurance schemes embodied in CHIP and the Ullman bill would substantially increase out-of-pocket costs to BCHS users. Costs to current beneficiaries would either drop or remain unchanged under the Kennedy-Corman legislation.

Table 17-3 depicts the changes in net cost for families using community health centers. The annual cost for each of 500 families (average size 3.32 persons) was estimated to be $1,222. We checked this estimate by comparing it to estimates from the Social Security Administration's Research and Statistics Notes (Worthington 1975) indicating that per capita health expenditures in 1973 were $405, exclusive of government public health services and other indirect subsidies. Multiplying this figure by the average family size of 3.32 yielded a cost of $1,334.60. This indicates that if our simulation erred in depicting the cost of care, it did so on the conservative side.

The community health center user family was estimated to currently pay $116 per year out of pocket for all health services, or 9 percent of the total cost of services. Under CHIP, the simulation projected that this family would pay $621 per year, or 51 percent of the total cost. The user dollar cost would increase by more than 500 percent. One explanation for this increase is that CHC families use services in a way that dictated that they would have to pay most or all of the deductible and a significant copayment before CHIP benefits would commence. CHIP, then, was "last dollar coverage."

Under the Kennedy-Corman bill, cost sharing was estimated to be virtually the same as under current provisions, with the NHI fund incurring the costs previously charged to BCHS or third-party payers. The Ullman bill would have boosted user cost to an estimated $513, a 450 percent increase.

Table 17-4 shows that the current cost of services to families using migrant health projects (MHP) was $1,310 per year. For 500 simulated families, the current annual out-of-pocket expense was estimated at $77. Under CHIP the annual user out-of-pocket expense was estimated to rise to $640. Families would be responsible

Table 17-3 Community Health Center: Estimated Annual Costs*

Income Level: All	*Now in BCHS Community Health Centers*	*Under NHI*		
		CHIP	*Kennedy-Corman*	*Ullman*
Total cost of care	$1,222	$1,222	$1,222	$1,222
Payments by BCHS	643	—	—	—
Payments by third parties (Medicaid, Medicare, private insurance)	470	8	0	10
Payments by NHI	—	685	1,197	768
Premiums paid by user	7	92	56	69
Deductibles, copayment	109	529	25	444
Total costs to user	116	621	81	513

*500 families simulated.

for 49 percent of the cost of care, an increase from the current 6 percent liability. The $229 premium, where virtually none existed before, stems from a CHIP stipulation that requires employment for a specified period (400 hours in a year) before an employer would pay any portion of the premium.

Estimates for the other two NHI bills were lower. The Ullman bill would increase costs to $460. Costs under the Kennedy-Corman bill would remain at approximately current levels or decline slightly, because premium increases would be offset by decreases in deductibles and copayments. In addition, the Kennedy-Corman bill would fully cover some services that BCHS users paid out of pocket.

Table 17-5 shows that under CHIP, costs of care for families in maternal and infant care programs would increase by more than 325 percent from current levels, largely the result of to significant additions to premiums, deductibles, and copayments. The Ullman bill

Table 17-4 Migrant Health Programs: Estimated Annual Costs*

Income Level: All	*Now in BCHS Community Health Centers*	*Under NHI*		
		CHIP	*Kennedy-Corman*	*Ullman*
Total cost of care	$1,310	$1,310	$1,310	$1,310
Payments by BCHS	774	—	—	—
Payments by third parties (Medicaid, Medicare, private insurance)	460	4	0	7
Payments by NHI	—	895	1,297	993
Premiums paid by user	1	229	59	150
Deductibles, copayment	76	411	13	310
Total costs to user	77	640	72	460

*500 families simulated.

would double premiums and increase deductibles and copayments by 240 percent. On the other hand, the Kennedy-Corman bill would reduce out-of-pocket expenses to one-third of current levels.

Users of children and youth programs could expect their present annual cost, $180 per family, to grow to $631 under CHIP or $442 under the Ullman bill. Costs would decline to $86 under the Kennedy-Corman bill. (See Table 17-6.)

Financial obligation estimates for users of family planning programs would greatly increase under CHIP and Ullman bills, while they would fall substantially under the Kennedy-Corman bill. Some cells in Table 17-7 contain two estimates: the first is costs of all health services for family planning program users; the second is costs of family planning services only.

Table 17-5 Maternal and Infant Care Programs: Estimated Annual Costs*

Income Level: All	*Now in BCHS Community Health Centers*	*Under NHI*		
		CHIP	*Kennedy-Corman*	*Ullman*
Total cost of care	$2,384	$2,384	$2,384	$2,384
Payments by BCHS	1,139	—	—	—
Payments by third parties (Medicaid, Medicare, private insurance)	1,000	22	0	7
Payments by NHI	—	1,621	2,350	993
Premiums paid by user	25	105	60	150
Deductibles, copayment	245	741	34	310
Total costs to user	290	963	94	460

*500 families simulated.

Discussion

This study has important methodological and health policy implications. In terms of methodology, the marriage of two key decision aids—simulation and subjective probability—can have significant benefits. Although simulation is a powerful research tool, its utility has been limited by the absence of a strong data base. This study suggests that subjective estimates from respected experts can be effective surrogates for solid empirical data. The accuracy of subjective probability estimation for generating data has been demonstrated by several researchers, as discussed in Chapter 8.

From the standpoint of policy, this study casts doubt on some expectations about the benefits of NHI. Although NHI is expected to improve access to health care for the poor, our results indicate

Table 17-6 Children and Youth Programs: Estimated Annual Costs*

Income Level: All	*Now in BCHS Community Health Centers*	*Under NHI*		
		CHIP	*Kennedy-Corman*	*Ullman*
Total cost of care	$1,599	$1,599	$1,599	$1,599
Payments by BCHS	906	—	—	—
Payments by third parties (Medicaid, Medicare, private insurance)	512	1	0	—
Payments by NHI	—	1,051	1,567	1,198
Premiums paid by user	1	84	54	41
Deductibles, copayment	179	547	32	401
Total costs to user	180	631	86	442

*500 families simulated.

that CHIP and the Ullman bill (as formulated in 1975) would raise barriers to access, not remove them, at least for several segments of the poor.

Several issues should be considered while interpreting these results. First, the study focused on a narrowly defined and special population, a few groups with access to health care through federal programs that were created to meet their requirements. There is no reason to assume that accessibility for all the poor would be similarly affected by NHI. Second, the study does not consider the possibility of significant changes in the financial burden on taxpayers who would subsidize NHI. This could be an important factor, especially in determining the effects of the Kennedy-Corman legislation which we expect would reduce economic barriers to health care access among the poor. The removal of such barriers could well result in a

Table 17-7 Family Planning Costs: Estimated Annual Costs*

Income Level: All	*Now in BCHS Community Health Centers*	*Under NHI*		
		CHIP	*Kennedy-Corman*	*Ullman*
Total cost of care	$1,584/58	$1,584	$1,584	$1,584
Payments by BCHS	571/45	—	—	—
Payments by third parties (Medicaid, Medicare, private insurance)	578/13	10	—	15
Payments by NHI	—	968	1,579	1,005
Premiums paid by user	3	177	89	109
Deductibles, copayment	435	606	5	564
Total costs to user	437	782	94	673

*500 families simulated.

surge in demand with consequent increases in costs to the NHI fund. The simulation study assumes no change in demand after NHI begins, but this will not, of course, be the case. Economic theories of price and demand elasticity provide clues to these changes. The increased levels of cost-sharing stipulated by CHIP, for instance, would likely decrease demand for preventive and perhaps other services. Assuming no demand change is a safe starting point for analysis if we plan to enter estimates of demand fluctuation in the simulation at some point. A third issue would be the shift in financial burden from the poor to other segments, particularly the middle class. The effects of increased financial liability on the middle class and the concomitant effects on the health delivery system are unclear at this time.

The primary point of this chapter is that we have three options if existing data are inadequate for a policy analysis: to acknowledge weaknesses in the data but use them anyway; to collect new data, which is reasonable if time and resources allow; or to assemble the best data available, point out their strengths and weaknesses, and revise the data according to the judgment of expert panels. We were initially surprised by the extent to which experts revised the data, because we had expected data to intimidate experts. In fact, experts looking at data on populations outside our study sometimes discounted them substantially, but they almost always considered the data a useful starting point.

We encourage you to look skeptically at analyses that use unadulterated empirical data from outside the population of interest, when a minimal investment in panels of experts could transform those data into a powerful resource. We say this with conviction because we have, since this study, convened several panels of experts to make subjective probability estimates on relationships between risk factors, such as family history of drug abuse, and outcomes, such as problem use of marijuana. In a few cases our data were quite good for the population, and our panel's estimates matched them very well. But when our data were poor, the panels' estimates greatly differed from them. The implication is that high-quality experts can be a powerful source of data.

References

Community Health Survey for Winnebago and Boone Counties. 1973. Report to Comprehensive Health Planning of Northwest Illinois Inc., Office of Community Health Research, Rockford School of Medicine.

Bureau of Community Health Services (BCHS). 1975. "Comparisons of 1974 and 1975 National Health Insurance Proposals." Legislative Services Branch, unpublished report.

Gustafson, D., R. Shukla, A. Delbecq, and W. Walster. 1973. "Comparative Study of Differences in Subjective Likelihood Estimates, Individuals, Interacting, Delphi and Nominal Groups." *Organizational Behavior and Human Performance* 9 (2).

Hester, J., and I. Levenson. 1974. "The Health Insurance Study: A Critical Appraisal." *Inquiry* 11 (March): 53.

Ludke, R. L., F. F. Stauss, and D. H. Gustafson. 1977. "Comparison of Five Methods for Estimating Subjective Probability Distributions." *Organizational Behavior and Human Performance* 19: 162.

Mental Health Registry. 1973–1974. *Reports of Utilization of Mental Health Services.* Mental Health Registry of Monroe County, NY, University of Rochester.

Newhouse, J. P. "A Design for a Health Insurance Experiment." *Inquiry* 11: 5–27.

RHIHSR. 1973. "Use of Physician Services in Rhode Island." SEARCH Reports, Rhode Island Health Services Research Inc.

U.S. Congress. 1974. *National Health Insurance Resource Book.* Staff of the Committee on Ways and Means. 93rd Congress. Washington, D.C.: GPO.

U.S. Department of Health, Education, and Welfare. 1975. "An Economic Analysis of HSA Program Potential, 1976–1981." Prepared for the Office of Planning, Evaluation, and Legislation, Health Services Administration by Geomet Inc. Contract HSA 105–74–7, HSA 105–74–68.

———. 1971. "Deep Coverage in Health Insurance and Treatment of Catastrophic Illness." Preliminary report by Arthur D. Little Inc. Contract HSM 110–71–197.

U.S. Social Security Administration. 1976. *National Health Insurance Proposals: Provisions of Bills Introduced to the 94th Congress as of February, 1976.* Compiled by Saul Waldman. Office of Research and Statistics.

Worthington, N. 1975. "National Health Expenditures, Calendar Years 1929–73." Research and Statistics Note No. 1. Office of Research and Statistics, U.S. Social Security Administration.

18

Conclusion: Decision Support Technologies and Public Policy

Throughout this book, we have focused on information—one factor among the many that affect how people formulate, implement, evaluate, and modify policy. We were concerned mainly with the role of information and information systems within the policymaking process. Although we understand the importance of other aspects of the policy process, we feel that information needs have been somewhat ignored in recent literature.

We have looked at information in several ways and have stressed that policymakers' information needs change partly in relation to the stage in the policymaking process. We have described that process in terms of the issue life cycle and developed several tools to understand the varying information needs at each stage of the cycle.

While we have dwelled on the use of these tools, we are more concerned with explicating the principles that guided us in their development. We hope readers will find our thought processes and the examples of their application useful in understanding how policy information systems based on decision theory and group processes can effectively support the formation and implementation of policy.

Our work has also indicated new roles for information systems within the policy- and decision-making process. We have shown how decision models can promote improved implementation planning and decision making. We recognize that other aspects of policymaking, such as problem structuring, are also amenable to improvement with these tools, and feel that further opportunities of this type remain to be addressed. We hope this text will cause a new type of policy information system to develop that can meet the information needs of policymakers and policy analysts. But we also recognize that the models we propose have only begun to be tested, and that other studies are needed to test and refine them.

We believe rationality is essential for formulating public policy. By rationality, we mean using analytical tools to aid a policymaker's intuition and prevent a policymaker's biases from interfering with decisions more than necessary. For instance, the models we present can help policymakers anticipate which issues will become important so they have time to deal with them. If, as we assert, issues largely conform to our issue life cycle model, policymakers can use methods presented here to prepare for issues likely to develop in the next few months. This will allow policy analysis resources to be allocated more effectively and give policy analysts the time they need to do an excellent job of supporting the policymaker.

As modern life has gained complexity, the nature of policymaking has changed. One person or group can no longer know enough to make intelligent policy decisions on many topics. Consumers, citizens, and special interest groups have become more skillful in pressing their issues and influencing policy. While systems can now be analyzed in greater complexity, change occurs faster than ever. The increasing need for policy analysis, coupled with its greater power, has allowed it to become an academic discipline. In addition, many government agencies and private sector organizations have devoted entire offices to policy analysis and planning. But given the increasing instability of the health and social service environments, policymakers and analysts need an approach that reduces noise in the system and reconciles conflicting positions. Tired of constantly fighting fires, policymakers want to regain control of their work.

Even though it has become increasingly evident that decisions should be made deliberately, in the real world confusion and disarray remain the rule, not the exception. Typically, a policymaker shuffles a dozen call-back messages from constituents while trying to reorganize the department and write next year's budget before the fiscal year commences. When a decision analyst brings a ream of computer printouts and a confusing mathematical model to a meeting, the decision maker and the decision analyst must find common ground for communication. We hope this book provides guidelines and models that can help policy analysts understand the needs of their customers (the policymakers and ultimately the consumers of, and payers for, health and social service programs) and develop support systems to meet those needs more effectively.

Unfortunately, designers of decision analysis and information systems have relied on a number of myths. The first was that if poor decisions arose from a paucity of information, decision analysts could remedy the situation merely by providing more information. This might have worked if the policy analysts had understood what information was actually needed, but they usually provided too much information and too little filtering, overloading the users and preventing them from "hearing" what they received. The second myth was that information is objective and numbers are inherently valid. This led policy analysts to provide uninterpreted facts and figures which they considered self-explanatory, even though the same information does not lead everybody to the same conclusion. We contend instead that information always is interpreted according to subjective values and that the key job of the analyst is understanding and interpreting information in light of these values.

The third myth was that information systems should be portable, so a system designed for the Bureau of Health Planning in Wisconsin should function equally well in Texas or Oregon, whether for health planning or a similar function. This myth led information analysts to design information systems in a vacuum, with insufficient contact with their real-world users. The attempt to make systems portable again assumed that information needs are consistent across cultures, whereas we realize that the character of both the subject matter and the organization play vital roles in determining the type of information needed. This recognition leads us to caution the reader not to blindly adopt our models. Models for information needs analysis and implementation analysis should be reformed in the context of the application environment to ensure that they fit its unique needs.

These myths also prevented decision makers from getting usable information and caused them to grow frustrated with policy analysts. The relationship was further strained because policymakers have enormous information needs, but policy analysts are not providing suitable information. If they are providing it, they do so in a form that does not match the setting, problem, or personality. As a result, policy analysts have belatedly recognized that it is not just pieces of information that influence policy formation, but also values and subjective judgments that play key roles.

This book discusses a three-part process to formally integrate quantitative information and values in a way that effectively promotes policy analysis. The first draws on mental models of a variety of processes, including problem solving (and process improvement), organizational change, and issue life cycle. These mental models, long the focus of study by social scientists, provide an important foundation for our work by giving us a logical way to view the task of policy formation. But mental models can be even more powerful if their concepts can be quantified and thus made amenable to formal analysis.

This integrative mechanism is provided by the second process covered herein: quantitative modeling. Our process relies on two modeling strategies, one dealing with values and another dealing with uncertainty. We limit our focus to two models because they fit within the decision science construct we adhere to and because we have found them helpful to a wide variety of management philosophies. They allow a "rational" policymaker to quantify values and uncertainties in a way that permits the application of optimization processes. They offer an administrative decision maker the consistency and rigor needed to ensure uniform application to policy and permit political policymakers to capture the values and uncertainties of their constituents so they can effectively be considered in the policy formation process.

The third process we build on is an understanding of how people function as decision makers and problem solvers, both alone and in groups. By understanding the strengths and weaknesses of individual and group decision makers, we can build models and support systems that supplement, not replace, their strengths. We provide group process techniques that can reach excellent resolutions to difficult issues. We show how models can overcome the weaknesses inherent in the human tendency to process complex information inappropriately. While we disaggregate the policy process into manageable components, we also recognize that values and constituencies are fundamental to policy formation, so we present strategies and examples for eliciting and using them in the disaggregated analysis.

These processes depend on a thorough debunking of the myths mentioned above. First, we need not collect all possible information; we need only examine the crucial information, and for this step we can use techniques such as the information analysis tool, which indicates the kinds of information needed on the basis of actual events. Second, once the critical information has been collected, the policy analyst must decide how to present it. It would be nice to know exactly what weaknesses our policymaker "customer"has so we could provide compensating information, but unfortunately the theory of human information processing has not advanced far enough for us to do this. Rather, we offer a design of a policy support system that is flexible enough to respond to different information needs so policymakers can draw on a wide variety of support. As in other applications of information systems, we expect policymakers will be unable to resist opportunities to access and process information in ways they never dreamed of.

So, in addition to detailing a process to support rational decisions, we are also offering a prototype policy support system. This system provides several types of information to policymakers in diverse types of organizations. First, this system tells us which issues are about to ripen and which will be particularly embattled. This can be done by analyzing public information from newspapers, court decisions, legislation, and special interest groups with our "issue anticipator." Policy support systems can use information analysis to yield information to help estimate the impact an important issue might have on a system. Note that when an issue is controversial, we can learn a great deal about the value systems and hypotheses of key stakeholders; this information can point toward possible compromise down the line. If we create a policy information management system flexible enough to respond to the varying needs of different policymakers, we can effectively reduce policymaker uncertainty. Finally, a policy information system can evaluate projects with the implementation analysis tool, which allows us to determine modifications to an implementation plan to give it the greatest potential for success.

The goal of a policy support system is to help policymakers manage the overwhelming amounts of information available to them. We hope the conceptual and quantitative models, the insights into human information processing, and the methods of capturing values and uncertainties will give useful guidance to those who seek to improve health and social service policy.

Index

About the Authors

David H. Gustafson, Ph.D., is Professor of Industrial Engineering and Preventive Medicine at the University of Wisconsin in Madison. He is a past chairman of the industrial engineering department there and founder of the Center for Health Systems Research and Analysis (CHSRA), a multidisciplinary research center employing systems analysis, decision science, and decision support technologies to address patient care and health policy problems.

Gustafson holds bachelor's, master's, and doctoral degrees in industrial engineering from the University of Michigan, where he was a W. K. Kellogg Fellow in Health Administration. His research interests focus on advancing the ability of decision science, decision support systems, and systems science to improve the delivery of health and human services. His research on computer-based health promotion systems received the 1989 American Medical Association Award of Excellence for Education and Prevention.

Gustafson has served on the Commission on Education in Health Administration, the National Health Council's Committee on Health Education, the Health Manpower Data Task Force of the National Institutes of Mental Health, the National Demonstration Project on Hospital Quality, the Hospital Productivity Committee of the National Commission on Productivity, the Institute of Medicine's Health Advisory Committee, and the Health Care Technology Study Section of the National Center for Health Services Research.

In addition to numerous journal articles and book chapters, Gustafson is the coauthor, with Andre Delbecq and Andrew Van de Ven, of *Group Techniques for Program Planning: A Guide to Nominal Group and Delphi Processes.*

William L. Cats-Baril, Ph.D., is Associate Professor of Information and Decision Sciences at the University of Vermont School of Business. He has been a research fellow at the London School of Economics, and has been a visiting faculty member at the European Institute of Business Administration (INSEAD) in Fountainebleau, the International Management Center in Budapest, the China-Europe Management Institute in Beijing, and the University of Puerto Rico-Mayaguez. He has taught decision analysis and information systems in executive programs in Asia, Europe, and Latin America.

Cats-Baril was a Principal Investigator in the Vermont Rehabilitation Engineering Center (VREC) where he developed a predictive risk model of low back pain disability. His research at VREC won the Eastern Orthopaedic Association Award for Spinal Research in 1988. He was appointed to the National Advisory Board for the Multi-Disciplinary Study of Low Back Pain and has served on the outcome studies task forces of the American Academy of Orthopaedic Surgeons and the North American Spine Society. In addition, Cats-Baril was Special Assistant for Systems Analysis and Policy to the Chief Executive Officer and Chairman of the Board of the University Health Center in Burlington, Vermont.

Cats-Baril is the founder and President of Devanda Corporation, a company that develops expert systems in the diagnosis and treatment of low back pain. He is a consultant on consensus building, strategy, and information needs assessment. His current research interests focus on developing a methodology to identify and monitor strategic performance measures of organizations. He has published numerous articles and book chapters on strategic information needs assessment and systems design, conflict resolution and negotiation, predictive risk and cost-effectiveness models.

Farrokh Alemi, Ph.D., is a decision analyst working on the design, quality, and financing of health care systems. He is the chair of the Health Application Section of the Operations Research Society of America, and teaches in the health administration program at Cleveland State University. In addition, he has taught health care managers at Tulane University.

Alemi is the founder of Interpractice, a health maintenance organization that uses delayed communications and computerized referrals to reduce outpatient costs. In addition, he is the creator of AVIVA, a voice-interactive software program for assessing health risks, and VODO, a software program to teach sexual decision making to teenagers.

Alemi has done ground-breaking work on comparing the various systems for assessing hospital quality and on methods of assessing quality of care provided to AIDS patients. He is currently working on a telephone-based system for managing and educating cocaine-abusing pregnant women. His research has been funded by organizations such as the Robert Wood Johnson Foundation, the Health Care Financing Administration, the National Institute of Drug Abuse, and the Jewish Endowment of New Orleans. He has published articles in numerous journals, including *Medical Care* and *Interfaces*.